CAMBRIDGE LIBRARY COLLECTION

Books of enduring scholarly value

Physical Sciences

From ancient times, humans have tried to understand the workings of
the world around them. The roots of modern physical science go back to
the very earliest mechanical devices such as levers and rollers, the mixing
of paints and dyes, and the importance of the heavenly bodies in early
religious observance and navigation. The physical sciences as we know them
today began to emerge as independent academic subjects during the early
modern period, in the work of Newton and other 'natural philosophers',
and numerous sub-disciplines developed during the centuries that followed.
This part of the Cambridge Library Collection is devoted to landmark
publications in this area which will be of interest to historians of science
concerned with individual scientists, particular discoveries, and advances in
scientific method, or with the establishment and development of scientific
institutions around the world.

Memoir and Scientific Correspondence of the Late Sir George Gabriel Stokes, Bart.

Lucasian Professor of Mathematics at Cambridge and President of the
Royal Society, Sir George Gabriel Stokes (1819–1904) made substantial
contributions to the fields of fluid dynamics, optics, physics, and geodesy,
in which numerous discoveries still bear his name. The *Memoir and
Scientific Correspondence of the Late Sir George Gabriel Stokes*, edited by
Joseph Larmor, offers rare insight into this capacious scientific mind, with
letters attesting to the careful, engaged experimentation that earned him
international acclaim. Volume II (1907) includes important professional
correspondence with James Clerk Maxwell, James Prescott Joule, and many
others, with particular attention given to Stokes' activities with the British
Meteorological Society. Many of his foundational innovations in optics are
also explicated in these letters, serving in place of the authoritative volume he
unfortunately never had the opportunity to complete.

Cambridge University Press has long been a pioneer in the reissuing of out-of-print titles from its own backlist, producing digital reprints of books that are still sought after by scholars and students but could not be reprinted economically using traditional technology. The Cambridge Library Collection extends this activity to a wider range of books which are still of importance to researchers and professionals, either for the source material they contain, or as landmarks in the history of their academic discipline.

Drawing from the world-renowned collections in the Cambridge University Library, and guided by the advice of experts in each subject area, Cambridge University Press is using state-of-the-art scanning machines in its own Printing House to capture the content of each book selected for inclusion. The files are processed to give a consistently clear, crisp image, and the books finished to the high quality standard for which the Press is recognised around the world. The latest print-on-demand technology ensures that the books will remain available indefinitely, and that orders for single or multiple copies can quickly be supplied.

The Cambridge Library Collection will bring back to life books of enduring scholarly value (including out-of-copyright works originally issued by other publishers) across a wide range of disciplines in the humanities and social sciences and in science and technology.

Memoir and Scientific Correspondence of the Late Sir George Gabriel Stokes, Bart.

Selected and Arranged by Joseph Larmor

<small>VOLUME 2</small>

<small>GEORGE GABRIEL STOKES</small>
<small>EDITED BY JOSEPH LARMOR</small>

CAMBRIDGE UNIVERSITY PRESS

Cambridge, New York, Melbourne, Madrid, Cape Town, Singapore,
São Paolo, Delhi, Dubai, Tokyo

Published in the United States of America by Cambridge University Press, New York

www.cambridge.org
Information on this title: www.cambridge.org/9781108008921

© in this compilation Cambridge University Press 2010

This edition first published 1907
This digitally printed version 2010

ISBN 978-1-108-00892-1 Paperback

SIR GEORGE GABRIEL STOKES

MEMOIR

AND

SCIENTIFIC CORRESPONDENCE

Presentation Bust (1899) in the Hall of
Pembroke College.

MEMOIR

AND

SCIENTIFIC CORRESPONDENCE

OF THE LATE

Sir GEORGE GABRIEL STOKES, Bart.,

Sc.D., LL.D., D.C.L., Past Pres. R.S.,

KT PRUSSIAN ORDER *POUR LE MÉRITE*, FOR. ASSOC. INSTITUTE OF FRANCE, *ETC.*
MASTER OF PEMBROKE COLLEGE AND LUCASIAN PROFESSOR OF MATHEMATICS
IN THE UNIVERSITY OF CAMBRIDGE.

SELECTED AND ARRANGED BY

JOSEPH LARMOR, D.Sc., LL.D., Sec. R.S.,

FELLOW OF ST JOHN'S COLLEGE AND LUCASIAN PROFESSOR OF MATHEMATICS.

VOLUME II.

CAMBRIDGE:
AT THE UNIVERSITY PRESS.

1907

𝕮𝖆𝖒𝖇𝖗𝖎𝖉𝖌𝖊:

PRINTED BY JOHN CLAY, M.A.

AT THE UNIVERSITY PRESS.

CONTENTS OF VOL. II.

CORRIGENDA.

Vol. ii. p. 362, line 5, *for* ramified *read* rarefied ;
p. 375 headline, *read* sensitive.

SECTION III.

SPECIAL SCIENTIFIC CORRESPONDENCE.

LETTERS FROM JAMES CLERK MAXWELL.

TRINITY COLLEGE, *Oct. 3rd*, 1853*.

DEAR SIR,

    What I said about chlorophyll was merely a vague remembrance of one out of a number of facts in Chemistry which Professor Gregory mentioned to me last Winter during a visit to his laboratory. I am very sure that he told me of its isolation but less sure about its colour. He gave me no reference to any source of information and at that time I had no intention of asking for any.

    I have written to him on the subject, but I have not yet received an answer.

    With respect to my conjecture about Haidinger's brushes, your statement of it is quite accurate and expresses all that deserves the name even of conjecture†. In August 1850 I noted down some experiments and deductions from them, some of which are the same with those which I have since seen described in the fourth part of Moigno's *Répertoire d'Optique*, particularly the experiment with a plate of quartz.

    I have published nothing on the subject. In fact I have had no opportunity of obtaining sufficient foundation for a theory, and a conjecture is best corroborated by the existence of identical or similar conjectures springing up independently.

Yours faithfully,

JAS. CLERK MAXWELL.

---

* Prof. Clerk Maxwell graduated B.A. in the Mathematical Tripos of Jan. 1854.

† Cf. *Scientific Papers*, vol. i. p. 242, read at Brit. Assoc., Cheltenham, 1856.

TRINITY COLLEGE, *Feb.* 16, 1856.

DEAR SIR,

Professor Forbes writes to me that the chair of Natural Philosophy at Aberdeen (Marischal College) is vacant, and suggests that I should become a candidate.

I have no hesitation in acting on that suggestion at once but I would like to know what sort of men go in, and what chance there is of success, because I have no wish to have what I cannot get.

Now as several of your Cambridge acquaintances have obtained such situations in Scotland & Ireland, I thought you might both know the style of man wanted and whether any such were in the field and might also have learned the outline of the proper method of making application.

I find the presentation belongs to the Crown, that is, I suppose, equivalent to the fact that the Home Secretary & Lord Advocate are the proper persons to apply to.

I hope to hear the details from someone who knows; meanwhile I have no doubt that anything you could say in my favour would have its weight in a certificate or testimonial, so as soon as I am better acquainted with the matter I will write and ask you for one, if it is wanted.

I have been trying to simplify the theory of compound Optical Instruments and I have been reading up all that I can find on that subject*. I find a great deal in Smith's *Opticks* that is good, and also investigations by Euler, Lagrange, & Gauss. Do you know of any other researches on this subject?

I have greatly simplified the calculation of the foci of an instrument as far as first approximations by considering the "principal foci" and "focal centres" of the instrument. In this way I can begin with two axioms about the *possibility* of conjugate foci, undistorted images, and thence immediately deduce the law of conjugate foci and the magnitudes of images in any instrument. I thus render the theory of instruments independent of the particular optical apparatus used in the instrument and make no reference to the propositions about refraction at spherical surfaces.

---

* *Scientific Papers*, vol. i. pp. 238—240, read at Camb. Phil. Soc., May 12, 1856; *ibid.* pp. 271—285 from *Quar. Journ. Math.*, Jan. 1858.

The geometry is very simple and short and all that is necessary to be remembered is easily expressed in two equations, thus—

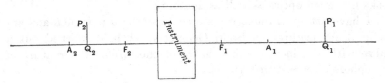

Let $F_1 F_2$ be the principal foci, $A_1 A_2$ focal centres, $P_1 P_2$ conjugate foci, $P_1 Q_1$, $P_2 Q_2$ perp. to axis,

$$\frac{Q_1 F_1}{F_1 A_1} = \frac{F_2 A_2}{F_2 Q_2} = \frac{P_1 Q_1}{P_2 Q_2}.$$

There are several other things in Geometrical Optics that I find worth examining.

A modification of the same method seems applicable to the case of small oblique & excentric pencils.

I hope I do not trouble you in the first part of this letter. You must understand that I do not wish to trouble you now, but merely to inform you of my design and to intimate that if I went through with it I might find it necessary to write to you for your certificate as Professor.

Yours truly,

J. C. MAXWELL.

TRIN. COLL., *Feb. 22nd*, 1856.

DEAR SIR,

I have received your two letters. I acted upon your advice with regard to Lord Aberdeen by asking Alex. Herschel to get his father to tell Lord Aberdeen that my case was worth attending to. Two of my friends are writing to two friends of the Lord Advocate. So much for private influence. With respect to testimony, I should think my friends are all saying their utmost in my favour judging by the specimens I have received. I am obliged to you for promising me a testimonial. As I am told it is important to be early I hope you will send it soon.

I have received two applications myself for testimonials. I intend to write out my estimate of the men and leave it to them to consider whether they will use it at once or wait till another time.

I find the nearest thing to my optical propositions in Smith's *Optics*. You will find a good description of a stereoscopic experi-

ment in vol. ii. page 388 and in the "Remarks," page 108. Miller
has lent me a paper by Listing on the Dioptrics of the Eye, which
looks like good optics as well as anatomy.

I have set up a reflecting stereoscope* on a new plan and am
using it for lectures on Solid Geometry. I have worked out a
curve with the locus of the centre of the circle of curvature of
the sphere of curvature; the curve is

$$x = 4a \cos^3 \theta, \quad y = 4a \sin^3 \theta, \quad z = 3a \cos 2\theta.$$

It is the best curve I know to explain everything on, for all the
results have simple expressions.

I read the second part of my paper on Faraday's "lines of
force†" at the Philosophical some weeks ago. It is an examination
of the "Electrotonic state" and an application of the mathematical
expression of it to the solution of problems. I have not got it
into shape owing to other things on hand, Aberdeen among the
rest; but the first part, which is entirely elementary, I consider
intelligible without professional mathematics, and I would like to
know if a somewhat long investigation of the strict theory of the
lines of force would be acceptable in the *Phil. Mag.*‡ I want to
get Tyndall and Faraday to read it. Thomson has read it. It
requires attention and imagination but no calculation, at least
that is all done by me and put in the form of diagrams of the
lines. I have no more time just now.

Yours truly,
J. C. MAXWELL.

Is it true that Cayley is going in?

129, Union Street, Aberdeen.
27th Jan., 1857.

My dear Stokes,

Although I am not likely to be much in London, I am
fully sensible of the honour and advantages of belonging to the
Royal Society, and therefore I am much flattered with your
remembrance of me in connection with that body. I do not
know whether your office precludes you from proposing a candidate.
If not, then in the event of my coming forward I should be glad
of your paternity.

---

* See *Scientific Papers*, vol. ii. p. 148 (1867) footnote. The curve in the
reference following has not been traced.

† *Scientific Papers*, vol. i. pp. 155—229.

‡ *Phil. Mag.* 1861-2: *Scientific Papers*, vol. i. pp. 451—513. The formal
memoir is in *Phil. Trans.*, Oct. 27, 1864; *Scientific Papers*, vol. i. pp. 526—597.

With respect to the composition, I have no immediate prospect of anything in the way of a paper for you. I have made preparations for an analysis of the colours of the spectrum with respect to the sensations they produce, but it is doubtful whether anything presentable will be produced this summer*.

I constructed an instrument 2 feet long which acts very well as a portable instrument of illustration, but I must make great progress before I come into the field with such men as Helmholtz. Have you seen his physiology of the eye in *Karsten's Encyc.*? I hope he will give his last words on colour and sensation in the next fasciculus.

So my conclusion is that I am desirous of joining, but not yet. I would rather wait a year if you think the same as I do.

My fingers are frozen and the post is ready.

Yours truly,
J. C. MAXWELL.

GLENLAIR, SPRINGHOLM, DUMFRIES,
8th May, 1857.

DEAR STOKES,

In conversing with Prof. Smyth of Edinburgh about his excursion to Teneriffe and his experiments on the solar spectrum, it occurred to me that the method which I arranged last year for exhibiting the mixture of the colours of the spectrum might be made useful in experiments on the spectrum by travellers and out-of-door observers.

My portable instrument consists of a wooden box 24 × 6 × 4 inch inside measure. The light enters at slits at the end of the box and falls on a prism at the other end. After refraction it is reflected by a concave mirror back through the same faces of the prism, and finally forms a pure spectrum near the place of entrance.

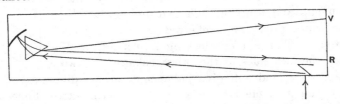

---

* For his earliest work, at Cambridge, Nov. 1854, with historical references, see *Scientific Papers*, vol. i. pp. 119—125 under date Jan. 4, 1855. This is followed by a memoir in *Trans. Roy. Soc. Edin.*, and a note in *Phil. Mag.* June, 1857

In my arrangement the light enters at the side and is reflected by a small plane mirror, the prism is equilateral and of glass, and the mirror is a lens silvered at the back of about 1 foot focal length by *reflexion*. For the refrangible light, the light should enter directly at the end of the box, the prism should be quartz, and the mirror metallic with a tool for repolishing.

I found this arrangement give a very large and distinct spectrum. The prism should be cleaned well and the mirror adjusted to make the reflexion in the plane of refraction. Then the focussing is done by turning the prism on its axis, which gives a very long range of focal lengths, and different parts of the spectrum are examined by turning the mirror and prism (which are on the same frame) together.

There should be a finder outside the box in the fashion of an alidade or better a spectacle glass at one end and a mark for the sun's image at the other.

Fluorescent substances may be viewed by looking down from the side of the box, and photographic materials might be slid into the same place.

The main advantage is the getting rid of the long rays inclined at an angle and the boxes to hold them. They are all doubled up and put in one box. I am going to try a new plan for comparative trials of different mixtures of light.

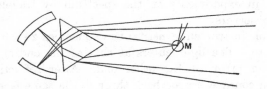

The various light is to be admitted by slits as in other experiments and sent through two different angles of the same prism and reflected by two mirrors separately. Thus an eye looking into a mirror M will see the two sides of the prism coloured in two different ways, and will give judgement on the difference.

Can you give me references to the experiments of Foucault, Fizeau, and others if any? 1st, on the velocity of light in air, 2nd, comparative velocity in air and water, and 3rd, comparative velocity in water running opposite ways.

Moigno's *Répertoire* & his *Cosmos* are the only authorities I know of.

I have had a dynamical top* of brass made at Aberdeen and have been simplifying the theory of the motion of the "invariable axis" (normal to invariable plane) in the body. The extremity of this axis describes spherical ellipses about the greatest or least principal axes, and the areas (projected on coordinate planes) described about these axes are proportional to the number of revolutions of the body about that axis (not to the absolute time). I have made some very delicate experiments with the top, to exhibit these movements, and it appears to be very little disturbed by accidental causes even when the difference of principal axes is less than $\frac{1}{300}$. I have also succeeded in setting the axis on the path leading *to* the mean axis so that the axis *approached* the mean axis for a considerable time before it left it for ever.

Yours truly,
J. C. MAXWELL.

Is it nitrate of strontian which becomes doubly absorbent when crystallized out of logwood infusion?

GLENLAIR, SPRINGHOLM, DUMFRIES,
Sept. 7th, 1858.

DEAR STOKES,

I am just finishing my essay on Saturn's Rings†, and there is a problem in it about a set of fluid rings revolving each with its proper velocity with friction against its neighbours. Now I am in want of the *coefficient* of *internal friction* in water and in air, in order to condescend to numbers. I have your paper on pendulums, but it is locked up at Aberdeen, so I have written to you rather than wait till I go.

It is the coefficient B of your paper on elasticity and fluid friction, that is the number of units of force acting tangentially on unit of area sliding with unit of relative velocity past a fluid plane at unit of distance, stating units of force and length.

I think you have got it somewhere at hand. I have not.

It is not long since I began to do mathematics after a somewhat long rest. I have set up a neat model for showing the

---

* *Scientific Papers*, vol. i. pp. 245—262, from *Brit. Assoc. Report*, 1856; and *Trans. Roy. Soc. Edin.* 1857, with application to Earth's *free* precession.

† Adams Prize Essay for period 1855-7; *Scientific Papers*, vol. i. pp. 286—376.

disturbances among a set of satellites forming a ring. It was made by Ramage of Aberdeen, who made my dynamical top.

I have done no optics this summer, but have been vegetating in the country with success.

Yours truly,

JAMES CLERK MAXWELL.

GLENLAIR, SPRINGHOLM, DUMFRIES,

30 May, 1859.

DEAR STOKES,

I saw in the *Philosophical Magazine* of February, '59, a paper by Clausius on the "mean length of path of a particle of air or gas between consecutive collisions," on the hypothesis of the elasticity of gas being due to the velocity of its particles and of their paths being rectilinear except when they come into close proximity to each other, which event may be called a collision.

The result arrived at by Clausius is that of N particles $N\epsilon^l$ reach a distance greater than nl where l is the "mean path" and that

$$\frac{1}{l} = \tfrac{4}{3}\pi s^2 N$$

where s is the radius of the "sphere of action" of a particle, and N the number of particles in unit of volume.

Note.—I find $\qquad \dfrac{1}{l} = \sqrt{2}\pi s^2 N*.$

As we know nothing about either s or N, I thought that it might be worth while examining the hypothesis of free particles acting by impact and comparing it with phenomena which seem to depend on this "mean path." I have therefore begun at the beginning and drawn up the theory of the motions and collisions of free particles acting only by impact, applying it to internal friction of gases, diffusion of gases, and conduction of heat through a gas (without radiation). Here is the theory of gaseous friction with its results

---

* Cf. *Scientific Papers*, vol. i. Editorial notes to pp. 387, 392, 405.

Divide the gas into layers on each side of the plane in which you measure the friction, and suppose the tangential motion uniform in each layer, but varying from one to another. Then particles from a layer on one side of the plane will always be darting about, and some of them will strike the particles belonging to a layer on the other side of the plane, moving with a different mean velocity of translation.

Now though the velocities of these particles are very great and in all directions, their mean velocity, that of their centre of gravity, is that of the layer from which they started, so that the other layer will receive so many particles per second, having a different velocity from its own.

Taking the action of all the layers on one side and all the layers on the other side I find for the force on unit of area

$$F = \tfrac{1}{3} MNlv \frac{du}{dr}$$

where $M =$ mass of a particle, N no. of particles in unit of volume, $l =$ mean path, and $v =$ mean molecular velocity, and $\frac{du}{dr}$ velocity of slipping. Now you put

$$F = \mu \frac{du}{dr}$$

$$\therefore \quad \mu = MNlv = \rho lv.$$

Now
$$v = \sqrt{\frac{8k}{\pi}},$$

and if we take

$$\sqrt{k} = 930 \text{ feet per second}, \quad v = 1505 \text{ feet per second},$$

and if
$$\sqrt{\frac{\mu}{\rho}} = \cdot 116 \text{ in inches \& seconds},$$

we find
$$l = \frac{1}{447000} \text{ of an inch},$$

and the number of collisions per second
$$8,077,000,000,$$
for each particle.

The rate of diffusion of gases depends on more particles passing a given plane in one direction than another, owing to one gas being denser towards the one side, and the other towards the other.

It appears that l should be the same [?] for all pure gases at the same pressure and temperature, but I have found only a very few experiments by Prof. Graham on diffusion through measurable apertures, and these seem to give values of l much larger than that derived from friction.

I should think it would be very difficult to make experiments on the conductivity of a gas so as to eliminate the effect of radiation from the sides of the containing vessel, so that this would not give so good a method of determining l.

I do not know how far such speculations may be found to agree with facts, even if they do not it is well to know that Clausius' (or rather Herapath's) theory is wrong*, and at any rate as I found myself able and willing to deduce the laws of motion of systems of particles acting on each other only by impact, I have done so as an exercise in mechanics. Now do you think there is any so complete a refutation of this theory of gases as would make it absurd to investigate it further so as to found arguments upon measurements of strictly "molecular" quantities before we know whether there be any molecules ? One curious result is that μ is independent of the density, for

$$\mu = MNlv = \frac{Mv}{\sqrt{2}\pi s^2}.$$

This is certainly very unexpected, that the friction should be as great in a rare as in a dense gas.   The reason is, that in the rare gas the mean path is greater, so that the frictional action extends to greater distances.

Have you the means of refuting this result of the hypothesis ?

Of course my particles have not all the same velocity, but the velocities are distributed according to the same formula as the errors are distributed in the theory of least squares.

If two sets of particles act on each other the mean *vis viva* of a particle will become the same for both, which implies, that equal volumes of gases at same press. and temp. have the same number of particles, that is, are chemical equivalents. This is one satisfactory result at least.

I have been rather diffuse on gases but I have taken to the subject for mathematical work lately and I am getting fond of it and require to be snubbed a little by experiments, and I have

---

* *i.e.* inadequate.

only a few of Prof. Graham's, quoted by Herapath, on diffusion so that I am tolerably high-minded still.

I had a colour-blind student last season who was a good student and a prizeman so I got pretty good observations out of him. Any four colours can be arranged, three against one, or two against two, so as to form an equation to a colour-blind eye. I took 6 colours, Red, Blue, Green, Yellow, White, Black, and formed the 15 sets of four of them and tried 14 of these by the help of my student, making no suggestions of course to him, after I had instructed him in the things to be observed.

I then found the most probable values of the 3 equations, which ought to contain the whole 15, and then deducing the 15 equations from the 3 I found them very near the observed ones, the average error being 2 hundredths of the circle on each colour observed, and much less where the colours were very decided to a colour-blind eye. I hope to see him again next year.

I suppose we shall hear something of the theory of absorption of light in crystals from you in due time. I hope Cambridge takes an interest in light in May.

Yours truly,
JAMES CLERK MAXWELL.

GLENLAIR, SPRINGHOLM, DUMFRIES,
1859, Oct. 8.

DEAR STOKES,

I received your letter some days ago. I intend to arrange my propositions about the motions of elastic spheres in a manner independent of the speculations about gases, and I shall probably send them to the *Phil. Magazine**, which publishes a good deal about the dynamical theories of matter & heat. I have not done much to it since I wrote to you, as all my optical observations must be done when I have sun and that is getting lower every day. But I think I have made as much optical hay as will make a small bundle for you as Secretary R. S., some time in winter if you can tell me under what forms it should be sent in. I would

---

* *Phil. Mag.* Jan. and July, 1860, read at Brit. Assoc. Aberdeen, Sept. 1859; *Scientific Papers*, vol. i. pp. 377—409. Formal memoirs on gas-theory, *Phil. Trans.* 1865-6; *Scientific Papers*, vol. ii. pp. 1—78.

call it On the Relations of the Colours of the Spectrum as seen by the Human Eye.

As I have only lately reduced some of my observations I cannot say what more results I may come to, but I shall tell you the present state of affairs that you may judge whether it is better to show it up at once or to keep it a year till the experiments are more numerically formidable.

My experiments have been of this nature. I mix 3 colours of the spectrum so as to produce a white equal to a given white, thus

$$18\cdot4\,(24) + 30\cdot7\,(44) + 30\,(68) = W$$

where the numbers in brackets denote the position of the colour in the spectrum and the coefficient denotes the quantity of that colour. (24) (44) & (68) are the 3 standards of colour which I have taken.

red   green   blue

Then I take any other colour say (28) of my scale and mix with 44 & 68 to produce the same white in quality and intensity as before. I get—

$$16\,(28) + 25\cdot2\,(44) + 29\cdot7\,(68) = W.$$

Subtracting one from the other

$$16\,(28) = 18\cdot4\,(24) + 5\cdot5\,(44) + 0\cdot3\,(68)$$

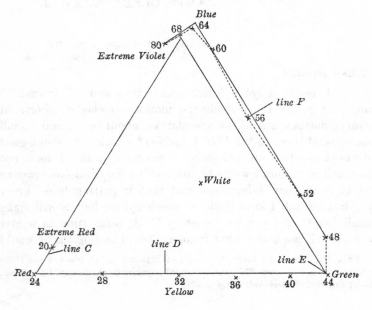

which gives the relation of (28) to the 3 standard colours, and from this I can place it in Newton's diagram when I have fixed the three standards anywhere. In this way I have laid down 16 points of the spectrum with great care from observations by myself and two others, and 15 intermediate points from observations by myself, and I find,

1st, the spectrum forms a curve which is not a circle as Newton imagined but two sides of a triangle with doubtful portions at each end of the third side. This shows that all the colours can be made of three, which are marked (24) (46) & (64). These being the purest colours in the spectrum and therefore in nature may be reckoned standard colours, although they may not coincide with those assumed at first to guide the experiments.

(The position of these colours is different for different observers, and I find *real differences* of large amount between the results of different eyes in certain parts of the spectrum. The figure is from the results of a very good observer.)

I determined the wave-length in air for the different divisions of the spectrum by observing the bands seen by analysing light reflected from a thin plate of air giving a retardation of 41 undulations for the line D, and comparing with Fraunhofer's measures. I thus found for the different divisions

	Wave-length
20	2450
24	2328
44	1951
46	1924
64	1721
68	1688

and so on.

I propose to give a short geometrical theory of Newton's diagram of colours, and to lay down the spectrum upon it, and so to establish a method of ascertaining the colour produced by any mixture of any number of colours in the spectrum. I have come on great differences between different observers but they all agree in having 3 primary colours, though they differ in defining them. I can make a colour which, placed beside white, looks green to one and red to another; the reason being, that the space just before the line F is dim to all but much dimmer to one than another, so that if I mix this with red it will be neutral red and green according to the degree of obscurity of this part of the spectrum. If you think any of these results proper aliment for the Royal Society pray tell me how I may administer it in due form, as I do not purpose to be in London.

The subject is one about which we have had conversation before, so that I have not thought it necessary to do more than indicate new results as you know by what method they must be obtained and how they are expressed in Newton's method.

I remain, Yours truly,

J. CLERK MAXWELL.

After the 25th Oct. address Aberdeen.

MARISCHAL COLLEGE,
ABERDEEN, 1859, *Dec.* 24.

DEAR STOKES,

I have addressed to you as Secretary of the Royal Society an abstract of my paper on the Colours of the Spectrum. I have the paper itself in a presentable form and can send it also if required, but I do not wish it to be considered as finished yet, for I am getting more apparatus and expect to do more before long. If my present paper could be considered as Part I. of a more complete one it would answer best for me. I suppose the Society will send back the MS. of the paper if I send it, so that I may refer to it in preparing a second part, for some of the tables in it are the result of calculations which I do not wish to go over again without good cause, and I have not yet taken copies.

An optician here is making me a reflecting colour apparatus with which I intend to get colour-blind equations among the prismatic colours which will determine accurately the missing colour.

I intend to produce whites with one fixed colour and one variable one, and so find the neutral point of the spectrum and the two points of greatest intensity of colour as seen by my observer. If afterwards Mr Pole should wish to try experiments I have no doubt the instrument could be sent up to him, as I am making it strong, so as to be portable, it is

3 feet $6 \times 9\frac{1}{2}$ in. $\times 5$ in. in dimensions.

I should think a quartz prism of 35° or so with one side ground convex, radius 20 or 30 inches and well silvered, would make a good instrument for analysing the more refrangible rays.

I remain, Yours truly,

J. C. MAXWELL.

MARISCHAL COLLEGE,
ABERDEEN, 1860, *Jan.* 3.

DEAR STOKES,

I have sent you my paper on the Colours of the Spectrum, which is as complete as I hope to make it*. If I do anything more worth publishing it will be upon dichromatic vision, with reference to the spectrum, and I shall not have light for it before March, even if I get my instrument into good order, and find a convenient time for the observer and myself. I have written out the theory of Newton's diagram, and its relation to the method of representing colours by lines, drawn from a point in various directions. I have described the method of observation and its results as given by two observers, and explained what I believe to be the reason of the differences between them, which are very far beyond any errors of observation, but appear to me to arise from unequal absorption of certain rays by media which are more absorptive in some eyes than others, and not from any radical difference in the sensations excited by the different rays.

Though I expect to get better observations by taking more pains, I believe the results which I send you to be not far from the truth, and certainly I think I have evidence that the spectrum on Newton's diagram is a triangular figure, with one side broken out. So that if the R. S. is willing, I am ready to put my name to the paper and have it printed. I have copied out all that I shall need in the way of tables, &c., so I shall not want it back. I have addressed it to Cambridge as I do not know whether the Royal Society meets about this time.

I remain, Yours truly,

J. CLERK MAXWELL.

* Bakerian Lecture read March 22, 1860; *Scientific Papers*, vol. i. pp. 410—444. Also Roy. Inst. lecture, May, 1861.

MARISCHAL COLLEGE,
ABERDEEN, 1860, *Feb.* 25.

DEAR STOKES,

I have received yours of yesterday and am much honoured by the Council of the Royal Society in being appointed Bakerian Lecturer.

I am not aware of the constitution or laws of the Bakerian Lectureship, but from what I have read under the title of Bakerian Lectures I suppose that the Lecturer reads in the ordinary way the paper which the Council have honoured with the title of Lecture, and that his duties are then over.

I cannot leave Aberdeen till April, but after that I shall be able to go to London.

I should myself prefer the 19th April if that would suit the Society.

The instrument I used in summer is not at all portable but I have now got a reflecting one,

3 ft. 6 in. × 11 in. × 4,

which is warranted to be safe when turned upside down and moderately jolted. I have not had much sun but I find that a certain uniform cloud which we have sometimes is not bad for observations, and I have had one set of observations from a colour-blind gentleman who spent a day in Aberdeen and two sets from a colour-blind student from whom I expect about 3 sets a week till the degree examinations interfere with optics.

They are both perfect cases like Mr Pole's, and the student is a good observer. Both agree that the neutral point of the spectrum is near *F* and the student sees a dark cloud in the axis of vision when he looks at this light just as I do myself.

All less refrangible light is yellow, all more refrangible blue, and by equating white to selected pairs of colours from each side of the spectrum, I propose to determine the brightness and blueness and yellowness of every point of the spectrum, in terms of two selected colours. Before April I shall have probably ascertained the accuracy of the observations and got some normal observations with the same instrument to compare with them.

If I should be able to arrive at results and to communicate them to the Society before I go up, would it be lawful to append them (with proper date) to the former paper ?

I suppose I had better prepare one or two diagrams and bring the instrument with me.

If I have mistaken what I have to do pray let me know, and if the 19th April will not do, then any day after that will do for me.

I remain, Yours very truly,

J. C. MAXWELL.

MARISCHAL COLLEGE,
ABERDEEN, 1860, *March* 12.

DEAR STOKES,

If it is found necessary for the convenience of anyone to have the day changed from the 19th April, I am willing to come on the 26th, but I cannot go to London at all at any later time. I should prefer the 19th to the 26th if it should turn out equally convenient for Dr Sharpey to have his lecture at a different time, but if the 19th secures advantages to the Croonian Lecture which are not to be had afterwards (such as the presence of a man like Kölliker) then do not scruple to change my day to the 26th as my reasons are all private. After the 26th I should prefer to have it read in absence by deputy, for I do not think I can go up after that time.

What are the rules about one who is not a fellow reading a paper? I see the Bakerian lecturer is called one of the fellows in the foundation, but I suppose from what you say that an extraneous person may deliver it.

I have seen a large diagram of the spectrum with the fixed lines in various lecture rooms. Do you think I could get one in London to point to during lecture in order to indicate the colours to be mixed?

I hope to have the triangle of colour ready on a large scale in time for the lecture.

My colour-blind student has been ill some time and I have got no more observations.

I remain, Yours truly,

J. CLERK MAXWELL.

I have had a letter from a colour-blind man in Manchester and expect to hear more of his case in answer to my reply.

Care of J. MacCunn, Esq.
Ardhallow, Dunoon.
23 *April*, 1860.

Dear Stokes,

I had hopes when I last wrote to you that I should have got a considerable number of colour-blind observations of the spectrum, from a last year's student. Unfortunately he fell ill just when the bright weather came and only recovered at the end of the session, so I got his observations for two days only, but as these seem pretty consistent perhaps they might be worth reducing and publishing till some one finds a better opportunity to observe. The results are very similar to those of another gentleman whom I tried on a cloudy day. Here are the results.

Taking a green at the line E and a blue $\frac{2}{3}$ from F towards G as arbitrary standard colours, the first appears "yellow" the second "blue" to the colour-blind. The red end of the spectrum is visible as far to them as to us, using a blue glass to cut off the bright part of the spectrum, but the colour is "yellow" apparently of the same kind as that produced by a very narrow aperture at E in the green. In order to apply my ordinary method we must have enough of the colour to produce the standard white when combined with blue, and this cannot be done without taking all the light from the end of the spectrum to the line D.

Now this quantity of light which to our eyes is much brighter than the standard white, requires an addition of blue to make it equal to white to the colour-blind, showing that the red end is very feeble in its effects upon them.

The quality of the colour of this end of the spectrum, as shown by the quantity of blue required to produce the standard white, seems to be nearly uniform from the extremity as far as a point half-way between E & F and is what the colour-blind call "yellow." The quantity or intensity as shown by the amount of the given colour required to neutralise the blue is very small up to D, is greatest at $\frac{2}{3}$ from D to E, and then diminishes.

This part of the spectrum therefore is nearly uniform in colour, and any part may be taken as a representative of "yellow" by a colour-blind person.

From $\frac{1}{2}$ between E & F to $\frac{1}{4}$ from F towards G the colour is to them a mixture of blue and yellow, being neutral close to F.

Near F the intensity is feeble, so that three times the breadth of slit is required to produce white when compared with the breadth at the standards. Also in viewing a uniform field of this colour the "yellow spot" of the retina is seen dark on a light ground.

From the point $\frac{1}{4}$ from F to G to the end of the spectrum the colour seems quite pure blue, increasing in intensity to $\frac{2}{3}$ from F to G and then diminishing, but being visible as far as we see it and apparently as bright.

Comparing my observations with these I find

	Red	Green	Blue
J.C.M.	$22\cdot6 +$	$26\cdot0 +$	$37\cdot4 = W$
J.S.		$33\cdot7$	$33\cdot1 = W$

$$D = 22\cdot6\,(R) - 7\cdot7\,(G) + 4\cdot3\,(B).$$

The colour which if it could be exhibited we would see and they would not is a red with a little blue, and purged of that green (or "yellow") which renders it still visible to them.

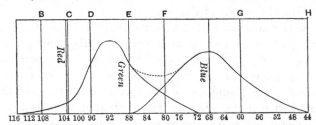

I send you a plan of the spectrum according to my student, showing the intensity of blue & of yellow at different points of the spectrum. I find my new instrument is not altered by being carried about so that if I find any more subjects I can compare them wherever they are.

I remain, Yours truly,

J. CLERK MAXWELL.

After the 30th April my address will be Glenlair by Dumfries.

The curve on the right represents the amount of the "blue" element and that on the left the "yellow" element of colour in different parts of the spectrum as seen by the colour-blind. The dotted line is the sum of the elements.

GLENLAIR, SPRINGHOLM, DUMFRIES,
May 7, 1860.

DEAR STOKES,

I sent you this morning an account of observations by
Mr James Simpson, on the colours of the spectrum, and also
a comparison of his observations on coloured papers with mine.
It is not very easy to find colour-blind persons who know what is
required to make good observations. If Mr Simpson had not
been in ill health, and by no means strong when he observed the
spectrum, I would have been able to send you colour-blind equations
better than most ordinary ones. However they are really very
good and certainly correct in the main features. I still believe
that the great difficulty in comparing eyes of either the dichromic
or the trichromic type lies in the absorptive power of the media
of the eye, especially the yellow spot, the seat of " Haidinger's
brushes."

Observations of colours near the line F are extremely unsatisfactory, as the brightness and even colour of the field vary with
the direction of the axis of the eye.

It is very remarkable that this peculiarity should be confined
to a small portion of the spectrum while neighbouring parts, *very
similar in colour*, show nothing of the kind. I expect that if
I could get enough of the yellow material of the spot to try it,
it would cut a dark band out of the spectrum near F, and that in
a very marked manner. Do you know whether there is any other
more abundant yellow substance of the same composition in the
animal kingdom ? for if it is peculiar to a small spot on the
thinnest part of the human retina, my conscience recoils from
the amount of slaughter necessary to verify my conjecture.

With respect to comparisons of brightness of dissimilar colours.
I find all my observers decline to say which of two very different
colours is brightest. I have made attempts to compare white
with pure colours, and the result is that if I make red (*e.g.*) and
white equal in brightness (that is = in impressing my eye with
a sense of something being there), I find that the red requires
green and blue to be added to it to neutralise it and still is not
brighter than the white with which it was compared; thus if

$$R = W$$

& $$R + G + B < W$$

something must be wrong. I think positive colours have more justice done them by the eye than white light.

I was some time ago doubtful about presenting myself as a candidate to the R. S. If you could get me placed as a candidate for admission according to the rules of the society I should be much obliged to you, in addition to the thanks I owe you for your advice as to my paper and other optical cooperation.

Yours truly,

J. CLERK MAXWELL.

GLENLAIR, DALBEATTIE, DUMFRIES.

DEAR STOKES,

I have just received your letter of the 19th May which went astray owing to its being addressed to Aberdeen. My safest address is always as above but I have written to Aberdeen to put them right.

I wrote an abstract of my postscript and sent it to Dr Sharpey.

In my paper I avoided making any definite statement as to the view taken by Sir D. Brewster of the composition of light. I think he regards the three elements of colour as objective realities or actual substances of which the light is compounded. I consider them as properties of homogeneous rays having reference to the human organ of vision whether dichromic or trichromic.

On the one theory it is conceivable that absorption may alter the proportions of the elements in a ray of the pure spectrum. On the other theory it can only have the same effect as weakening the light in any other way.

This is the field of a controversy into which I have not entered, as I have not yet sufficient experiments to prove that absorption always acts in the same way as simple weakening of the light.

I also avoid stating Sir D. Brewster's method of drawing curves, as he is not very distinct about them, and in cases of this sort I think that his methods of observation are sometimes much better than his description seems to indicate, so that if you go by the description merely you will find that you are under-rating the forces of the enemy and may require to modify a statement.

I shall endeavour to state the difference of the two theories by making one very distinct and stating the other in its own words only.

II. J and K are two different observers, I shall try to distinguish them properly when I get proofs.

III.   Unless the curve of the spectrum has angular points no colour in it represents a primary.

I fear oxen and sheep have no yellow spot, at least I never found it.   They have plenty of glittering tapetum but I must consult anatomists to see what animals have the yellow spot.

I have not been in Cambridge since you left Pembroke, so I am not sure whether your residence derives its name from the lenticular form of its enclosure, and if I address right under that impression.

I think it would be worth while to select some portions of the spectrum and determine their heating power, also their action on various chemicals, and their red, green and blue luminous elements and their fluorescent effects on various things, so as to prove that if the wave-length and the heating power remain the same everything else will be the same, whatever modifications the light may have undergone.

Yours truly,

J. C. MAXWELL.

GLENLAIR, DALBEATTIE,

1862, Sept. 10.

DEAR STOKES,

I send you reports on Dr Robinson's papers.   I did not know that the reciprocating discharge in a closed glass vessel had been tried and found to exhibit the lines of the constituents of glass.

I have been comparing the results of colour observations by different eyes in order to see whether they can be reduced to a single diagram by altering the mode of projection, that is taking different units of the standard colours.

I find that I get consistent results in the three cases I have tried but that the numbers by which the coordinates must be multiplied vary from one to three, so that if we suppose one person to see blue three times stronger relative to green than another person, we get consistent results. I find that the centre and circumference of my retina differ, so that white appears bluer to the retina generally than to the centre and therefore the colour complementary to red is of higher refrangibility for the retina than for its centre. I cannot perceive any difference in merely looking at white paper, but I suppose any constant difference in different parts of the retina can be discovered only by reason as it would not be an object of perception. By choosing a standard eye, or taking an average, I could express the facts of colour vision in two diagrams :

1st, a triangle of colour giving relations in the quality of colours and the same for all eyes.

2nd, a curve representing the intensity of colour at each point of the spectrum. This curve is different for different eyes and for different parts of the same eye. These differences may arise from absorption of certain rays before they reach the retina. Irregularities in the curve extending over small spaces are probably of this sort. There may also be differences in the sensibility of the nerves to particular sensations. This would account for inequalities extending over entire regions of colour in a regular manner.

The fact that all colours lie nearly in two straight lines makes it impossible to assign the exact position of the sensation defective in the colour blind. It lies in the production of the line joining green with the extreme red, but its exact position differs according to the normal person with whom the colour-blind eye is compared, and colour-blind eyes differ as much with regard to intensity of different colours as normal eyes do.

Yours truly,

JAMES CLERK MAXWELL.

8 PALACE GARDENS TERRACE, LONDON, W.
1864, *May* 6.

DEAR STOKES*,

I have your letter and my paper. Fizeau does not lean with confidence on any theory of the dependence of the rotation

* Maxwell's memoir on the Electrodynamic Field was communicated to the Royal Society on October 27, 1864; cf. next following letter. In a letter to Sir W. Huggins of date 1867, printed in *Phil. Trans.* 1868, p. 532, he considers the general problem of the influence of the Earth's motion on optical phenomena. For further references cf. Larmor, *Aether and Matter*, p. 14.

of the plane of polarization on the velocity-ratio of propagation in the media.

The rotation of course depends on the ratios of *intensity* of rays polarized in different planes and passing through the inclined plate.

You have proved that on Fresnel's theory the *direction* of the refracted ray is not affected by the motion of the aether.

Have you or Jamin or anyone else done anything towards the determination of the intensity ?

Fizeau rests his calculation on experiments with plates of different indices of refraction.

I am not inclined and I do not think I am able to do the dynamical theory of reflexion and refraction on different hypotheses: and unless I see some good in getting it up, I would rather gather the result from men who have gone into the subject.

If Fizeau has really found a phenomenon related to the earth's motion in space or rather in the luminiferous medium, a great deal may be founded upon it independently of a good optical theory.

I think M. Faye corroborates Fizeau's experiment. If the experiment is good it would be worth the while of an eminent astronomical observer to have the instrument properly mounted, and to work for a year determining frequently the direction and (proportional) velocity of the rush of aether through his observatory, and so to have a log book of the earth's motion.

I have been reading Fraunhofer on the spectrum, and making estimates of the intensity of the pure colours by bringing each to an equality with my standard white light. I used to try this by making an adjustment and then observing, then making another adjustment and so on, and I got very inconsistent results. Now the adjustment is made while I observe, and I find the results not very inconsistent, not worse I think than Fraunhofer's. Has any one else repeated such measurements ? I can do them quickly, and mean to get a good many to compare.

Dove believes in the comparison of intensity of light of different colours by his photometer with a photographic print.

I mean to try it with two half-square prisms joined

with Canada Balsam laid on in narrow streaks so that certain portions are transparent and others reflecting*.

Yours truly,

J. CLERK MAXWELL.

8 PALACE GARDENS TERRACE, W.
1864, *Oct.* 15.

DEAR STOKES,

I have been reading Jamin's note on the Theory of Reflexion and Refraction, *Ann. de Ch.* 1860, pt. i. p. 413.

I am not yet able to satisfy myself about the conditions to be fulfilled at the surface except of course the condition of conservation of energy.

Jamin insists on the equality of the motion both horizontal and vertical in the two media. I do not see the necessity for equality of motion; but I think action and reaction must be equal between the media, provided the media pure and simple vibrate and nothing along with them.

If the gross matter in each medium does not vibrate, or has a different phase and amplitude from the ether, then there will be six relations between the four quantities :—two portions of ether and two kinds of gross matter.

Have you written anything about the rival theories of reflexion? or can you tell me of any thing you agree with or eminently differ from on that subject? I think you once told me that the subject was a stiff one to the best skilled in undulations.

Jamin deduces (p. 422) from his conditions of equality of motion in the two media for vibrations in the plane of incidence that the density of the medium is the same in all substances.

That is to say he gets this by pure mathematics without any experiment.

Or according to him no such vibrations could exist in the media unless they were of equal density.

This I think simply disproves his original assumption of the equality of the displacements in the two media.

* This is the principle of the Swan-Brodhun photometer.

In fact the equality of displacements combined with the equality of energy involves the equality of density.

Therefore the general theory, which ought to be able to explain the case of media of unequal density (even if there were none such) must not assume equality of displacements of contiguous particles on each side of the surface.

I suppose if two media of different density were glued hard together and large vibrations sent through them they would be separated at their common surface.

But there is nothing in the surface of separation of two media analogous to this gluing together that I can detect.

I have now got materials for calculating the velocity of transmission of a magnetic disturbance through air founded on experimental evidence, without any hypothesis about the structure of the medium or any mechanical explanation of electricity or magnetism.

The result is that only transverse disturbances can be propagated and that the velocity is that found by Weber and Kohlrausch which is nearly that of light. This is the velocity with which such slow disturbances as we can make would be propagated. If the same law holds for rapid ones, then there is no difference between polarized light and rapid electromagnetic disturbances in one plane.

I have written out so much of the theory as does not involve the conditions at bounding surfaces, and will send it to the R. S. in a week.

I am trying to understand the conditions at a surface for reflexion and refraction, but they may not be the same for the period of vibration of light and for experiments made at leisure.

We are devising methods to determine this velocity = electromagnetic $\div$ electrostatic unit of electricity. Thomson is going to weigh an electromotive force. Jenkin and I are going to measure the capacity of a conductor both ways and I have a plan of direct equilibrium between an electromagnetic repulsion and electrostatic attraction.

Yours truly,

J. C. MAXWELL.

8 Palace Gardens Terrace, W.
1866, *Dec.* 18.

Dear Stokes,

I enclose a report on Mr Tarn's paper and Thomson's report on mine*. Thomson's theory of matter being continuous but much more dense at one place than another is a molecular theory, for the dense portions constitute a finite number of molecules: and I do not assert nor does any one since Lucretius that the space between the molecules is absolutely empty.

I take Statics to be the theory of systems of forces without considering what they act on, Kinetics to be the theory of the motions of systems without regard to the forces in action, Dynamics to be the theory of the motion of bodies as the result of given forces, and Energetics to be the theory of the communication of energy from one body to another. I therefore call the theory a dynamical theory because it considers the motions of bodies as produced by certain forces.

I think I have stated plainly enough the difference between my molecules and pure centres of force. I do not profess to have a dynamical theory of the molecule itself.

I have considered the equilibrium of heat in a vertical column and find I made a mistake in equation 143. I now make the temperature the same throughout, and have inserted (on two pages) an addition to that effect which I mean to be instead of p. 60. I have also amended equation 54 in which the effect of external forces was omitted by mistake in writing out and copying from equation 47.

The result is a corroboration of the theory of the distribution of velocity as the errors are in the theory of " Least Squares "; for we require to have

$$\overline{\xi^4} = 3\,\overline{\xi^2} \cdot \overline{\xi^2},$$

which agrees with that theory.

I think something might be done to the statical theory of elasticity with centres of force whose displacement is a function of the index of the molecule as well as of its initial coordinates,

----
* "On the Dynamical Theory of Gases," *Phil. Trans.* May, 1866.

the number of kinds of indices being small and the groups of n particles being originally of the same form, but it would require an investigation to itself *.

I remain, Yours truly,

J. CLERK MAXWELL.

I shall be at Cambridge from 26 Dec. to 26 Jan. at Trinity.

8 PALACE GARDENS TERRACE, W.

1867, *Feb.* 27.

DEAR SIR,

I find that I have made two mistakes in a paper I have sent to the R. S.† about Siemens' and Wheatstone's machines.

It is all right about the *first* kind of commutator.

In finding the effect of the second kind of commutator which breaks one of the circuits I have omitted to consider the effect of the stoppage on the other coil by induction. I send you the result taking the inductive effect into account.

The second mistake was in the case of two currents always disconnected. Such a system cannot maintain currents by its motions.

The result I came to arose from neglecting the inductive effect of stopping the current.

In fact a closed circuit cannot have a current always in one direction produced by any inductive action.

The nearest approach to it is to take a long flexible magnet and wind it on the coil like as a silk is wound to cover a wire. During the winding a current will go in one direction and during the unwinding it will go in the other direction.

The part relating to two independent coils is therefore all wrong.

If you can allow that part to be cut out, and the enclosed to be substituted for what relates to commutators of the second kind, it will prevent a false statement from going beyond you and me.

Yours truly,

J. CLERK MAXWELL.

---

* See *infra*, p. 36.   † *Roy. Soc. Proc.* 1867.

GLENLAIR, DALBEATTIE.

June 26, 1869.

DEAR SIR,

As you have studied series of periodic functions I should like to know from you whether there is any method of finding the coefficients of such a series

$$S = A_0 + A_1 \cos \theta + A_2 \cos 2\theta + \ldots + A_n \cos n\theta$$
$$+ B_1 \sin \theta + B_2 \sin 2\theta + \ldots + B_n \sin n\theta,$$

so that between the limits $\theta = 0$ and $\theta = \alpha$ the series S shall have values given in terms of θ, and between the limits $\theta = \alpha$ and $\theta = 2\pi$ some other series depending on S such as $dS/d\theta$ shall have values given in terms of θ.

For instance, can we find the values of the coefficients A in the series such that $A_0 + A_1 \cos \theta + A_2 \cos 2\theta + \ldots + A_n \cos n\theta$ shall be constant, $= C$, for values of θ of the form $\theta = 2\pi n \pm \phi$, when ϕ is less than α and at the same time $A_1 \cos \theta + 2A_2 \cos 2\theta \ldots + nA_n \cos n\theta$ shall be zero for values of θ not within the above limits ?

The question arises from the application of Fourier's theorem to the determination of the potential of an electrified grating consisting of a number of parallel strips all in one plane, the breadth of each being $2a\alpha$, and the intervals between them being

$$2 (\pi - \alpha) a.$$

The potential is of the form

$$V = A_0 + A_1 e^{-\frac{y}{a}} \cos \frac{x}{a} + \&\text{c.} + A_n e^{-n\frac{y}{a}} \cos \frac{nx}{a},$$

with the conditions $V = C$ on the strips and $dV/dy = 0$ in the intermediate spaces.

I have succeeded by a different method (that of transformation by conjugate functions) in finding the effect of interposing a grating of fine parallel wires at potential zero between a plane at potential zero and an electrified parallel plane.

If E be the electrification produced through the grating and E' that produced when the grating is away then

$$\frac{E'}{E} = 1 + \frac{\dfrac{1}{\alpha}}{\dfrac{1}{b_1} + \dfrac{1}{b_2}},$$

where b_1 and b_2 are the distances between the grating and the two planes and α is a line equal to $\dfrac{a}{2\pi}\log_e\left(\dfrac{1}{2}\operatorname{cosec}\dfrac{\pi c}{a}\right)$, a being the distance between the axes of the wires and c their radius. This is only true when a is much greater than c and when b_1 and b_2 are each much greater than a. I have since got a method of finding the complete solution for cylindrical wires in a series of the form $A_0 F + A_1\dfrac{dF}{dy} + A_2\dfrac{d^2F}{dy^2} + $ &c., where

$$F = \log\left(e^y + e^{-y} - 2\cos x\right).$$

Can you tell me the value of the infinite series

$$\frac{1}{1\,.\,(1+\alpha)} + \frac{1}{2\,.\,(2+\alpha)} + \frac{1}{3\,.\,(3+\alpha)} + \text{\&c.,}$$

where α is any fraction?

Excuse my troubling you with these questions*, I wish you to answer them only if you can do so without trouble.

Yours truly,

J. CLERK MAXWELL.

Glenlair, Dalbeattie.

8 July 1869.

Dear Sir,

I am very much obliged to you for your information about Fourier's series, &c.

I am afraid I shall not be able to attend the meeting of the B. A. at Exeter† which I hope will be a great success. Jenkin is I believe with the Great Eastern and will probably be busy for some time. Balfour-Stewart knows most about the resistance experiments, and Dr Joule has got the whole subject up independently in a most thorough manner so as to correct several numerical mistakes in the Report.

I know how to reduce the problem I sent you to a case of integration, by a method which Thomson has applied to the complete solution of the electrification of a segment of a spherical shell.

---

* See *Electricity and Magnetism*, vol. i. (1873), part i. chapters xi. xii.

† British Association Meeting, under Prof. Stokes' presidency.

I first find by this method the distribution on an arc of a cylindric surface, and then putting this in cylindric coordinates r and θ, the functions are periodic in θ. Then putting $r = ce^{\frac{2\pi x}{a}}$ and $\theta = \dfrac{2\pi y}{a}$, the functions become suited for rectangular coordinates and periodic in y.

The curves on the opposite page* represent the equipotential surfaces and lines of force due to an infinite series of *wires*, of which one is given at A, placed parallel to a plane on the left hand which has an equal and opposite charge on its opposed surface.

Yours truly,

J. CLERK MAXWELL.

GLENLAIR, DALBEATTIE.
11 Jan. 1871.

MY DEAR STOKES,

I received the copy of the Adams Prize Essay some time ago, and have been gradually getting into the subject.

Did not you set the theorem about the surface integral

$$\left(\left(\frac{d\gamma}{dy} - \frac{d\beta}{dz} \right) dy\,dz + \dots + \right)$$

over a surface bounded by the curve s being equal to

$$\left(\alpha \frac{dx}{ds} + \beta \frac{dy}{ds} + \gamma \frac{dz}{ds} \right) ds?$$

I have had some difficulty in tracing the history of this theorem†. Can you tell me anything about it?

Yours truly,

J. CLERK MAXWELL.

I hope you saw the eclipse well.

---

* Not reproduced : given in *Electricity and Magnetism*, vol. i. plate xiii.

† In *Electricity and Magnetism*, vol. i. p. 16, this theorem, fundamental in regard to the vectors of physics, is traced to Smith's Prize Examination Papers for 1854, and the result has since been known as Stokes' Theorem. It occurs however also in a letter from Lord Kelvin to Prof. Stokes of date 1850 : see Stokes' *Math. and Phys. Papers*, vol. v. p. 320.

GLENLAIR, DALBEATTIE.
8 *Jan.* 1872.

DEAR PROFESSOR STOKES,

I send you a paper for the Royal Society in which a peculiar kind of image of an electromagnetic system is shown to be formed by a plane sheet of conducting matter.

If a man on board ship were to drop into the sea every second a piece of lead and a piece of cork attached to each other by a string of such a length that, in sinking, the cork is always one second in the rear of the lead, the series of plummets and corks would form a kind of trail in the rear of the vessel which is always sinking with uniform velocity while it retains its shape as a whole.

This gives a mental image of the trail of images of an electromagnet in a conducting sheet. The plummets are positive images and the corks negative images.

N.B. The paper is not offered for the *Transactions*, as it is to be put into a slightly different form in a separate book *.

Yours very truly,

J. CLERK MAXWELL.

11 SCROOPE TERRACE, CAMBRIDGE.
12 *Feb.* 1872.

MY DEAR STOKES,

I send you a note on my paper on induction in a plate, making mention of the researches of Felici and Jochmann, which I could not refer to in the country.

The mutual induction of the induced currents must not be neglected when the relative velocity of the magnet and plate is comparable with V, which, for a copper plate 1 mm. thick is about 25 metres per second and for a thicker plate is less; so that the secondary phenomena described in the paper ought to occur, as indeed they do with a thick plate and a high speed, *i.e.*

Primary phenomenon,	Tangential dragging,
Secondary „	Repulsion from disk,
Tertiary „	Attraction towards axis.

Yours truly,

J. CLERK MAXWELL.

* *Electricity and Magnetism*, vol. ii.

GLENLAIR, DALBEATTIE.
8 *July*, 1872.

MY DEAR STOKES,

The figures in Mr Latimer Clark's paper are better than anything I could do. If therefore I can obtain the consent of the Society and of Mr Clark, I should like to obtain clichés of the engravings of the Electrodynamometer and its parts*, and in doing so I have supposed it best to begin with the Society, for though I am acquainted with Mr Clark, I have no direct relation with him in the capacity of referee of his paper.

I had already drawn figures, not so good as Mr Clark's, but they are not yet engraved, so that if there is any objection to my getting either the loan of the blocks or plates, or impressions of them, my work will not be delayed.

As some of the figures are already engraved I should think it likely that Mr Clark means to use them in some work of his own, and that they have been engraved independently of the R. S. though they may first appear as illustrations in the *Phil. Trans.*

With respect to the time at which I should have to send my last figures to the engraver, I think the end of September is about the limit of safety.

Yours sincerely,

J. CLERK MAXWELL.

GLENLAIR, DALBEATTIE.
3 *August* 1875.

DEAR STOKES,

I have done nothing about the reflective power of gold, silver, or platinum, but am willing to continue my labours if so required.

Browning is making a divided circle to carry a Jellett's prism, a $\frac{1}{4}$ undulation plate or a Babinet's compensator, any or all of them, and it is to be capable of being fastened to an inclined arm so as to examine the ray reflected from a liquid.

Yours very truly,

J. CLERK MAXWELL.

* For the *Electricity and Magnetism*.

I have been thinking about Dr Andrews' experiments on mixtures of N and CO_2.

I find the conditions of equilibrium between two mixtures in different states to be*:—

Let e = energy of unit of mass of mixture ($e = f(v, \phi, q_1, \ldots q_{n-1})$),

v = volume of do.

ϕ = entropy of do.,

$q_1, q_2 \ldots q_{n-1}$ the masses of the substances $_1, _2 \ldots _{n-1}$ in unit of mass of the mixture (there being n constituents).   Then

$$\text{pressure} = p = -\frac{de}{dv},$$

$$\text{temperature} = \theta = \frac{de}{d\phi},$$

$$\text{"reaction,"} = r_1 = \frac{de}{dq_1},$$

$$r_2 = \frac{de}{dq_2},$$

$$\&\text{c.,}$$

and the conditions of equilibrium between two mixtures, one of which is distinguished by an accent, are

$$p = p',$$
$$\theta = \theta',$$
$$r_1 = r_1',$$
$$r_2 = r_2',$$
$$\&\text{c.,}$$

and

$$e + pv - \theta\phi - r_1q_1 - r_2q_2 - \&\text{c.} = e' + pv' - \theta\phi' - r_1q_1' - r_2q_2' - \&\text{c.} \ldots$$

* These are precisely the general analytical conditions for the equilibrium of coexistent heterogeneous substances that were formulated by Willard Gibbs in his epoch-making memoir, of which the first part was communicated to the Connecticut Academy in October 1875. Prof. Gibbs employed the term 'potential' of a constituent substance for what is here provisionally named 'reaction.' The theory was probably elaborated by both Gibbs and Maxwell from Gibbs' Thermodynamic Surface, of which an account was inserted by Maxwell in the fourth edition (1875) of his *Theory of Heat* (pp. 195—208), giving the method he employed in constructing the surface to scale for water-substance. On March 8, 1876, Maxwell, then President of the Cambridge Philosophical Society, communicated to the Society an account of Gibbs' methods "which seem to me to throw a new light upon Thermodynamics" (*Proceedings*, vol. ii. pp. 427—430). The special problem which here suggested to Maxwell this general investigation, has been developed with much success by van der Waals and his pupils on the same lines as those indicated above.

These conditions seem to me to be all right. The difficulty is to conceive clearly the energy as a function of volume, entropy, and composition, say in liquid CO_2 saturated with absorbed N in contact with a gaseous mixture of CO_2 and N.

But it seems plain that if the thermodynamic surface (co-ordinates, volume entropy and energy) has a valley or hollow place which works itself out at the critical point, after which the surface is convexo-convex, and if the surface for N is everywhere convex, then the effect of mixing N with CO_2 will be to make the head of the valley more convex, that is it will work itself out sooner, or the critical point will be lowered.

But the difficulty is to see how to form the surface for a mixture when those of the constituents are given.

However, I must interpret my conditions in this case, and in that of two liquid mixtures, as of benzol and (alcohol and water), and in solutions of salts (supersaturated &c.), and in solutions of gases (supersaturated &c.).

GLENLAIR, DALBEATTIE.

9 *July* 1875.

DEAR STOKES,

I have no objection to Sir W. Thomson seeing my report.

What I meant about the third paper was that the quarto paper on the Electrodynamic Qualities of Metals would suffer greatly if the new facts about very soft iron were not worked up and inserted into it. If the magnetic molecules have got a slight preponderance of set in one direction, then a strain of extension in that direction would (in a homogeneous medium) tend to increase that set. But this cannot be the explanation for the "unital extension" is small compared with the percentage increment of magnetization.

We arrived here on 25th June breaking the journey at Carlisle. Mrs Maxwell stood the journey very well, and is making steady progress. We have had a week of hot weather, but to-day the rain is heavy and constant.

Yours sincerely,

J. CLERK MAXWELL.

GLENLAIR, DALBEATTIE.
25 *Sept.* 1875.

MY DEAR STOKES,

I quite concur with you that Mr Gore's paper should now be printed *in extenso* in the proceedings.

Mrs Maxwell has been keeping pretty well and has been able to get out on her pony.

Would you agree with the following statements about elasticity and viscosity, as related to a molecular theory?

When after being strained the groups of molecules in a body tend to return to the same stable configuration as when unstrained, the body is elastic. If at corresponding stages of the straining and restitution the stresses are the same, the body is perfectly elastic. If the stress during restitution is less than that during deformation the elasticity is imperfect.

If, when the strain exceeds a certain value, complete restitution does not occur, this value is called the limiting strain and the stress the limiting stress of elasticity.

If, when the stress is removed, the body does not completely return to its original form the body is said to be plastic or viscous. A viscous body, if kept strained long enough, loses all tendency to change its form. A plastic body does not permanently change its form unless the stress exceeds a certain value, and if kept strained it never loses all stress.

THEORY.

In an elastic solid the thermal agitation of the molecules does not carry them beyond the limits of oscillating about stable configurations.

But as the thermal agitation increases so many molecules per second are thrown out and oscillate about a new configuration, the nature of which is determined by the present form of the body and not by its unstrained form*.

The greater the strain it is probable that more molecules will be so thrown out; but the number is not proportional to the strain, but varies very little for small strains, and suddenly rises enormously for breaking strains.

* Cf. Maxwell's article " Constitution of Bodies," *Ency. Brit.* 9th ed., *Collected Papers*, vol. ii. p. 624, referred to by Rayleigh in connexion with Ewing's model of magnetic hysteresis; the next page gives a development of his ideas.

If $1/l$ of the molecules are thrown out in unit of time from a state of strain represented by e, f, g, a, b, c, see Thomson and Tait, and if these molecules enter into the state $\frac{1}{3}\theta, \frac{1}{3}\theta, \frac{1}{3}\theta, 0, 0, 0$, where
$$\theta = e + f + g,$$
then*
$$\frac{de}{dt} = -\frac{1}{l}\left(e - \frac{1}{3}\theta\right) + \frac{du}{dx},$$
$$\frac{da}{dt} = -\frac{1}{l}a + \frac{dv}{dz} + \frac{dw}{dy};$$
and if k and n are the coefficients of elasticity so that
$$P = (k - \tfrac{2}{3}n)\,\theta + 2ne \quad \text{and} \quad S = na,$$
the equations of motion will be of the form
$$\frac{1}{\dfrac{1}{l} + \dfrac{d}{dt}}\left\{\left[\frac{k}{l} + \left(k + \frac{n}{3}\right)\frac{d}{dt}\right]\frac{d\theta}{dx} - {}^{\cdot}n\nabla^2 u\right\} + X - \rho\frac{Du}{Dt} = 0,$$
with 2 others and
$$\frac{d\theta}{dt} = \frac{du}{dx} + \frac{dv}{dy} + \frac{dw}{dz}.$$
If l is small we may put
$$p = C - k\theta \quad \text{and} \quad nl = \mu$$
and the equation becomes
$$-\frac{dp}{dx} - \mu\nabla^2 u + X - \rho\frac{Du}{Dt} = 0.$$

Yours very truly,
J. CLERK MAXWELL.

[1875.]

DEAR STOKES,

.... The exploring of a liquid electrolyte appears to have been original and there was plenty of good work all through, but it was of the kind which does most good to the worker himself, except in so far as it encourages other people to work also.

I would say the same of all experimental verifications of mathematical formulae founded on Ohm's law except (1st) experiments to determine specific resistances, intensities of polarization &c., or in short to evaluate physical quantities, and (2nd) experiments to test whether or not Ohm's law is true to 10^{-8} as we know it is to 10^{-4}.

---

* Here (u, v, w) is velocity. The signs of the terms involving $\nabla^2 u$ below should be changed, unless $\nabla^2 u$ has itself the quaternionic sign.

Scotch, a verb or a termination as in Butterscotch, an English delicacy the nature of which I do not know. Perhaps the last t should be r.

Scots, native form.

Scottish, English form.

Yours very truly,

J. CLERK MAXWELL.

GLENLAIR, DALBEATTIE.

7 *July* 1876.

MY DEAR STOKES,

I think I read the abstract of the paper on the dragging of air by water in the *Proc. R. S.*

I agree with you that the experiments are good experiments, but that the plan of experiment is not such as to lead to any result of great scientific importance. I remember verifying ——'s formula as given in the abstract, and finding it right. The chief results of value are

1. That it is possible to have a regular jet of water down the axis of a tube. This may turn out useful for experiments.

2. That air sticks to water as much as to glass (within the degree of accuracy of the experiments).

Now Kundt and Warburg have measured the amount of slipping between air and glass, and find it insensible at ordinary pressures but measurable at very small pressures.

In calculating the effect of a slight breeze in raising ripples on water we may therefore suppose the relative velocity of air and water at the surface to be zero.

If the experiments are put forth as researches into the nature of fluids then I think the plan is not good. If they are experiments to illustrate the correctness of the formula as given then their value consists in the development of the skill of the author, which we hope to hear of again in some more important research.

I have not seen the paper in full so I can say nothing about its suitableness for the *Transactions*, but from all I have heard I believe the author is better than his subject, and would do well to wait till he can bring up a heavier piece of ordnance to storm the Committee of Papers withal.

GLENLAIR, DALBEATTIE.
26 *July*, 1876.

DEAR STOKES,

Many thanks for your information ± about the theory of light.

With respect to —— if he can explain how he measures the temperature which he calls 750° I would delete the (?). If he does not I would leave it as you have marked it. But I would cut out any statement of what would take place at temperatures below − 300° C.

There are many things which a man may make a note of for himself, but which Foster or I may think it desirable not to send (of our own motion) to the R. S. For instance, an author may think it a matter of principle to prune his calculated results down to such a number of decimal places as he considers true beyond the possibility of contradiction by future investigators.

But the future investigator may wish to know whether his results agree with those of the earlier author as far as that author carried them.

If ——'s results had been glaringly inconsistent with each other in an early place of decimals it would be absurd to print a long row of figures, but if they are decently consistent and if the author himself is not afraid to write them down, the question of printing them at length or docking them seems one of expense not of principle.

I think there is a short account of some experiments on the *rate* of conduction of heat in the different bars which shows that —— has not studied his Fourier very deeply. His aim is to determine *the* velocity with which heat or the heat wave travels along the bar. If he has profited by his experiments so much as to disbelieve in the existence of such a velocity, the experiments may have contributed to convert a man of science to the truth, but they have not much value to others.

I am now copying out for press the "laboratory book" of Henry Cavendish. If he had published his electrical experiments he would not probably have published the details of the experiments. But as he did not do so, and only wrote out (for publication ?) an account of the earlier ones, but made a full index of the

later ones, I think it right to give the whole details, as they are
quite intelligible, and as they are unique of their kind even if
the date were the corresponding years of this century instead
of 1771–2–3.

The accuracy of his measurements of capacity is astonishing,
considering that his electrometer was a pair of pith balls which he
caused to diverge to a measured distance.

If he had had a heterostatic electrometer of any kind, he
might have got on much quicker.   But he compares the formulae

$$\frac{A_1 L}{\log L - \log D}, \quad \frac{A_2 L}{\log 2L - D}, \quad \frac{A_3 L}{\log 4L - D},$$

for the capacity of a long cylinder whose length is L and diameter
D, and his experiments are sufficient to make him decide in
favour of the second, which we now know is right.

Mrs Maxwell desires me to say that her health has greatly
improved by the quiet of her Scottish home.   She joins me in
kind regards to yourself and Mrs Stokes.

Yours very truly,

J. CLERK MAXWELL.

Remarks on Mr G. F. FitzGerald's paper "On the Electromagnetic Theory of the Reflexion and Refraction of Light*."

Cavendish Laboratory, Cambridge.
6 <i>Feb.</i> 1879.</div>

...At the same time I think it desirable, and even necessary
to the completeness of the paper, that the author should make the
several statements of his assumptions more explicit.

What he says about the interior of a medium can generally be
interpreted by comparison with my book on electricity, but his
statement of the boundary conditions (p. 4) can only be interpreted
by working backwards from his results.

He says "the values of the elements of the integrals must be
the same for the two media," but what these elements are depends

---

* <i>Phil. Trans.</i> 1880.

on the set of variables which we adopt to express these integrals by *.

It appears from the subsequent work that the boundary conditions adopted are

(1) The components of the intensity of the electric force parallel to the surface are continuous in passing from one medium to the other.

(2) The corresponding components of the magnetic force are also continuous.

The normal component of the electric force is not continuous, if there is surface-electrification. The normal component of the magnetic *induction* (as defined in my book as the product of the magnetic force into the "permeability") is continuous, and is necessarily so.

Mr FitzGerald finds in the course of his work that the normal component of the magnetic force is continuous, but this is because he supposes the "permeability" the same in both media. See p. 7, line 3.

This part of the theory, which I did not touch in my book, has been taken up by Dr H. A. Lorentz of Leiden in his academic dissertation "Over de theorie der terugkaatsing en breking van het licht," Arnhem, 1875.

It is satisfactory to find that Lorentz adopts the same boundary conditions (though he states them somewhat more explicitly), and arrives at the same results when the passage from one medium to the other is supposed to be abrupt, namely, Fresnel's values of the intensities of the reflected and refracted rays, the plane of polarization being assumed to be that of the *magnetic* disturbance and perpendicular to the *electric* disturbance.

Lorentz has attacked the theory of the change of phase at reflexion, which, as Mr FitzGerald remarks, is of a higher order of difficulty, but he has not attempted the theory of a magnetized medium.

It is in this that Mr FitzGerald has broken new ground, and it is here that I think it especially desirable that he should write out his assumptions in a more explicit form.

---

* Prof. FitzGerald's procedure is explained by Larmor, Report on " The Action of Magnetism on Light," *British Association*, 1893, see § 8. It was extended to the theory of electrons in *Phil. Trans.* 1894–5–7; see especially *Aether and Matter*, 1900, chapter vi.

In order to form the equations for a magnetized medium, Mr FitzGerald professes to borrow an expression from my book. The expression he writes down seems to serve his purpose very well, and it may possibly be capable of reconciliation with mine, or it may possess merits altogether its own, but I have failed to trace any correspondence between them, especially in those points in which I think my own theory faulty.

In my book I did not attempt to discuss reflexion at all*. I found that the propagation of light in a magnetized medium was a hard enough subject.

At first sight it might seem very natural to expect that the corkscrew relation between electric and magnetic vectors would introduce terms into the equations of wave-propagation, of the forms which MacCullagh and Airy have shown to be required in order to explain Faraday's discovery of the rotation of the plane of polarization.

But if we take the *pure* electromagnetic theory, in which the only phenomena considered are electric and magnetic disturbances, and in which no bodily motion of a medium is introduced—the theory, in fact, which has been taken up by Lorentz, FitzGerald and myself—we find that the terms which seem to lead up to a rotation of the plane of polarization, though they enter into the preliminary equations, disappear from the final ones, so that Faraday's phenomenon remains unexplained by the pure electro-magnetic theory.

I have not stated in my book the various hypotheses which I formed and rejected in seeking for a theory of this phenomenon. The one which is found in my book is a hybrid theory, in which bodily motion of the medium is made to cooperate with electric current, so that there is a term in the expression for the kinetic energy which involves the product of a matter-velocity and an electricity-velocity, two kinds of motion between which in other parts of my book I had endeavoured to show there was no demonstrable connexion.

---

* Maxwell has already mentioned (p. 26 *supra*) in this correspondence that the agreement or disagreement of the index of refraction with the square root of the ordinary dielectric constant would be a criterion as to whether the sympathetic vibrations of the molecules would have to be included in the theory of optics for material media. It was possibly their disagreement in general that led him to postpone the treatment of optical reflexion.

Hence the expression which I gave for the kinetic energy in a moving magnetized medium differs from those given long ago by MacCullagh and Airy only in being ostensibly based on a physical hypothesis which I have stated as plausibly as I could, while I pointed out that the fundamental assumption—that matter-velocity and electric current are quantities capable of interaction—is not warranted by any other known phenomenon.

Now in Mr FitzGerald's work I cannot find that he ever introduces the bodily motion of the medium. At least the symbols he uses are all defined as electric or magnetic quantities, and no change of meaning is pointed out.

If he has succeeded in explaining Kerr's phenomena as well as Faraday's by a purely electromagnetic hypothesis*, the fact that he has done so ought to be clearly made out and stated, for it would be a very important step in science; and even if his expression for the kinetic energy is not founded on a dynamical hypothesis, but is merely a mathematical expression from which certain observed facts may be deduced, its value would be very greatly increased by such an interpretation as I am sure Mr FitzGerald could easily give.

J. CLERK MAXWELL.

P.S. Though I have no skill in Quaternion notation, I think that the operator

$$\frac{d}{d\theta} = \alpha \frac{d}{dx} + \beta \frac{d}{dy} + \gamma \frac{d}{dz},$$

where α, β, γ are the components of $\dot{R}$ should be written $- S\dot{R}\nabla$.

GLENLAIR, DALBEATTIE.
21 August 1879.

MY DEAR STOKES,

I have not been able for work of any kind for some time, so that it is with difficulty that I can answer your letter, though it is about my own paper, and you have been taking so much trouble to make it somewhat more like a paper.

---

* FitzGerald's analysis, suitably extended to metallic media, is in fact the usually accepted theory. Cf. Leathem, *Phil. Trans.* 1897 A.

At p. 3 *I would read*—

Hence the normal pressure will be greater on the convex surface than on the concave surface, and if we were to neglect the tangential pressures we might think this an explanation of the motion of Mr Crookes' cups.

At p. 6.   Note added May 1879] should be put at p. 7 in italics before the square bracket].

At p. 12 transfer Added May 1879] to the end of line 12 of p. 13 *in italics* with square bracket thus

$$\frac{2}{3}\frac{p}{\mu}.\quad Added\ May\ 1879]$$

At p. 18 transfer Note added June 1879 to foot of page and put it in italics.

At foot of p. 26 delete]

in p. 18, line 6 from bottom, *for* $\mu^4/\rho^2 p\theta$ *read* $\mu^4 \div \rho^2 p\theta$.

,,	7	,,	,,	$\mu^2/\rho\theta$	,,	$\mu^2 \div \rho\theta$.
,,	9	,,	,,	$\mu^3/\rho p$	,,	$\mu^2 \div \rho p$.
,,	13	,,	,,	μ/p	,,	$\mu \div p$.

I hope the R. S. will not send me any papers to report on for I could not do it.

My wife got caught in the rain on the 13th, and on Monday I was afraid of bronchitis, but the doctor thinks that it has taken a better turn now, but she is very much distressed with neuralgia in the face and with toothache.   She was much pleased by getting a letter from Mrs Stokes this morning.

Among the subscribers to Dr Smith's *Optics*, 1738, appears the name of "Mr Gabriel Stocks." I dare say however your optical studies were already somewhat advanced (from a heredity point of view) in 1738.

Yours very truly,

J. CLERK MAXWELL.

GLENLAIR.

2 *Oct.* [1879].

MY DEAR MRS STOKES,

We had 3 Doctors yesterday and they all agreed in one thing, that Mr Maxwell must go at once to Dr Paget, who is celebrated in what they think this complaint. Dr Saunders I liked very much. We hope to leave to-day, when we can have an invalid carriage to go direct to Cambridge without stopping anywhere. Wonderful from being so weak all summer I can do everything, order the packing &c. With kindest regards to you all,

Ever y^r affec^{te},

K. M. CLERK MAXWELL.

We will likely leave for Cambridge on Monday.

[*In Prof. Maxwell's handwriting.*]

A small but heavy box arrived this morning from Cambridge directed to Arthur. We shall take it back to Cambridge, as Arthur will probably be soon back at King's.

I am a little stronger to-day*.

* Prof. Maxwell died on Nov. 5: cf. *Life* by Campbell and Garnett, ch. xiii.

Letters from JAMES PRESCOTT JOULE.

New Bailey St., Salford, Manchester.
July 24, 1847.

Dear Mr Stokes,

I feel greatly obliged for your kind communication which will afford me a great deal of matter for reflection and study. Your prediction respecting the success of the mercurial apparatus* is fully verified by the only experiment hitherto made with it. The apparatus is similar to that exhibited at Oxford, only smaller and of iron. It contains 205625 grs. of mercury and weighs of itself 39615 grs., which altogether equals the capacity for heat of 11142 grs. of water. By expending on it a force equal to 5230 lbs. through a foot, it was heated 4°·20 which equals 6°·685 per lb. of water: hence the equivalent from this experiment comes out 782, which is wonderfully close to 775 the equivalent already deduced, considering that the corrections are not yet fully estimated.

Believe me Dear Mr Stokes with my best thanks for your obliging and highly interesting communication,

Ever most truly yours,

JAMES P. JOULE.

* See the classical memoir " On the Mechanical Equivalent of Heat," *Phil. Trans.*, communicated to the Royal Society June 21, 1849 by Faraday : reprinted in Joule's *Scientific Papers*, vol. i. see p. 312.

ACTON SQ., SALFORD.
Dec. 10, 1852.

MY DEAR STOKES,

I was unable to visit the Polytechnic on last Wednesday week—and therefore send the enclosed by post. I have felt excessively interested in your recent discoveries, which must be considered as giving the " coup de grace " to the old unmechanical hypothesis of light. I hope you are actively pushing forward a research which promises to yield such beautiful and important truths.

April 12, 1853.

I was much gratified in receiving the very important and interesting papers you were so kind as to send me.

The copies for the Lit. and Phil. Society will be presented at its next meeting and duly acknowledged. I perceive that you are employing quartz instead of glass for your prisms &c.

Melloni in a paper in the *Comptes Rendus*, vol. xviii. p. 42 states that by decomposing the solar rays by "sel gemme" which is equally permeable to every kind of calorific radiation (constituting so to speak the white glass for heat) the maximum of heat will be found in the obscure space beyond the red. The above quotation from Melloni will I doubt not be interesting to you unless as I have little doubt you are already well acquainted with it. Now it naturally occurs to the mind that if quartz be so much superior to glass the limit of perfection may not be attained even with quartz, and I should imagine that an examination of all the transparent substances in your reach would repay your trouble. I have an idea that even elastic fluids might be used as prisms without any solid medium to envelop them. This I think might be accomplished by forcing them through triangular orifices, under pressure.

Thomson and I are getting on as quickly as we can with the air experiments. We have found what Thomson anticipated, viz. the opposite effects of cold and heat in the same stream of air rushing at high pressure. The thermometer held immediately over the orifice and close to it shows 40° of cold: but if held at a distance of $\frac{1}{2}$ an inch so that the air rushes between the bulb and some material surrounding it so as to leave a narrow passage, the thermometer rises 100°: of course these numbers are not exact but they suffice to show that we have found the main theoretical principles to be correct. The air rushing through the porous plug shows 2° Fahr. of cold at a pressure of about 4 atmospheres, a result similar to that attained with the smaller apparatus previously used.

November 12, '53.

I am glad to hear that an institution is about to be established which will doubtless be of the greatest value to Cambridge. I have thought a good deal on the subject of your queries, and in the first place I do not think that for experiments on heat it is absolutely essential to have an uniform temperature, although it would be desirable to obviate sudden and uncontrolled changes. If a room were built without windows it would be rendered useless for many purposes; and I think that for experiments on heat it would be desirable to use a room of considerable size such as might at times be used for different purposes, such as experiments on light, and chemical experiments, probably also for magnetic experiments. I would therefore recommend abundance of light from windows, which might be so furnished with shutters as to shut out light and the effect of external temperature when required. My impression is that if there were two shutters, one opening inside, the other outside of the window, and both well constructed of well-seasoned wood, say 2 inches thick, the window would resist the passage of heat as well as the wall, and thus by closing the shutters all the advantages of a room without windows would be secured.

This arrangement would as you will observe render the room available for experiments on light, for the outer shutter and the window being thrown open, the inner shutter (furnished with proper slits &c.) might be closed: and to facilitate this object it would be desirable that the window should be folding, being hung by hinges to the sides of the opening: by this means it will be possible to open the entire space between the walls instead of only a part as is the case in ordinary windows.

In the accompanying sketch I have given my ideas as to the proper arrangement of the room. I would have an adjoining apartment to contain a steam boiler and if necessary a steam engine. In very many experiments on heat it would be desirable to have the means of conveying heat by steam, which would be best done from an adjoining apartment. A chimney rising between the two rooms would convey the smoke from the boiler flue and from a fire-place or stove in the large room.

I would have the entrance to the large room in which the experiments are conducted through a doorway from the boiler room, which arrangement would prevent the sudden admission of a large quantity of air of different temperature, on any person entering the room by a door from the outside.

For a great many experiments it is necessary to have the command of a great and easily adjustable heat, which would be most easily effected by means of a gas stove. I would therefore recommend a gas pipe of sufficient size, say $1\frac{1}{2}$ inch bore, to be conveyed into the room.

The shutters and windows might be made almost perfectly tight by means of a strip of india rubber placed round their edges.

It would be advisable to have the means of ventilation, which would be most conveniently effected by means of an opening near the ceiling into the chimney, of say 6 inches area, closed when not required or opened to the requisite extent.

The above are the ideas which strike one at present, and if of any use to you I shall be most glad.

A flagged floor would be preferable.

OAK FIELD, MOSS-SIDE, MANCHESTER.
Nov. 29, 1858.

You will see by the enclosed that I am engaged to read a paper to our Society tomorrow and thus am prevented enjoying a pleasure I had looked forward to, that of meeting my many friends at the Royal Society Annual Meeting. I see also that the same cause hinders me from attending the Council for the last time, of which I fear I shall be considered to have been a very indifferent member. The subject of sewage which has occupied my attention is one which, I dare say, it will be thought I have no business with; but it is one which sooner or later will force itself on every intelligent member of the community. I have searched through the Blue Book &c. and find that, except in one or two instances, viz. the appointing of Hofmann and Witt to make a report on a part of the subject, and a letter from Airy printed in small type, scientific men have not been consulted on a momentous subject involving questions which relate to several departments of Science. The whole matter lay in the hands of, I doubt not, very able engineers, but who in the absence of scientific data and information have come to conclusions diametrically opposite to one another. I hope to see the day when the Royal Society will be consulted as the representative of Science in England on all public measures in which Scientific questions are mixed up.

343, LOWER BROUGHTON ROAD, MANCHESTER.

12/10/72.

I fear it is out of my power to give any useful advice to the expedition in the respect of magnetic observations, because the apparatus on which I have worked long is not yet in such a forward condition as a portable and readily renewed instrument, as would be essential for the purposes of an instrument used by an expedition. I have no doubt whatever as to the superiority of the system of suspension I have adopted, but for a long expedition it would be hardly advisable to use instruments which have not been in use more generally at home. The principal difficulty I apprehend is the difficulty of replacing spider or silk filaments accidentally broken. I fully comprehend the immense difference in working at home and on a voyage. However, I am pushing these inquiries farther, and if there is an opportunity of serving the expedition I will not neglect it.

I suppose that there is no better plan for finding the dip at sea than that old one but useful one of Fox's.

However, it seems to me that an apparatus might be made of needles suspended on loops of silk that would give readily the number of vibrations of a needle, horizontally, and vertically or in the dip; whence the dip could be got on board a ship at sea. I am now thinking of such an apparatus as would be practical. I don't think it will involve much difficulty. If anything occur to me further I will write to you at the Royal Society.

Pray give my kind regards to Dr and Mrs Robinson.

MECHANICAL EQUIVALENT OF HEAT—HISTORY. BY PROF. STOKES*.

In a paper published in 1842, Mayer showed that he clearly conceived the convertibility of falling force, or of the *vis viva* which is its equivalent, or representative in visible motion, into heat, which again can disappear as heat by re-conversion into work or *vis viva* as the case may be. He pointed out the mechanical equivalent of heat as a fundamental datum, like the space through which a body falls in one second, to be obtained from experiment. He went further. When air is condensed by the application of pressure, heat, as is well known, is produced. Taking the heat so produced as the equivalent of the work done in compressing the air, Mayer obtained a numerical value of the mechanical equivalent of heat, which, when corrected by employing a more precise value of the specific heat of air than that accessible to Mayer, does not much differ from Joule's result.

This was undoubtedly a bold idea, and the numerical value of the mechanical equivalent of heat obtained by Mayer's method is, as we now know, very nearly correct. Nevertheless it must be observed that one essential condition in a trustworthy determination is wanting in Mayer's method; *the portion of matter operated on does not go through a cycle of changes.* Mayer reasons as if the production of heat were the sole effect of the work done in compressing air. But the volume of the air is changed at the same time, and it is quite impossible to say *a priori* whether this change may not involve what is analogous to the statical compression of a spring, in which a portion or even a large portion of the work done in compression may have been expended. In that case the numerical result given by Mayer's method would have been erroneous, and *might* have been even widely erroneous. Hence the practical correctness of the equivalent got by Mayer's method must not lead us to shut our eyes to the merit of our own countryman Joule in being the first to determine the mechanical equivalent of heat by methods which are un-exceptionable, as fulfilling the essential condition that no ultimate change of state is produced in the matter operated upon.

* A Royal Medal was awarded to Dr Joule in 1852 and the Copley Medal in 1870, for his researches on the mechanical equivalent of heat. The Copley Medal was awarded to J. R. Mayer in 1871 ; the statement printed above was prepared by Prof. Stokes for this occasion at the request of Sir Edward Sabine, Pres. R.S., and was included by him in his remarks in presenting the Medal.

LETTERS FROM W. HAIDINGER*.

VIENNA, 14 Nov. 1852.

DEAR SIR,

I beg leave to introduce myself, and to address you directly. I have read a notice on the interesting crystals with a metallic surface, you described at the Belfast meeting†. I enclose a small portion of crystals I examined some years ago, which I trust will be very interesting to you, as they almost exactly agree with those you examined, except in the point of transparency. The crystals I send are also allied to chinine. They were discovered by Prof. Wöhler, and named *Grünes Hydrochinon*. To him I have been indebted also for the small portion wrapt up here. The present crystals have evidently the same metallic aspect as those which you described, but they have a deep and beautiful violet blue colour by transmitted light. They are, however, but very faintly translucent, and in no direction they appear colourless or transparent, as it is said in the notice contained in Abbé Moigno's *Cosmos* 24 Octobre 1852, page 574.—There is also a mode of chemical preparation given there, but I failed in succeeding. Then I believe I can do nothing better, than asking you the favour yourself, in exchange for the small portion of hydrochinon, to send me also a very little quantity of the crystals you examined. Beside the one here included, I have examined a number of other crystals and substances shewing similar phenomena, and I lately drew up a catalogue of them, and presented it to our Academy. I sent a separate copy, inscribed to you, together with a number of others, to our mutual friend, Professor Miller, begging him to forward them to you. Since this publication I still examined a few more

---

* The Wilhelm Haidinger Medal of the Imperial Academy of Sciences of Vienna was awarded to Prof. Stokes in 1856. Dr Haidinger was for a long period Secretary of the Vienna Academy ; he was a For. Mem. R. S.

† *Math. and Phys. Papers*, vol. IV. pp. 18—21 (1852), 'On the optical properties of a recently discovered salt of quinine.' See note added at the end of the paper.

crystals. The occurrence of these real surface-colours is not over rare, but it well deserves the attention of scientific persons, and particularly your own, who are so eminently prepared for duly appreciating its value. Above all the double cyanurets of platinum and other substances, potash, magnesium, baryta, lithium, &c., are of the most splendid. I believe I gave Prof. Miller a few small crystals of the magnesium compound ($PtMgCy_2$ or) $Pt_5Mg_6Cy_{11}$, the latter being the formula by Quadrat and Redtenbacher.

I was very much delighted with your observations on the "brushes" in the Report of the Edinburgh meeting*. There can remain, I think, no doubt that the phenomenon is produced by diffraction, and your own results contributed much to ratify the explanation, which I proposed in the paper "Über das Interferenz-Schachbrettmuster &c†."

And now, my dear sir, I beg you will accept of my apology for thus intruding upon you, but believe me ever, with the greatest respect

Your obedient servant,

W. HAIDINGER.

26 *Jan.* 1853.

I am under very great obligations to you indeed for your kind letter of 15th Dec., and for the crystals enclosed of that beautiful and most remarkable substance, as also for the hints it contained, respecting the formation of them, which I immediately caused to be made in our laboratory, and most successfully indeed. The process given in Abbé Moigno's *Cosmos* could not succeed, because unhappily by an error of print, there was said *solution d'acide* instead of *solution d'iode* dans l'alcool. When I got your letter I only remembered that Prof. Wöhler already had written me on the substance, and actually sent some of the greenish metallic substance; but I omitted examining it with the microscope, only making use of my dichroscopic lens, and so I failed in my first examination. I also had polished some of it on a ground but not polished glass surface, but as it did not assume a bright green

* *Math. and Phys. Papers*, vol. II. pp. 362—4.

† But see Extracts from a letter by Prof. Stokes relating to optical subjects, communicated by Dr Haidinger to the Vienna Academy, reprinted *loc. cit.*, pp. 56—60. See also letter of April 12, 1854, printed *infra*, p. 58.

surface colour, but only a deep blue one, I again had laid it aside. Now upon subjecting those beautiful crystals, deposited on the mica, to a magnifying power of 90, all immediately was clear.

I was the more delighted with the sight I had, as also a great puzzle vanished, which had made me somewhat uneasy since the time I had read your communication, it is true not the one given at Belfast, but only the one in Abbé Moigno's *Cosmos*, and still more so, when I had read both Dr Herapath's papers, which I also compared. All the *green* surface colours, which I had observed up to that time, were coexisting with a *red* colour by transparency; now the new substance—for brevity's sake I have called it *Herapathite* in a notice lately given in our Academy—was described as being colourless or pale greenish polarized in the direction of the axis, in the principal section of the axis, while it was opake and *as black as midnight* polarized perpendicular to it. The green metallic surface colour you found likewise to be polarized perpendicularly to the axis, or to agree in the direction of polarization with the more absorbed ray of light. Very thin crystals of the Herapathite appear of a pale *reddish* colour, while thicker ones are merely pale *greenish*, but if a rhomb of Iceland Spar is placed upon the microscope, the thinner crystals shew the colour polarized in the direction of the axis very pale greenish white almost entirely colourless, so much so, that the very thinnest crystals are altogether invisible, while the colour polarized perpendicularly to the axis is a very beautiful more or less deep *red*. In the thicker crystals the ordinary ray is more or less inclining to green, while the extraordinary is so deep red, that it is actually opake, yet it is possible to follow up all the intermediate stages between red and perfectly black and opake. This observation also accounts for the *green* tint of the metallic reflected light, which is then the complementary tint to the *red* colour of the crystal by transparency. It is very remarkable that the very same tints of colour appear in the transparent andalusite from Brazil, a description of which you will find in one of the papers I sent, with the date of 1843, the same pale greenish colour, the same deep hyacinth red. Thin plates, possessing themselves a pale red colour, being laid upon one another in a parallel position appear of a pale greenish white, if they are crossed they shew the deep red. Andalusite could be used like tourmaline, it were almost preferable, being so very transparent, but it is unfortunately too rare, the herapathite would likewise be an excellent 'artificial tourmaline' as Dr Herapath says, but for the smallness of the crystals. It were a grand result, if it should be possible to incorporate it in an isomorphous crystal of another substance, to have larger crystals with corresponding properties, more easily handled.

Of those most interesting observations of your *degraded* light, I have not yet made any, if I except that beautiful general appearance of the solutions of quinine, or of the solution of the cyanuret of platinum and potash in thin films, as it is retained by capillarity between the crystals deposited on the sides and the glass vessels in which they are formed. But I have, myself, no command of a ray of sunlight in my dwelling, and can not leave it for seeking one anywhere else; but M. von Ettingshausen proposed to repeat them, when the sun shall be a little more favourable, winter being nearly as unfavourable with us as it can be in England. The fact that you found this property in the cyanurets, I should believe, indicates* an intimate connexion between the colours of *surface* and these phenomena, the 'blue' of quinine, of fluor, &c., being in fact dispersed in all directions, it seems to me internal dispersion indeed in regard to the whole mass of liquid, or the entire crystal, but in fact emanating from the surface of the smallest particles of the bodies. I am very anxious to hear of your farther results, when you shall have again set about the sun's rays.

Pray remember me kindly to Prof. and Mrs Miller, and believe me ever My Dear Sir most respectfully and faithfully yours,

W. HAIDINGER.

5 *May*, 1853.

I received your kind letter on the 27th March, and the really wonderful account of your great discovery on the 30th; I have so long deferred returning my thanks, till I should have read it all through, which I was prevented from immediately accomplishing. I perfectly agree with you in considering the internal dispersion, or light with lowered refraction, or fluorescence, as a phenomenon totally different from the relation of difference of the colours of surface and body. This is proved by your own experiments beyond any doubt. When I wrote my paper on the

* See next letter.

surface colours of several substances, I was anxious to look out for anything that should bear any similarity to the subject in the communications till then published. It seemed to me that I should not leave unnoticed the observations of Sir John Herschel or Sir D. Brewster, rather for pointing out that there were properties or phenomena worth the closest examination than for comprising them under the same head. Now that you have given the solution of the problem, there cannot remain the least doubt respecting the difference of the two. You have now really an immense field of discovery before you, and certainly no man is better prepared for extending it on all sides. I follow your labours with the greatest interest, and I shall be most happy to hear particularly of the results of your farther examination of the double cyanurets.

Allow me to express my thanks particularly also for the views on the polarized metallic colours and opacity or other phenomena*, which you were kindly pleased to call 'a subject which especially belongs to me.' I may be permitted to decline this designation. I was happy indeed to make some observations in this respect, but I think the field is everywhere free, as soon as the observation is published; and I was indeed in the habit of publishing what I myself thought very incomplete, in order to induce other persons, better prepared for the task, to take the inquiries in their hand. Moreover, your method of explanation, and phraseology, are far more scientific than what I could propose, as I stuck as close as possible to the observations, without entering much in any theoretical detail. I was greatly interested in seeing the way by which you were led to your discoveries. Carthamine I also examined, but only last summer; it is a very beautiful substance. Your observation on the not mathematically exact complementary state of surface and substance colour is very just; and the substance of it forms indeed the solution of what had appeared to me extraordinary in the case of herapathite, the black colour of transmitted light, and the green colour of reflected light, whereas the latter depends on the red colour of transparency of a very thin stratum contiguous to the surface. I also pointed out several substances in which, a deviation or exception from the general law

---

* The subject of these letters is further explained and elucidated, *Math. and Phys. Papers*, vol. IV. pp. 38—49 (Dec. 1853), 'On the metallic reflexion exhibited by certain non-metallic substances.'

seems to take place, but leaving the explanation to further inquiries, which, I do not doubt, will throw much light on the matter. I should like to call your attention also particularly to the classical chrysammate of potash. This most curious substance if spread with a knife on a hard body, for instance on glass or crystal with a dull polish, will shew a different colour if you examine it lengthways in the direction of the streak of the knife, or across the direction of the streak. I enclose a small quantity of it spread on paper.

Of all the expressions used for that beautiful phenomenon of the quinine, I may be permitted to say, I like best your own 'fluorescence,' being a single word not implying any theoretical view. It appears to me that the rays refracted beyond the usual spectrum are by no means rendered visible to the eye, but that as you have excellently expressed it, page 503, *the fluid or solid medium is self-luminous so long as it is under the influence of the active light.* I perfectly acknowledge also the explanation of the increase of the blue light, the latter not being lost in the crystals of the platino-cyanide of potash, when they are covered by a thin layer of the liquid solution; in the latter however there is certainly a difference in the phenomena observable on the terminal plane owing to cross-fracture, and the sides of the prism, the blue light being polarized perpendicularly to the incident ray in every azimuth in the former planes O, while it is *polarized perpendicularly to the axis* only, and not parallel to it in every one of the longitudinal planes M. I can not exactly reconcile this with the statement of your letter *that the blue light emitted by platino-cyanide of potassium is polarized in the plane of the axis when the axis is perpendicular to the ray entering the eye.* Perhaps the discrepancy originated only in the use of different words for expressing the same thing. I am in the use of observing with the dichroscopic lens in the manner represented in the diagram [*omitted*]. The ray AB entering the eye in C separated into the ordinary ray O and the extraordinary ray E, shews uniformly O, white from the ordinary superficial reflexion, polarized *in the plane of incidence,* and E blue, from the action of the crystallized substance itself on the colour, polarized *perpendicularly to the axis.* In the position above represented Fig. 1 perpendicular to the axis is at the same time *perpendicular to the incident* ray, but when a position of the crystal is chosen like Fig. 2 [*omitted*], then the blue ray polarized perpendicularly to the axis lies *in the plane of incidence*, and is no longer visible in the extraordinary image E, which in that case is without either lustre or colour, but it reaches the eye at the same time and with the same kind of polarization, as the light reflected and polarized in the ordinary way in O. I doubt not, the whole will find its natural explanation, whenever you shall have exposed your observations more circumstantially. Within these days I shall also send my observations on the herapathite, and the reconciliation of the red and green. And now again pray accept my most sincere thanks for your so highly valuable, interesting and important papers, as also for the kind letter you favoured me with. But the whole range of the inquiries is so vast, that it requires time, to comprise in all its variety. I beg leave to be remembered also to Prof. Miller, and believe me, My Dear Sir, yours ever most truly and respectfully,

W. HAIDINGER.

12 *April*, 1854.

I indeed should have answered your kind, and I well may add, grand letter of 1st Febr. immediately, so much pleasure I felt in unfolding it, but unfortunately I was rather unwell at the time I received it, so that I in fact did not read and study it all through, till about three weeks after. In the meantime I communicated it to my friend Professor von Ettingshausen, who likewise was very much gratified with it, and who urged me particularly, to give an account of it to our Academy, in which I heartily concur with him, and which I still propose to do, but was detained till now, as well as from answering and giving you my thanks for your most interesting communications, by various other avocations. As to the memoirs and books sent, I indeed learn from your letter, that there must have been some error committed in packing up and addressing the parcels. But probably this took place at our own dispatching office; and I am from time to time in the necessity of offering apologies, or sending numbers lost, as we have at present not less than 157 Societies or Institutions to whom our Memoirs are sent, and a great many more, between 600 and 650, to whom the *Annals* are regularly sent. The parcel for you was ordered to be enclosed to the Royal Society, as also that for Prof. Miller,

agreeably to the communication I had from the latter, while I had
the pleasure of seeing him in Vienna. To him was also enclosed
a set of our publications for the Cambridge Philosophical Society,
in order to endeavour to open a correspondence, and if the Society
were inclined to it, an exchange of our publications. If you happen
to see Prof. Miller, I should be much obliged to you, if you would
kindly ask him, whether or not that parcel has arrived. Perhaps
this also was mislaid or lost, and I should in that case immediately
send another set, taking care to have them regularly addressed
and enclosed to the Royal Society. I was delighted with the
copious details into which you entered on the different subjects.
As to the phenomenon of the 'Schachbrettmuster*,' I quite concur
with you in considering [it] as one of the more simple character
you claim, and not of any 'Interference' properly or rather more
strictly so called, consisting of the successive spectra distinct from
one another. But I took it in a more general sense, and then
a name was necessary to call the phenomenon by, and so the
one was chosen, and then it is difficult to change it afterwards.
Diffraction would have been a more suitable word. But when
I published my account I rather more thought of publishing the
few observations that had struck me, than of entering into a series
of investigations to settle the matter. I requested some younger
persons to follow up the study, particularly by projecting the
colours on a screen. I am happy to learn that indeed you have
done so, and added to it your explanatory observations, which are
very just in every respect. Particularly the observation of a
bright luminous point through a dark blue cobalt glass bears an
exact and as you call it 'striking' analogy. I likewise fell on that
observation and described it or at least alluded to it in a paper on
the 'Löwe'schen Ringe†.' I was endeavouring to observe the
red coloured 'brushes' in candle light or sun light, which you had
described in the Report of the Meeting in Edinburgh in 1850,
Notices and Abstracts, p. 20; and as it seemed to me, that the
colours of the fringes in the 'diffraction-chessboard' as you
would like to call it, exactly agree with the tints of the brushes,
the observation of a distant candle light, red in the centre, fringed
outside with blue, presented itself very naturally. Being brought
nearer the eye than the distance of distinct vision, a small luminous

---

* Footnote, *supra*, p. 53.
† *Sitzungsberichte der Kais. Akad. d. Wiss.* 1852, July, Bd 9. S. 240.

point is of a violet blue, but fringed outside with a deep red, the faint tints of yellow and pale blue of white light being heightened by means of the blue cobalt glass to the splendid blue and deep ruby red of the phenomenon. But as I had not met with a description of it, though it had been published by Dove and I believe also by Plateau, I just mentioned it without naming any author, but had afterwards an opportunity of acknowledging Dove's prior publication. Now I see that also yourself had long ago observed the same want of coincidence of the red and violet rays for distant objects, proceeding however from a different starting point.

I had great pleasure also in reading your account in Poggendorff's *Annals* on metallic reflections, particularly with regard to the carthamine, as well as also on the red colour of substance of the herapathite and the supermanganate of potash. I readily confess your observations to have been conducted with much greater accuracy, as I never made any use of instruments to see Fraunhofer's lines, which indeed is almost indispensable to convey a more deep insight into the nature of the phenomena. As to myself, I must content myself with stating what I can skim as it were from the surface, being drawn in so many different directions, and with but poor means to strive at greater perfection. This does not prevent me, however, from experiencing the greatest pleasure if other persons explain the same phenomena better than I was able to do. In giving an account of your kind letter to our Academy I shall have occasion to bespeak several of these points, and I shall not fail to send you the communications as soon as they are printed. I lately dwelt upon the fact of the absolute interception of every trace of light by two—say Nicol's prisms—plates polarized perpendicularly to each other, as a reason to consider the colour of the brushes as really produced by diffraction, the same as the penumbral tints of the 'chessboard,' the red side of the spectrum being extinguished by the blue, the blue side by the red, all of it referable to the eye not being achromatic. I am very anxious to see your explanation. As to any separate copies you would kindly destine to me, I shall be most glad to receive them, also the mathematical ones, notwithstanding I am far myself from being a consummate mathematician; but particularly my friend Prof. von Ettingshausen takes the greatest interest in all your observations, inquiries and results. You might kindly direct the parcels to be sent to me through the Geographical, or the Geological, or the Linnean, or the Chemical, or Palaeontographic Society, or the Government Geological Survey Office, from all of whom we receive publications. We send our publications also to the Royal Society, but hitherto without receiving any return. Or through Messrs Pamplin or Williams and Norgate of London, and Braumüller bookseller of Vienna. With the most sincere thanks for your kind, and comprehensive and most interesting letter, believe me ever My Dear Sir yours very sincerely,

W. HAIDINDER.

Pray give also my best compliments to Prof. Miller, Prof. Sedgwick, Prof. Whewell, all of whom I had formerly the pleasure of meeting somewhere or other.

LETTERS FROM JULIUS ROBERT PLÜCKER.

The following statement was made by General Sabine, President of the Royal Society, at the Anniversary Meeting, Nov. 30, 1866, in transmitting the Copley Medal to Professor Plücker.

To an audience not exclusively mathematical it is obviously impossible to enter into details of researches which deal with geometrical questions of no ordinary difficulty. Amongst these, however, may be indicated, as especially appreciated by those who are interested in the progress of analytical geometry, his theory of the singularities of plane curves as developed in the " Algebräische Curven," with its six equations connecting them with the order of the curves: the papers on point and line coordinates, and on the general use of symbols, may also be noticed as establishing his claim to a position in the department of abstract science which is attained by few even of those who give to it their undivided attention. But Professor Plücker has high merits in two other widely different fields of research, viz. in Magnetism and Spectrology: and to these I may more freely invite your attention.

Shortly after Faraday's discovery of the sensibility of bodies generally to the action of a magnet, and of diamagnetism, Professor Plücker, in repeating some of Faraday's experiments, was led to the discovery of magnecrystallic action,—that is, that a crystallized body behaves differently in the magnetic field according to the orientation of certain directions in the crystal. The crystals first examined were optically uniaxal, and it was found that the optic axis was driven into the equatorial position;

(that is, of course, assuming that the magnecrystallic action is not masked, in consequence of the external form of the body, by the paramagnetic or diamagnetic character of the substance). New facts, discovered both by Faraday and by Plücker himself, led him to a modification of this law, to the effect that the optic axis was impelled, according to the nature of the crystal, *either* into the equatorial or the axial position. This subject was afterwards followed out by Professor Plücker into the more complicated cases in which the conditions of crystalline symmetry are such as to leave the crystal optically biaxal; and after having recognized the insufficiency of a first empirical generalization of the law applicable to crystals of the rhombohedral or pyramidal system, and accordingly to uniaxal crystals, he was led to assimilate a crystal to an assemblage of small ellipsoids, capable of magnetic induction, having for their principal planes the planes of crystalline symmetry where such exist; and to apply Poisson's theory. The result of this investigation is contained in an elaborate paper read before the Royal Society in 1857, and published in the Philosophical Transactions for the following year. In this paper Professor Plücker has deduced from theory, and verified by careful experiments, the mathematical laws which regulate the magnecrystallic action. These laws have not necessarily involved in them the somewhat artificial hypothesis respecting the magnetic structure of a crystal from which they were deduced; and at the close of his memoir Professor Plücker recognizes the theory of Professor Sir William Thomson, with which he then first became acquainted, as a sound basis on which they might be established. The laws, however, remain identically the same in whichever way they may be derived.

Another subject to which Professor Plücker has paid much attention is the curious action of powerful magnets on the luminous electric discharge in glass tubes containing highly rarefied gas. In this case the luminous discharge is found to be concentrated along certain curved lines or surfaces. He has succeeded in obtaining the mathematical definition of these curved lines or surfaces, by a simple application of the known laws of electromagnetic action, regarding an element of the discharge as the element of an electric current. With regard to the blue negative light, for instance, starting from a point in the negative electrode, he has shown that there are two totally distinct paths, one or other of which, according to circumstances, it may take, going either within the enclosed space along a line of magnetic force, or else along the surface of the glass in what he calls an " epipolic curve," which is the locus of a point in which the inner surface of the vessel is touched by the line of magnetic force passing through that point.

Angström appears to have been the first to notice that the spectrum of the electric spark striking between metallic electrodes through air on another gas at ordinary pressures is a compound one, consisting of very bright lines varying with the metal, and others, usually less bright, depending only on the gas. Under the circumstances which presented themselves in his experiments, the latter can frequently be but ill observed; and the diffused light of a rarefied gas in a wide tube is but faint, and does not form very definite spectra. But Plücker found that by employing tubes which were capillary in one part, brilliant light and definite spectra were obtained in the narrow part. These spectra were observed by him with great care, and were found to be characteristic of the several gases and to indicate their chemical nature, though the gases might be present in such minute quantity as utterly to elude chemical research. It further appeared that compound gases of any kind were instantaneously, or almost instantaneously, decomposed; at least the spectra they offered were the spectra of their constituents.

In a recent memoir, which has only just been published in the Philosophical Transactions, Professor Plücker has investigated the two totally different spectra frequently afforded by the same elementary substance according as it is submitted to the instantaneous discharge of a Leyden jar charged by an induction-coil, or rendered incandescent by the simple discharge of the coil, or else, in some cases, by ordinary flames. The two spectra show a remarkable difference in *character*, and are not merely different in the number and position of the lines which they show. Some phenomena which he had previously noticed receive their explanation by this twofold spectrum.

This difference of spectra is attributed by Professor Plücker, with the greatest probability, to a difference in the temperature of the glowing gas when the two are respectively produced. The discovery opens up a new field of research, the exploration of which may throw much light on the correct interpretation of celestial phenomena, especially in relation to the physical condition of nebular and cometary matter.

BONN, den 28 *Sept.* 1858.

DEAR SIR,

Ihr ausführliches Schreiben kam in Bonn an demselben Tage an, als ich meine Ferienreise antrat, von der ich gestern zurückgekehrt bin. Ich muss noch eine zweite kleine Reise machen, aber nach dem 20 October werde ich Bonn nicht mehr verlassen und dann erst Musse finden, auf Ihr verehrtes Schreiben erschöpfend zu antworten. Jetzt aber schon einige vorläufige Worte. Der Eindruck den dieses Schreiben auf mich machte,

war dass ich einen grossen Fehler gemacht habe, wenn auch
unabsichtlich, dass ich in der eingesandten Abhandlung der
Arbeit des H. Thomson* nicht Erwähnung gethun habe, der, was
keinem Zweifel mehr unterliegt, zuerst die Theorie des Magnetis-
mus der Krystalle gegeben hat. Es war mir lieb nachträglich
zu sehen, dass diese Theorie mit den von mir mitgetheilten
Resultaten übereinstimmt †. Ich habe niemals diese Theorie
H. Poisson zuschreiben wollen. Von diesem und seinen Nach-
folgern habe ich nun die Theorie der Induction eines unendlich
weit entfernten Poles auf ein dreiaxiges Ellipsoid adoptirt und für
den Fall einer Ellipsoids aus Eisen verificirt. Die doppelte
Beobachtung dass unter der Voraussetzung paralleler Kraftlinien,
der ganze Krystall sich *einerseits* richtet wie jedes beliebige
Fragment desselben, und *andrerseits* wie ein Ellipsoid, das gegen
die Krystallform eine gegebene Lage hat—gestattete mir aus
dieser Theorie die Theorie der magnetischen Induction der
Krystalle unmittelbar abzuleiten, nachdem H. Thomson *früher
schon*, von mir unberücksichtigt, seine Theorie dieser Induction
gegeben hatte.

Mir scheint es hiernach am angemessensten, und Sie stimmen
mir hierin vielleicht bei, dass meine Abhandlung gedruckt und
ihr eine Note mit späterm Datum hinzugefügt würde, in der ich
mein Versehen redressirte und überhaupt einige kurze sachliche
Bemerkungen machte.

Ich sehe bisher nichts was einer vollständigen Uebereinstim-
mung unserer beiderseitigen Ansichten entgegen wäre ; doch
in einigen Wochen schreibe ich ausführlicher. Ich bedaure nur,
Ihnen so viele Mühe zu machen. Ich bediene mich in diesem
Briefe der deutschen Sprache, um nicht, durch eine ungebräuchliche
Anwendung eines Englischen Ausdrucks, vielleicht missverstanden
zu werden.

Meine beiden kleinen Abhandlungen werden Sie von London
aus richtig erhalten haben, deren Inhalt H. Tyndall so freundlich
war in dem *Phil. Mag.* mitzutheilen. Seitdem haben die schönen
Erscheinungen, von denen ich einige H. Miller bei Ruhmkorff
in Paris zu zeigen Gelegenheit fand, in einer einfachen mathema-

* 'A Mathematical Theory of Magnetism,' by Prof. W. Thomson, *Phil. Trans.*
1849 and 1850 ; reprinted with additions in *Papers on Electrostatics and Magnetism*,
pp. 344—592.
† 'On the magnetic induction of crystal,' by. Julius Plücker. *Phil. Trans.*
(Mar. 26) 1857, pp. 543—587. See note at end.

tischen Theorie ihre Erklärung erhalten. Die Abhandlung darüber ist bereits gedruckt. Eine zweite Abhandlung über die Spectra der verschiedenen Gase und die Analyse der Gase vermittelst ihrer Spectra ist unter der Presse (Von beiden eine Notiz im *Cosmos*). Ich werde mich beehren, die Abdrücke, sobald sie mir zugehen, einzusenden.

Hochachtungsvoll der Ihrige,

PLÜCKER.

Bonn, 4th of November, 1858.

Since my last letter I read with the greatest interest your communication and learnt from it, that all phenomena regarding magnecrystallic induction, as described in my paper, may be mathematically deduced from the hypothesis, that the induced particles don't sensibly act on each other. But this hypothesis, I think, does not permit us to take into consideration the exterior form of the induced body, and gives no information about the direction of an isotropic ellipsoid. If we intend to consider *both* kinds of phenomena, we can scarcely avoid to enter into considerations analogous to those communicated by Poisson and his successors. For this moment I am not able to judge, if Mr Thomson's method be simpler and more elegant; I am inclined to think so.

In regard to myself I beg you to remember that my intention was not at all to enter into the general problem of mathematic physics. I adopted the results then known to myself; the theory —deduced according to Poisson's principles—of the magnetisation of an ellipsoid induced by an infinitely distant pole. I partially verified this theory in the case of an ellipsoid of iron, and by means of a new hypothesis, suggested to me by mere experiment, I succeeded to get a mathematical law of the phenomena presented by magnetically induced crystals. To be sure, properly speaking, the theory of the magnetisation of crystals is not due to Poisson. I got it myself by an additional hypothesis, but I thought what I added myself rather too insignificant as not to attribute the whole theory to Poisson. If I had known Mr Thomson's theory I would have chosen my expressions more cautiously. Then there is no doubt, the theory of the magnetisation of crystals has been presented by this great philosopher before I myself deduced from Poisson's results a similar theory. This fact is to be restablished.

You are, dear sir, perfectly right in saying, that all results I obtained till now by observation and calculation can be deduced from the mere fact of the existence of the ellipsoid. We know of this ellipsoid the magnetic axes only, perpendicular to its circular sections, i.e. one homogeneous equation between the squares of all three axes. You arrive directly to these results. Similar results are obtained by substituting an ellipsoid of iron to the crystal, where the reciprocal action of particles is not to be neglected. There is also an ellipsoid of induction if the inducing pole is infinitely distant, which Poisson proved first. Should there be no physical connexion between both cases? I think there is one, without attributing any importance to the hypothesis I proposed.

The experimental researches regarding the magnetisation of isotropic ellipsoids offered to me such an interest that I prepared some series of new experiments, depending upon the ratio of any two axes of the ellipsoid of induction. For this purpose I procured two powerful cylindric electromagnets, exhibiting an exactly straight line as magnetic axis. I procured also from Birmingham two fine pieces of metallic nickel and cobalt, to be turned out into ellipsoids, quite similar to the iron one, examined by myself (cobalt seems to be too hard). I expect to find, in the case of nickel, the angle formed by the two magnetic axes greater than in the case of iron.—I know very well that the attraction of an infinitely distant pole on an isotropic ellipsoid disappears, but I hope to be able to determine approximately the ratio of two such attractions along two axes of the ellipsoid: this ratio, which theoretically takes the form 0/0, being the ratio of the corresponding axes of the ellipsoid of induction. The axes of this ellipsoid, according to Poisson's theory, may be deduced by means of elliptical functions from the axes of the given ellipsoid, &c., &c.

Not having at first found a sensibly different attraction exerted on a crystal along different directions, I was much retarded in my earlier theoretical researches. Prof. Tyndall proved the fact. In certain cases and under certain conditions the difference is very considerable; in one case the ratio of the attractions, along the magnetic axis and perpendicular to it, approached to 2 : 1. I don't know if the first observation must be referred to Faraday (Note of Thomson's paper*).

I know there are experiments made with flat poles in order to get parallel lines of force, but this case seems to me a rather complicated one. I intend to repeat these experiments after having determined between the two flat poles the lines of force, which now may be done in the most striking manner by adhibiting electric light.

I fully adopt that a body pointing like a crystal may be obtained as well by putting spheres in different distances from each other along different directions (Prof. Tyndall) as by giving to symmetrically placed particles a non-spherical form, an elliptical for instance.

I am very sorry, Dear Sir, that I gave to you so much trouble, highly regretting that I did not call on you at Cambridge before I left England. All would have been settled in half an hour. I agree if you would send my paper, as it stands, to the press; then I may, if convenient, add a few pages of notes, regarding especially the priority of Thomson's theory of magneto-crystallic action. You had the kindness to propose to me three different ways of correcting the proofs. I would be extremely obliged to you, if you should superintend the correction of the press. But at the same time you may, wherever convenient, dispose on my cooperation.

I send to you only one of the two papers announced in my last letter; the copies of the other one 'On the Electric Spectra of Gases' (the proofs have been corrected by myself nearly a month ago) are not yet in my hands. I follow up with greatest interest this new kind of chemical analysis. Since my letter I made a new series of experiments regarding the electric discharge. Formerly I had examined the light at the negative pole, which gave to me the most splendid phenomena of the illumination of the lines of magnetic force. Recently I got no less splendid appearances at the *positive* pole. In longer exhausted tubes the electric current, with its dark bands, is the mere continuation of this light, separated by a dark space from the negative light. By substituting to the tube a small ball (about two inches diameter) and by conducting through it the two platina electrodes, you will

* *Papers on Electrostatics and Magnetism*; see pp. 499, 504, 540.

observe in a most striking way the peculiar light of both electrodes at the same time. While the negative light is bent by the magnet into the finest surfaces, as described already by myself, the positive light generally moves along spirals, starting from the positive wire towards the magnetic surface which passes through the negative wire, but not joining this surface. A full account of all these phenomena will be published.

Excuse my bad English. Highly obliged to you,

Dear Sir, and most truly yours

PLÜCKER.

2/1/59.

In copying my paper I omitted by mistake (after p. 27, line 11) a *longer* passage regarding the correction of the numbers of the table. Thus my numbers become not intelligible. After having restored the omitted passage (with abbreviations) all my numbers will stand, except the values of θ_1^2, θ_{11}^2, θ_{111}^2 inadvertently written down. In repeating the calculation with the sufficient number of decimals I got the former numbers (only 4,301 instead of 4,300 and 5,649 instead of 5,647). I see from your numbers that the law of formula (30) is equally verified by the numbers of the incorrected table. The approximation being in both cases more than sufficient, I think a passage (p. 27, line 9–5 from bottom) rather insignificant and therefore suppressed it. Thus the corrections will give no trouble to the printer. In order to prevent any further mistake I copied the whole part of the paper.

I have not the least objection that my paper will appear at a later period. By then I will have found time to try new experiments on the same subject, since long prepared.

Excuse, dear sir, all the trouble I gave to you.

The questioned salt is the mineral, of *green* colour, commonly called with us Uranglimmer, therefore *phosphate of uranium and copper*.

My best thanks for all your kindness,

Most sincerely Yours,

PLÜCKER.

P.S. Among the great number of observations I made on
the electric discharge through different gasvacua, there are many
very beautiful too regarding fluorescent light, particularly of
glass, which I wished to show you. I got, for instance, spheres
of German glass, illuminated all over their surface with a beautiful
uniform green light, produced by a rather dark electric discharge
within*. By means of the magnet the fluorescent green light
is concentrated to determined curves, where the magnetic negative
light touches the interior surface of the glass; in certain cases
too we get very well designed spirals of fluorescent green light,
characterising the positive electric light. With English glass,
containing lead, the same phenomena will appear, there is no
doubt, in the finest blue light.

BONN, 3rd of July, 1865.

In reply to your kind letter of April 7th it will be desirable
for me to make a few remarks with respect to my geometrical
paper and its title†.

The geometrical conception of what I call a ' complex of rays '
has nothing to do, I think, with Mr Cayley's determination of
curves by means of intersected right lines. I presented the
fundamental idea in 1846, but afterwards it entirely disappeared
from my memory. The same idea recurred when by Mr Sylvester's
suggestion at the (Newcastle) Meeting§ I became desirous to resume
Geometry, entirely neglected by me so many years since. In
order to make its analytical development successful I introduced
a fifth coordinate ($r\sigma - s\rho$), depending upon r, s, ρ, σ. After its
introduction an immense field of geometrical researches was at
once opened. Here, instead of point or plane, the right line is
considered as the element of space. Under this supposition I
presented my Paper.

After a short indication of my general views I gave in the
Paper a complete theory of complexes of the *first* order. In its
second Section I gave an application of the most simple complexes
of the *second* order to double refraction. This digression was
ill calculated fully to explain my proper views. I had preferred

---

* The cathode steam.   § Of the British Association.

† 'On a New Geometry of Space' *Phil. Trans.* (Dec. 22), 1864, pp. 725—791.
There follows 'Fundamental views regarding Mechanics' *Phil. Trans.* (May 29),
1866, pp. 361—380 in which the idea of a *dyname* or wrench is introduced.

to give an outline of the general theory of complexes of the *second* order, but an attempt to do so appeared too long.

A *plane curve* is analytically determined by means of *two* coordinates, a *surface* by *three*, a *complex* by *five* (one of which depends upon the four remaining ones). A right line within a plane, a plane, a linear complex, is represented by equations of the first degree. Conics within a plane, surfaces of the *second* order, complexes of that order, are represented by the general equation of the *second* degree between two, three, five coordinates. Let me add a few words in order to explain what a complex of rays of the second order is.

A surface of the second order being given, there is in each intersecting plane a conic enveloped by rays tangent to the surface. All such conics, constituting the surface, as well as the surface itself, may be said to be represented by the equation of a complex of the second order. In the general case of a complex of that order, there is likewise in each plane a conic enveloped by its rays, and all such conics may likewise be said to be represented by its equation : but the different conics do *not* constitute a surface. On the other side each point of space is the centre of a cone of the second order, the generatrices of which are rays of the complex. The system of all such cones may be said to be represented by its equation, but, in the general case, the cones do not envelope a surface.

$$x = rz + \rho \qquad\qquad y = sz + \sigma$$

being the equations of a ray, the general equation of a complex of the second order involves 19 constants, which are reduced to 9 if it represents a surface. If the axes of the surface fall within the axes of coordinates, the number of constants is reduced to three. *Under the same conditions of symmetry* the equation of the complex in the general case is

$$Ar^2 + Bs^2 + C + D\rho^2 + E\sigma^2 + F(r\sigma - s\rho)^2 = 0.$$

In admitting the following two equations of condition between the constants

$$AD = BE = CF,$$

the distribution through space of the conics, indicated by the equation, is such that all belong to the same surface, which in making use of ordinary coordinates is represented by

$$Ax^2 + By^2 + Cz^2 + 1 = 0.$$

I have before me a long series of curious results concerning complexes of the second order; others will occur in treating their general equation in a way quite analogous to that followed in the discussion of conics and surfaces of the second order. But here the field to explore is a still more extensive one. In moments of leisure I intend to prepare a Paper for publication, containing some fragments.

I take the liberty to submit these few remarks. If after all you think the title unproper, I propose to replace it by the following: "On a new Mode of representing Space in Geometry."

When at first I communicated the contents of my Paper to Prof. Sylvester, he intimated that there might be some connection between my Paper and one of his, but as I have since heard nothing from him I conclude that this is not the case. If you think it desirable I will write myself to Prof. Cayley who has always been very friendly toward me.

I am, My Dear Sir, Yours truly,

PLÜCKER.

The sheet proof will reach me at Bonn till August 19.

BONN, 21*st of March*, 1865.

The necessity of communicating with Dr Hittorf of Münster had caused a delay in my reply to your letter.

As to your verbal corrections to my bad English I have only to express thanks and to regret for giving you so much trouble*.

I have called at Henry's. Five stones are printed; there remain five the printing of which will require two or three months, as some colours need a longer time. I was quite satisfied with what I saw. The stones will be preserved until the safe arrival of the plates is announced.......

(Page 19, art. 53, l. 7). You put the question "Is it true of Acetylene C_4H_2 which according to Berthelot is formed from its elements by the electric discharge?" The question is most decidedly to be answered in the affirmative. There is I think

* 'On the Spectra of ignited Gases and Vapours, with especial regard to the different spectra of the same elementary gaseous substance,' by J. Plücker and W. Hittorf, *Phil. Trans.* (Feb. 23), 1864, pp. 1—29.

no contradiction. But as the same question might be put by others I think it suitable to intercalate at the end of art. 54 the following :

"Acetylene, C_4H_2, though according to Berthelot and Morren formed from its elements when Davy's coal-light is produced within an atmosphere of Hydrogen, nevertheless when introduced into our tubes it is rapidly decomposed by the discharge and most incompletely recomposed after the discharge is passed. The inside of the tubes is instantly blackened; and in the first moment only, with the spectrum of Hydrogen, we perceive the groups of coal-lines seen in the case of Olefiant gas."

Mr Hittorf writes:—As we worked on the coal spectrum Berthelot's observation was rather new. I prepared the gas in the way commonly made use of by Berthelot, &c., &c.

(Page 20, end of art. 56). My mind is not quite settled on the question whether there be in some instances a spectrum of an undecomposed gas as A. Mitscherlich supposes. The reference to art. 61 concerns the mere fact that the absence of a gas is not proved by the absence of its spectrum.

With a former letter Dr Hittorf sent me an evacuated tube which does not permit the discharge neither of Ruhmkorff's smaller coil nor of a Leyden jar. I intended to send it to you when I had opportunity. In his last letter he writes: "I made progress in constructing evacuated tubes. I got one, with platina wires nearly in contact, not permitting the discharge of Ruhmkorff's *large* coil to pass if the length of the spark in open air does not exceed four inches."

Lately my attention has been engrossed by certain brilliant phenomena presented by calcareous spar, proper to be submitted to mathematical analysis. I gave an account of the results obtained in our Philosophical Society to which a short abstract refers in the *Cologne Gazette*. I take liberty to add it to my letter, though I fear on account of its brevity it may not be sufficient to give an idea of the subject, connected as well with the theory of luminous meteors as with the structure and genesis of crystals.......

My Dear Sir, Most truly Yours,

PLÜCKER.

From THOMAS GRAHAM.

4, MORETON SQUARE, W.C.
6 *July* 1861.

MY DEAR SIR,

I am much gratified by your attention to the condition
of the colloidal solutions described in my paper, and shall take an
early opportunity to forward a few specimens, as you desire, for
your more particular study.

After reflecting upon your remarks I am inclined to think
that you will obtain some light from the observation of silicic acid,
one of these solutions which has the advantage of being colourless.
The liquid condition of all such bodies, it is to be remembered, is
not permanent. Now in silicic acid the transition to the gelatinous
(solid) form is visibly preceded by a faint opalescence of the liquid.
This gradually increases during a few hours or even days, and is
sometimes very beautiful. But it is sure to end, sooner or later,
in the somewhat sudden solidification of the mass. The previous
opalescence may very well be due to suspended solid matter, like
Faraday's highly divided gold, as your theory supposes. It is
however the effect of an incalculably minute amount of suspended
matter, and does not touch the great mass of colloidal matter
present. In short the opalescence may be due to suspended
matter, although the colloid is truly liquid. Indeed one or two
per cent. of such substances generally produce a firm jelly, on
passing from the liquid condition, while the solutions operated
upon were perfectly limpid with from 3 to 10 per cent. of substance
in solution.

That colloids are really in solution appears also to follow from
the fact that they *are* diffusive. They possess the property, and
of several the rate has been accurately observed. They may be
twenty or fifty times less mobile than any crystalloids; but there
appears to be nothing liquid that is absolutely non-diffusive.
A liquid colloid also, such as caramel, is found to diffuse through
a jelly, as well as in pure water.

I have only further to add that the permanent or nearly permanent opalescence seen in ferrocyanide of copper is by no means a common property of colloidal liquids. Indeed the last named and the metaperoxide of iron are the only liquids of the class in which it has been yet observed. The property is particularly decided in the last solution, as was first observed by Peau de St Gilles. It seems hazardous, therefore, to infer much respecting the constitution of colloidal liquids from what appears to be an unessential character.

At the same time the solution of a colloid must be something very different from ordinary solution, as we know solution in the case of crystalloids. It would be interesting to know whether the difference in the two forms of solution is marked by any optical distinction between them.

Believe me, My Dear Sir,

Very faithfully yours,

THO. GRAHAM.

ON SPECTRUM ANALYSIS.

PROF. STOKES TO SIR J. LUBBOCK.

4, WINDSOR TERRACE, MALAHIDE, Co. DUBLIN.
4 *August*, 1881.

DEAR SIR JOHN LUBBOCK,

I got your proof here*, which was forwarded from Armagh. I expect to be going there tomorrow. I saw Miss Stokes the day before yesterday, and gave her your message.

I have read your proof, and have made some reference marks in the margin, numbers enclosed in circles. I also corrected a few misprints which I happened to see. It is no trouble to correct them, and they might be overlooked, especially by the author, who is apt to read a proof as it was meant to be.

In what follows, the numbers refer to the reference marks.

1. A few of the more conspicuous of the fixed lines were discovered by Wollaston near the beginning of the century, by

---

* See Sir J. Lubbock's (now Lord Avebury) Presidential Address to the British Association, Jubilee Meeting, York, 1881, *Report*, pp. 25—48.

viewing a slit through a prism applied to the naked eye. This observation however remained but little noticed.

2. Long ago, but I have not got the date, and have not reference books here, Fox Talbot showed that the red given to a flame by strontium and the red due to a salt of lithium might be at once distinguished by the prism. I think too, but I am not sure, that he dwelt on the delicacy of this test for the detection of lithium. I do not recollect whether this was before or after 1835 [Wheatstone's experiments]. It is to Kirchhoff and Bunsen that we owe the great impetus that has been given to spectrum analysis, and perhaps it would be proper to mention their names in this connexion; it rather looks as if they had merely applied what had been done by their predecessors. It is true however that it cannot be called their discovery.

3. The exact coincidence of position in the particular case of the double line D, between the dark line of the solar spectrum and the bright double line D seen in certain flames, had long before been demonstrated by Fraunhofer. It was not however then known that the bright line was due to the presence of a compound of sodium.

Foucault was the first to obtain the double *dark* line D artificially, by passing strong light through an electric arc, which itself alone gave out the bright line D. He did not apparently attempt to connect the presence of D with any particular substance. (*L'Institut*, 1848.)

I know I think what Sir William Thomson was alluding to*. I knew well, what was generally known, and is mentioned by Herschel in his treatise on Light, that the bright D seen in flames is specially produced when a salt of soda is introduced. I connected it in my own mind with the presence of sodium, and I suppose others did so too. The coincidence of position of the bright and dark D is too striking to allow us to regard it as fortuitous. In conversation with Thomson I explained the connexion of the dark and bright line by the analogy of a set of piano strings tuned to the same note, which if struck would give out that note, and also would be ready to sound it, to take it up in fact, if it were sounded in air. This would imply absorption of the aerial vibrations, as otherwise there would be a creation of energy. Accordingly I accounted for the presence of the dark D in the solar spectrum

* See *Math. and Phys. Papers*, vol. IV. Appendix.

by supposing that there was sodium in the atmosphere, capable of absorbing light of that particular refrangibility. He asked me if there were any other instances of such coincidences between bright and dark lines, and I said I thought that there was one mentioned by Brewster. He was much struck with this, and jumped to the conclusion that to find out what substances were in the stars we must compare the positions of the dark lines seen in their spectra with the spectra of metals etc.

On the strength of this conversation, and of his having mentioned the thing in his lectures to his class, he tried to make out that I was the first to point out the existence of sodium in the sun. I think he was quite wrong; for if a man's private conversations with his friends are to be brought in, there is an end to all evidence that such a man suggested or pointed out such a thing. I merely mentioned the circumstance to explain to you what Thomson had said to you; I think what you have written is quite right.

I should have said that I thought that Thomson was going too fast ahead, for my notion at the time was that though a few of the dark lines might be traced to elementary substances, sodium for one, probably potassium for another, yet the great bulk of them were probably due to compound vapours which, like peroxide of nitrogen and some other known compound gases, have the character of selective absorption.

4. I confess I thought Ångström's statement somewhat indefinite. I have it not by me now to refer to, and I am not now going to Armagh till Saturday. I look myself on Balfour Stewart and Kirchhoff as the two who independently of each other made the important extension of Prevost's Law of Exchanges from which the whole follows. I am disposed myself to give the priority to Stewart; for though Kirchhoff was the first to *publish* the thing with reference to *light,* Stewart had already some time before published it for radiant heat; and we have the strongest reason for regarding radiant heat and light as identical in nature; heating and illuminating being merely two different effects of one and the same kind of radiation, though according to the wave length one kind of radiation shows itself mainly or perhaps only by the heating effect, and another mainly by the illuminating effect.

5. Kirchhoff announced potassium in the sun on the strength of the supposed coincidence in position of the principal red bright

line of potassium and the dark line A of the solar spectrum. Afterwards by employing greater dispersion Kirchhoff found that the bright line of potassium in the red was distinctly less refrangible than the dark line A, so that the supposed evidence of the existence of potassium in the sun disappeared. Comparatively recently Lockyer has obtained evidence of its existence. His researches led him to perceive in what direction it was to be looked for, I mean as regards what particular lines, and so, guided by theory, he has established its existence.

Your statement as to what Kirchhoff thought he had done is quite correct as to potassium, but he afterwards corrected it.

6. I imagine that the jets come from the interior of the sun, and pierce the chromosphere; or rather perhaps that the ordinary chromosphere is nothing but an assemblage of little jets of a similar kind.

7. When the nebulae were supposed to be clusters of stars, they were imagined to be very distant. But now that the planetary and certain other nebulae are known to be gaseous, there is no reason to suppose them on the average more distant than the stars on the average. No attempt, so far as I know, has been made to determine their annual parallax, which could only I suppose be done by observing their stellar nuclei when they have such, as the nebulae themselves would be too vague to observe with sufficient precision. We have no evidence one way or other, and we have no reason *now* to conjecture that they are more remote than the stars.

8. I think, according to Lockyer's views at least, the greater simplicity of the spectra of the nebulae would indicate a *higher* temperature than that of the stars.

9. Doppler does not seem to have made even a rough calculation with actual numbers corresponding to any probable movements of the stars, and I don't know whether he would have been competent to make it; if he were, it seems very strange that he should not have done so. Doppler's proposal is really absurd when quantitatively considered in the roughest manner, the change of refrangibility being utterly inadequate to produce an effect on the colour which would be sensible. Had the change been such as to bring the matter within the possibilities of determination by a change of colour, the method would still have been open to the objection you mention. I have no objection to what you

say except that the serious entertaining of the objection you name seems to conceal the inherent absurdity—when quantitatively considered in the roughest way—of Doppler's proposal, and to give him more credit than he deserves for throwing out so crude an idea.

10. The word "suggested" seems to me merely to imply the throwing out of a new idea. From the time of Prout at any rate the idea was not new, and it must be in the minds of all chemists. It seems to me that "suggested" is giving small praise. I would suggest "Brodie's researches also naturally fell in with the supposition that the so-called......complex."

11. If you mean putting common salt or other compound of sodium on the wick, it requires a stronger heat than that of a spirit lamp to bring out the lines of sodium, other than the double line D. If you mean burning metallic sodium, it does not need a spirit lamp to keep it burning once it is set on fire. I do not happen to know whether burning sodium would show more than the double D. The way the other lines are brought out is most commonly by an electric discharge. It is likely enough, but I do not happen to know, that some of them at least might appear if a compound of sodium, or sodium itself, were introduced into an oxy-hydrogen jet.

12. It was in 1852 that my first paper [on fluorescence] was published. I recollect now for certain that it was, for a few months after it came out I gave an evening lecture at the meeting of the British Association in Belfast.

I had been to a certain extent anticipated, but I do not know that it is necessary to go into this. Herschel had noticed the extension of the spectrum when the spectrum is thrown on turmeric paper, but had misinterpreted the result. Ed. Becquerel had noticed a similar extension when a spectrum is thrown on certain solar phosphori, and had correctly interpreted the result; though he had not perceived the full bearing of his own observation, and had no notion that the solution of sulphate of quinine with which he was actually working, and of which he expressly mentions the "dichroism," owed its blue colour to a phenomenon of the same nature.

13. Was not one of the great improvements the discovery by Herschel of the hyposulphites, and of the peculiar properties of the alkaline hyposulphites with respect to oxide of silver, which

rendered it possible to fix the image, by removing the portion of the silver salt that had not been acted on, and so preventing the paper from being blackened all over?

14. I do not know what the more correct views refer to. If it was any contemporary of Plato's, or even any person for many centuries afterwards, I doubt if their views, interpreted by the light of modern knowledge, were more correct than the bold guess of Plato, as far as I can gather its meaning from the word you quote. Perhaps, however, Plato meant that fire and heat were substances produced by impact and friction.

15. I mentioned the thing in some lecture (or notice of some kind) but I am by no means the originator of it. I have not worked originally at the subject, and I would rather my name were not mentioned in connexion with it. I cannot say who first pointed out the thing, but I know that Thomson and Tait have both worked much and written much about it. If you cannot find who first showed the weak point, it would be better to say "as has been pointed out."

16. Is it not the distance between adjacent molecules? I don't see how we can possibly get at the size of the molecules, except as to assigning a superior limit.

17. I have not read Sorby's address, and I do not know on what evidence he bases the estimate of the number of molecules in a given volume. I suppose he refers to the speculations against reference mark 16.

18. There are *two* important colouring matters in madder, alizarin and purpurin. Alizarin is the more valuable. Would it not be well to say, the principal colouring matter of madder?

19. "The volumes......spaces." *Apropos* to what does this come in? As regards absorption, what we are concerned with theoretically is the natural times of vibration of the molecular structure, or rather what in a complex dynamical system most nearly answers to the natural times of vibration.

As I do not know whether you are at High Elms or in London, I think it safest to send this to Lombard Street.

Yours sincerely,

G. G. STOKES.

SIR JOHN LUBBOCK, Bart., M.P., F.R.S.

FROM PROF. W. H. MILLER.

DEAR PROFESSOR STOKES,

I made a great mistake this morning*. I got the prisms and telescope in 1836, and I believe I made the observations on the identity of position of the spirit lines with the solar double line D in 1837.

Very truly yours,

W. H. MILLER.

CORRESPONDENCE WITH PROF. H. E. ROSCOE.

OWENS COLLEGE, MANCHESTER.
Feb. 24, 1860.

MY DEAR MR STOKES,

I am sorry to find that you do not agree with the map of the spectrum I sent—I do not know how the difference can have arisen, for I certainly copied, as correctly as I could, the pencil map you lent me some time ago—which too I returned to you after I had copied it.

If you will allow me to keep your map for a day or two I will draw the map again†—keeping the distances the same—so as to correspond with your last.

I found no difficulty in recognising the lines as I drew them. I do not know what became of those between N_4 and Q, but I cannot help thinking that they were not contained in the copy you lent me formerly.

Have you seen in the last no. of the *Annales de Chimie et de Physique* a short note about Kirchhoff's discovery of the probable cause of the coincidence of the bands of light (produced for instance by a soda flame) and the dark lines of the spectrum?

Bunsen and he are working on the subject—one of the greatest interest.

Believe me very truly yours,

HENRY E. ROSCOE.

---

* March 8, 1854? See *Math. and Phys. Papers*, vol. IV, Appendix, p. 372, where the correspondence with Lord Kelvin on this subject is printed.

† Published in the memoir on photo-chemical intensity by Bunsen and Roscoe, *Phil. Trans.* 1859. See *Math. and Phys. Papers*, vol. IV. p. 206.

March 19, 1860.

I am sorry to have been so long in forwarding the copy of the Spectral actions for my paper and your map which I return. I could not find time before this.

Now you will, I hope, be able to recognise the lines—it is, however, difficult to draw them well. I hope that the engraver will not copy my blots and imperfectly ruled lines. The map does not pretend to be else than one by which to identify the position.

You said that in the *Phil. Trans.* for 1859 there is a map. I have not seen it. Perhaps if you have some copies of that as well as of your original paper (in 1852) you would be so kind as to give me one. I should be very much obliged.

I hear from Bunsen that he has detected Lithium in all the Potashes he has examined (by his and Kirchhoff's new Spectral method), also Lithium in 20 grms, of sea water. He mixed Mg, Ba, Sr, Ca, Li, Na, K salts altogether, put some of the mixture on the point of a pin—looked through a telescope and saw at one glance all the substances present! This is something like Qualitative Chemical Analysis!

Sept. 27, 1860.

I believe that I have not written to thank you for a letter which I received in Heidelberg from you respecting Bunsen and Kirchhoff's Paper.

I am extremely obliged by your kindness in looking over the Proof Sheets. I translated the paper in a great hurry on the two last evenings of my stay in England and I daresay you may have found many mistakes.

I saw all the experiments at Heidelberg. The sight of the *Solar* and superimposed *Iron* spectrum is one of the most beautiful and striking things I ever saw. All the bright iron lines (perhaps 50 or 100 in the field at once) exactly coinciding with dark lines in the solar spectrum. They have now found in the Sun's atmosphere Sodium, Iron, Magnesium (one of Fraunhofer's lines *b*), Potassium, Chromium, and Manganese, probably ere this many other bodies are discovered.

Steinheil of Munich is now making the instruments, but the first trial one is not yet finished.

You of course will have heard that Clifton is our new Professor. We are fortunate to get so good a man and I shall be very glad to have a Physicist near me.

Feb. 6, 1862.

I am about to give three lectures on "Spectrum Analysis" at the Royal Institution in Albemarle Street, and in one of these I must enter into the subject of Solar Chemistry. My aim will be to express the true history of these discoveries as accurately as possible, and I should feel greatly obliged to you if you would kindly inform me of the facts concerning the part you have taken in this portion of the subject. Of course I have read the letter you wrote to the *Phil. Mag.* calling attention to Foucault's experiment of the production of the absorption sodium lines, but I can find no other statements made by you on the subject before that time. I am induced to ask you for further information upon the point from a letter in the Feb. *Phil. Mag.* from Prof. W. Thomson, in which he speaks of "Stokes's *Theory of Solar and Stellar Chemistry*" having been taught for 6 to 8 years in his Classes at Glasgow, and I cannot find any references to the Theory except the letter above referred to.

You are doubtless aware that in 1855 Ångström expressed his opinion that the "explanation of the dark lines of the Solar spectrum involves that of the bright electric metallic lines," and that "when taken altogether the two spectra present the appearance of one being the reverse of the other." (*Pogg. Ann.* Bd. 94, p. 145.) But he does not express any opinion as to the composition of the Solar atmosphere.

I see that Crookes, the Editor of the *Chemical News*, wishes to claim for W. Allen Miller the discovery of the power of luminous gases to absorb light of the same kind as they emit, but I cannot believe (from the paper as published) that Miller could think of claiming this point. His sentence on this subject is not only vague, but, as I read it, positively incorrect.

May I refer also to your experiments showing the existence of metallic rays of very high refrangibility which I saw at Cambridge ?

LENSFIELD COTTAGE, CAMBRIDGE.
Feb. 7, 1862.

DEAR MR ROSCOE,

My share in the history of solar chemistry, I look upon it, is simply *nil*; for I never published anything on the subject, and if a man's conversations with his friends are to enter into the history of a subject there is pretty nearly an end of attaching any mention or discovery to any individual.

As well as I recollect, what passed between Thomson and myself about the lines was something of this nature. I mentioned to him the repetition by Miller of Cambridge of Fraunhofer's observation of the coincidence of the dark line D of the solar spectrum with the bright line D of certain artificial flames, for example a spirit lamp with a salted wick. Miller had used such an extended spectrum that the 2 lines of D were seen widely apart, with 6 intermediate lines, and had made the observation with the greatest care, and had found the most perfect coincidence *. Thomson remarked that such a coincidence could not be fortuitous, and asked me how I accounted for it. I used the mechanical illustration of vibrating strings which I recently published in the *Phil. Mag.* in connexion with Foucault's experiment. Knowing that the bright line D was specially characteristic of soda, and knowing too what an almost infinitesimal amount suffices to give the bright line, I always, I think, connected it with soda. I told Thomson I believed there was vapour of sodium in the sun's atmosphere. What led me to think it was sodium rather than soda, chloride of sodium, &c. was the knowledge that gases that absorb (so far as my experience went) yield solutions that absorb in the same *general* way, but without the *rapid* alternations of transparency and opacity. Now if the absorption were due to vapour of chloride of sodium we should expect that chloride of sodium and its solution would exercise a general absorption of the yellow part of the spectrum, which is not the case. Thomson asked if there were any other instances of the coincidence of bright and dark lines, and I referred to an observation of Brewster's relative to the coincidence of certain red lines in the spectrum of burning potassium and the lines of the group a of Fraunhofer. I am nearly sure this is in a volume of the Reports of the British Association, being analogous

* Cf. *supra*, p. 80.

to but not identical with Brewster's observation in the Report for 1842, pt. 2, p. 15. (Since I wrote this I have looked through the indices of the volumes of the Reports of the British Association and do not find it there.) Thomson with his usual eagerness said, " Oh, then, we must find what metals produce bright lines agreeing in position with the fixed dark lines of the spectrum," or something to that effect. I was, I believe, rather disposed to rein him in as going too fast, knowing that there were terrestrial lines (seen when the sun is low) which evidently take their origin in terrestrial atmospheric absorption, where metals are out of the question, and thinking it probable that a large number of lines in the solar spectrum might owe their origin to gaseous absorption of a similar character in the solar atmosphere. Even now I think it likely that some of the non-terrestrial lines in the solar spectrum may be of this character, though after what Bunsen and Kirchhoff have done I think it probable that they are a minority.

The idea of connecting the bright and dark lines *by the theory of exchanges* had never occurred to me, and I was greatly struck with it when I first saw it, which was in a paper of Balfour Stewart's read before the Royal Society and printed in the *Proceedings*. I was wrong in saying *lines*, for B. Stewart considers only solids, the spectra of which don't present such abrupt changes. Stewart's paper was independent of, but a little subsequent to, Kirchhoff's, though the same idea with reference to radiant heat occurs in two papers of his printed in the *Edin. Phil. Trans.* and much anterior to Kirchhoff's paper. These papers I was not acquainted with at the time when Stewart's paper on light came before the Royal Society.

We can by no means affirm from the theory of exchanges that every dark line in the solar spectrum must be capable of reversion. For it may be due to absorption by a compound gas which is incapable of existing undecomposed at the temperature requisite for becoming luminous, or which though not decomposed might yet have its mode of absorption completely changed, as we know that even a small elevation of temperature is sufficient materially to alter the absorption of light by NO_4 gas.

As to the mention of the metallic lines in the invisible region, I own to feeling a wish that the subject may be novel when I bring it forward, and yet I can't help feeling that that is mere selfishness and that I have no business to keep it bottled up.

When are your lectures to be given? If not for some time perhaps I may draw up a note for the Royal Society for publication in the *Proceedings* which would of course set you quite free *.

Yours very truly,

G. G. STOKES.

P.S. Do you know anything of Mr C. H. Akin, Ph.D., a Hungarian, who studied at Heidelberg and has seen you there? He has come to Cambridge for the purpose of study.

To Mr T. L. PHIPSON.

SPRINGFIELD, TIVERTON,
22 *Sept.*, 1869.

DEAR SIR,

Absorption bands are certainly not a phenomenon of *interference*. In interference there is *no loss of light* but only a *different distribution of illumination*. Thus in the case of interference mirrors or an interference prism, if we call 1 the intensity of either stream, if the two streams simply mixed without interference we should have a uniform intensity of $1+1$ or 2, but with interference we have a series of bands bright and dark with intensities 4 and 0. The mean intensity is 2 the same as before.

All that we can regard as positively settled as to the nature of absorption is that the incident vibrations are spent in producing molecular disturbances. What the nature of these disturbances is, is a matter of speculation. I don't think that in solids and liquids they have anything like the regularity of the vibrations of a pendulum, and consequently we cannot strictly speak of their periodic time. In the gaseous forms of matter we have evidence (in the bright spectral lines) of an approach to this regularity. I say approach, because the lines are not in general *mere* lines, but narrow bands. The same gas incandescent never different, but in different parts of the spectrum may show what are nearly

---

* "On the Long Spectrum of the Electric Light," published in *Phil. Trans.* (July 19), 1862; *Math. and Phys. Papers*, vol. iv. pp. 203—237.

bright *lines* and what are *bands* of very decided breadth. Different kinds of vibrations can go on simultaneously, and in the case of bright merely-lines would produce nearly a pendulum-like regularity. To the extent of approximation to which we may speak of narrow bands as mere lines, to that extent will the corresponding vibrations be regular, of a definite period, and the periodic time of the light absorbed will agree with that of the light emitted. But in the case of those molecular agitations which are not of that regular character the precise relations between the incident vibrations absorbed—used up—and the agitations of the ponderable molecules thereby produced are not I apprehend at present known. I have some theoretical views on the subject, but they are at present in an immature state.

I am, Dear Sir, Yours very truly,

G. G. STOKES.

P.S.   Thank you for the papers which I duly received.

EXPERIMENTS ON OPTICAL GLASSES.

LETTERS TO REV. W. V. VERNON HARCOURT, F.R.S.
(Selection.)

(From Obituary Notice of Rev. W. V. V. Harcourt, *Roy. Soc. Proc.,* Vol. 20, 1872.   *By Prof. Phillips ?)*

Accustomed to the use of the gas-furnace, Mr Harcourt turned it to experiments on transparent compounds of fusion, which might be made to have refractive indices beyond the ordinary ranges, combined with scales of dispersion more favourable to achromaticity.   In this he was guided by the trials of Faraday to prepare glass for optical purposes. Many years since, the writer, who was often helpful in this way, ground one of the earliest of the Harcourt glasses into a lens, and found it indeed a highly refractive clear substance, but too much traversed by striæ to be of practical use.

When, some years since, Mr Harcourt removed his residence to the family seat at Nuneham, near Oxford, he constructed furnaces of a different kind for the carrying on of these experiments, and followed them with the zeal, resolution, and patience which had always characterized his firm and well-regulated mind.

At an age when most men cease from continuous literary and scientific work, he with failing sight, but perfect memory, was indefatigable in training an assistant and superintending his work; making many new combinations with substances untried before, and now selected for quality of fusion, resistance to atmospheric vicissitudes, range of refraction and specific action on different rays of the spectrum. Thus it was hoped finally to acquire glasses of definite and mutually compensative dispersions, so as to make perfectly achromatic combinations. After innumerable trials, and the production of glass of extremely various quality, Mr Harcourt, continuing his inspection to the last, had the satisfaction of believing that, though he could not remain to witness it, a good result had been assured, and that Professor Stokes, to whom all the specimens were submitted for scrutiny, would be able to construct a lens of sufficient size to be fairly tried, and thus crown the long-continued labour with a permanent benefit to science.

It was hoped that a full account of these experiments might have been prepared by the author of them, for his strong mind felt little of the weight of eighty years and overruled the bodily infirmities of age. But it was a character of the man never to cease experimental or literary research till he was satisfied; resolute to contend with difficulties till all were overcome, and too truly a lover of knowledge, with faith in its progress, to be hasty in publishing views on account of their novelty, which might be made valuable by proofs of their truth. Prof. Stokes has already presented to the British Association a notice of these researches[*], and to him we must now look for further records of a work to which he has cheerfully contributed a large amount of valuable aid.

(*From Obituary Notice of* Sir J. F. W. Herschel, *same vol.*,
by Dr Romney Robinson.)

His contributions to optics rank next to his astronomical in importance and number; but we shall only mention two, which gave a great impulse to the progress among us of this branch of physics. The first is a remarkable memoir on the aberrations of compound lenses and object-glasses, which appeared in our *Transactions* for 1821. Before it opticians (at least in this country) corrected the spherical aberration of their object-glasses by empirical rules, and its theory was given in rude and unsymmetrical forms. By a happy choice of symbols and an elegant

---

[*] Read to the British Association Meeting in Edinburgh, August, 1871. [*Math. and Phys. Papers*, iv. pp. 339—343 ; also a later account of an actual achromatic telescope made with the new glasses, presented to the British Association at Belfast, 1874, *loc. cit.* pp. 356, 357 ; also further tests on glasses, *loc. cit.* pp. 358—360.]

analysis he presents the theory of aberration in all its generality, and in as simple a manner as the nature of the question admits. He gives examples of the application of his theory to the construction of aplanatic doublets, and then to that of object-glasses. In discussing this he considers the dispersive power as composed of several terms, of which the first only is taken into account by opticians, and the rest constitute the irrationality of the spectra. This defect cannot be removed in a double object-glass unless the dispersive powers of higher orders are proportional to the first; and he recommends that the attention of future inquirers should be directed to the discovery of such a medium (a result which there is reason to believe has at last been obtained by the combined investigation of the late W. Vernon Harcourt and Professor Stokes).

The condition which he assumes for correcting the spherical aberration is, that the compound shall be aplanatic for near as well as remote objects. He tabulates the results for the various crown and flint glasses then used in England with a completeness which leaves nothing to be desired. It may be feared, however (notwithstanding the popular explanation of his method which he afterwards published in the sixth volume of the *Edinburgh Philosophical Journal*), that this paper has had little influence on the practice of British workmen, who then and now are far behind the requirements of the age in the knowledge of geometry.

CAMBRIDGE,

6 *Aug.* 1863.

On laying down on paper the deviations for the four omitted prisms of which I have got the angles, those on No. 7 seemed not so bad considering the great waviness of the glass, and No. 29 seemed tolerable if "air" (respecting which there is some enormous error) be struck out. So I calculated the indices and send the results. No. 27 was not thus cleared up, and must be remeasured. In No. 7 I have taken the means of v_1 and v_2 for the superior limit.

I send also a table of dispersive powers thus calculated. I take the widest interval of the spectrum for which the observations can be trusted, and the corresponding interval in the other, and find the ratio $\dfrac{\Delta\mu}{\Delta\mu_0}$, μ_0 referring to the standard. I then regard

$\dfrac{\Delta\mu}{\Delta\mu_0} \times (\mu_{0_H} - \mu_{0_A})$ as sensibly equal to what $\mu_H - \mu_A$ would have been if it could have been observed, and take $\dfrac{\mu_H - \mu_A}{\mu_D - 1}$ as the dispersive power. The dispersive power estimated as from B to H may be got from this with very little error by multiplying by $\dfrac{\mu_{0_H} - \mu_{0_B}}{\mu_{0_H} - \mu_{0_A}}$ or $1 - \dfrac{\mu_{0_B} - \mu_{0_A}}{\mu_{0_H} - \mu_{0_A}} = 1 - \dfrac{36}{462}$ or $1 - \cdot 078$. The greatest error from irrationality which this would entail (supposing the $\dfrac{\mu_H - \mu_A}{\mu_D - 1}$ were exact) would be in the case of the low dispersive glasses, and even then would be only about ·0003, by which the $\mu_H - \mu_B$ would be made too small.

Sir David Brewster remarked long ago that in *most* cases the more dispersive of two media is also the more purple-refracting*. Hence green refracting power does not *alone* hold out much prospect of obtaining a substitute for flint glass which shall compensate crown without leaving a secondary spectrum. We must enquire whether the medium is *more* green-refracting than corresponds to its position in the scale of dispersive power.

I have tried a large number (not far short of 100, a good deal more than 50 at any rate) of compensations, chiefly with the better prisms (those which showed the dark lines and a few of the others which I was able to manage), and *almost* without exception brought the prisms out in the same order as that in which they are arranged by their dispersive powers. No. 24 seemed rather purple-refracting for its d. power. Nos. 2, 13, 14, 32, come out not much different from one another, the order appearing to be

$$\text{purple,} \quad \left.\begin{array}{c} 14 \\ 2 \end{array}\right\} \text{ nearly equal, 13, 32, ---, green;}$$

but 2 is much lower in refractive power. Hence either 2 is purple for its d. p., or the group 14, 13, 32 green considering its d. p. A Munich crown prism was distinctly green to 2 and distinctly purple to your crowns 16, 18, 19. Your crowns and Nos. 20, 22, 27, 25 were all nearly equal as to greenness. It struck me that No. 25 might have been expected to be greener than it seemed, considering its low d. p.

I send a graphical table of compensations. The result leaves

---

* "A treatise on new philosophical instruments." The results are noted in Herschel's "Light," *Ency. Metrop.*

a poor prospect of destroying the secondary spectrum or doing more than diminishing it. With fluids Dr Blair had a wider range of dissimilarity at command than is I suppose to be had with glasses.

Is there anything remarkable in the composition of Nos. 13, 14, 32? They seem to be somewhat greener than belongs to their place in the scale of dispersive powers.

I must pit your prisms against flint glasses to see if on the whole they come out more green-refracting for equal dispersive powers. If so they may be substituted with advantage for flint glass (provided of course they can be prepared in a homogeneous state, *i.e.* free from striæ, &c.), though a diminished secondary spectrum would still remain.

20th Oct. 1870.

MY DEAR MR HARCOURT,

The correction for achromatism can be effected, so far as irrationality permits of its being effected at all, by means of spherical surfaces, and the deviation from the spherical form if made would not help out the correction. It is the correction for spherical aberration, not achromatism, which admits of being made by deviating from the spherical form. Dr Pritchard must have fallen into some strange confusion on this point if it were not a mere slip of the tongue.

After making this correction, the question raised in your letter comes to this—Why cannot the correction for spherical aberration be made by means of spherical surfaces?

The answer is, It *can* be made (to the lowest order of approximation at least, to which alone the thing is sensible with the ratio of aperture to focal length to which the defect of irrationality compels opticians to limit themselves) by spherical surfaces; but in order that it should be so made two things are required; first the ratio of the spheres must be accurately calculated, from data furnished by accurate observations of refractive indices; and secondly the surfaces so calculated must be accurately executed, both as to the truly spherical form of the surfaces, which in the calculation are deemed to be spherical, and as to the agreement of the radii of the executed surfaces with the radii assumed as the result of calculation.

Let us see what the first of the two latter requirements demands. Take the case of an object-glass of 15 inches aperture and 15 feet focal length. If the rays of any one colour, instead of being brought to a point, were brought so as to form a minute caustic surface such that a section perpendicular to the axis was, for edge rays, a circle of the $\frac{1}{100}$th inch radius, I suppose the spherical aberration would be sensible when the glass was tested as to this point. Suppose that the aberration would vanish if the 4 surfaces were all truly spherical, and exists in consequence of an error of one of the crown-glass surfaces, such that it deviates from a sphere in the direction of a prolate or oblate spheroid. What amount of error is implied in the actual surface? The edge rays evidently deviate from their proper direction by the small angle $\dfrac{\text{rad. of small circle}}{\text{focal length}} = \dfrac{\cdot 01 \text{ inch}}{15 \times 12 \text{ inches}} = \frac{1}{10000}$ in circular measure. Assuming $\mu = 1\cdot5$ for the crown, this would imply an angular deviation in the direction of the surface at the edge of $\frac{1}{9000}$. From this we get by an easy calculation for the distance between the spheroid and its sphere of curvature, at the edge, where it is greatest, $\frac{1}{8}$ aperture × angle $= \frac{15}{8} \frac{1}{9000}$ inch $= \frac{1}{4800}$ or in round numbers the $\frac{1}{5000}$th of an inch. We see with what amazing accuracy the surfaces must be executed spherical, be they calculated ever so accurately, and be the executed radii ever so near the calculated radii, in order that the spherical aberration may be destroyed independently of trial and correction.

As to the first requirement, few I imagine, if any, of our English practical opticians possess the requisite mathematical knowledge. By trial, aided by calculations more or less rough, they get pretty near the mark, and leave the rest to trial and correction.

As to the mode of correction, I imagine from hints which have dropped out that most of our practical opticians in making a large glass, after they have got the spherical aberration moderately small, and satisfied themselves as to the destruction of chromatic aberration, correct for the residual spherical aberration by polishing the glass in zones by trial. This is a bungling process compared with that which Grubb employs—less scientific and incomparably more difficult.

Next as to the statement that "Grubb had committed a serious error in regard to spherical aberration in a glass made,

I think, for the Royal Society in consequence of the object by which he judged his work being at too short a distance."

I happen to know the circumstance in which I have no doubt this statement originated. Grubb had been commissioned to make a 12-inch object-glass, but before the work was complete the order was withdrawn, in consequence, I think, of an object-glass having been bequeathed to the party ordering, compensation I presume being made to Grubb in some way or other. He still has this glass in his hands. Some years ago Howard Grubb, the son, who was coming over to Cambridge, brought the glass with him, that it might be tried here, for it happened to be of very nearly the same focal length as the Northumberland telescope here, so that the same fittings served for both. Before sending it over they tested it for spherical aberration, using for some reason (I forget whether it was badness of weather or occupation of ground) an object at a less distance than this had been used to, I think only 7 or 8 focal lengths, and made it right for that. On being tested in Cambridge on stars, the object-glass, though otherwise good, was found to have a very sensible negative spherical aberration. The object-glass therefore in which this defect was noticed was not one which they had turned out of their hands, but merely hastily corrected for a temporary trial. It is likely enough that this circumstance made Grubb more keenly alive to the necessity of attending to the distance of the test object; but that *now at any rate* he is fully alive to it is proved by the following passage in a letter from him to me dated Feb. 10, 1870. He is writing with reference to a refracting telescope ordered for the Royal Society of 15 inches aperture and 15 feet focal length. "We have had more difficulty about the spherical aberration, consequent on the day objects being necessarily at a finite distance. We have two distances of these, say about 150 and 300 feet. The difference of correction for the shorter and that for parallel rays is considerable, and even for the longer distance very sensible." In a later letter he told me that they had got the correction for spherical aberration so complete that one could not tell which way the error lay.

I should feel obliged by your informing Dr Pritchard of these circumstances, because it is not fair that statements based on imperfect information as to the facts of the case should gain currency to the detriment of an optician in whom all that I have

seen leads me to place a very high degree of confidence, as I know
Mr Robinson does, who knows him much better, and has besides
far more practical acquaintance with instruments than I can
pretend to.

Believe me, Yours sincerely,

G. G. STOKES.

17th Dec. 1870.

At last I got sunlight to-day and I was able to measure the
prisms. I send you the results.

I had fancied 103 and 104, which are a little milky, were
molybdic prisms; but from the fact of the violet not being
absorbed, and from the magnitude of $\theta - \theta_c$, I conclude they are
titanic.

Two of the prisms hardly showed H, partly in consequence
of atmospheric absorption, which much weakens the violet when
the sun is so low as it is at this time of year and the air too somewhat hazy. The measures, however, for H agree very well with
the others.

The disk and...prism have not yet arrived from Darker.

20th Jan. 1871.

I found your letter on my return from London to-day. I felt
confident that there could be no mistake about my having taken
up some other prism and measured it when I thought I was
measuring H 107, but I looked over all the titanic prisms as to
their indices. Only in three cases was there a chance of agreement even with a very liberal allowance for errors of observation,
and then if the indices roughly agreed, the double deviations did
not, in consequence of the angles differing from one another
sufficiently completely to forbid the supposition that it was one
of them I had measured. I am confident there is no mistake
about the identification.

But I don't think the difference between H 107 and the
prisms of nearly similar composition inexplicable from the composition. The most striking difference lies in its great purple
refraction. There was introduced into the prism 107 (if I under-

stand rightly) 2 per cent. of nitre *more than* into the disk 107$^+$ making $2 + 1\cdot9 = 3\cdot9$ per cent. (I don't feel quite sure that it may not be 2 per cent. into 107 against 1·9 per cent. into 107$^+$, but I understand it to be $2 + 1\cdot9 = 3\cdot9$ per cent.) This would be equivalent, apart from its influence in keeping up oxidation, to 1·82 of ANHYDROUS potash. Now acids, especially sulphuric and phosphoric, are remarkable for green refraction. In a table of the order of sequence of 89 substances given by Brewster and quoted by Herschel (*Encyc. Metrop.*, art. Light, § 443) sulphuric acid stands at the head of the list, phosphoric acid second, water 6th, crown-glass 28th, flint-glass 40th. The *anhydrous* acids would probably be still more remarkable for green refraction.

On the other hand I found that solutions of salts, on being compared with the corresponding acids mixed with water, were strongly purple-refracting, the strengths being adjusted so as to give the same total dispersion in hollow prisms of 45°. The difference consists in the substitution of base for water.

Hence, putting the two things together, it is probable that anhydrous potash would be *very strongly* purple-refracting to anhydrous phosphoric acid. A phosphoric glass would be intermediate, and hence the introduction of anhydrous base (actually introduced in the shape of nitre) would tend to purple refraction.

But furthermore the long-continued firing may have acted by expelling water, which adheres so obstinately to phosphoric acid. The analyses of glacial acetic acid quoted by Gmelin are extremely discrepant. While theory requires 11·19 per cent. water, H. Rose got 7·3 to 9·48, Peligot 12·55, Dulong 17·08, Berthollet about 25. Peligot's was heated to redness. Rose's was heated so much as to leave a mixture of metaphosphoric acid and anhydrous acid. The long firing of H107 may have expelled water, which possibly may have raised the purple refraction more than corresponds to the elimination of the dispersion.

The condition of stability of a glass requires the acid and base to be adjusted within very close limits, and so prevents the green refraction or purple refraction due to an excess of acid or base from being much visible. Yet long ago I thought I perceived its effect in the case of the antimony prisms. The earlier cases were yellow, and unstable from excess of alkali, and these showed purple refraction; but when by the addition of a little more acid they were got colourless and stable, the purple refraction disappeared. And the most *determinately* purple-refracting of the molybdic prisms seemed to me to be the early brownish ones, which were unstable from excess of alkali.

21*st Jan.* 1871.

Since I finished my letter this morning I have calculated from Fraunhofer's indices the place of water (as to green or purple refraction) in the crown-flint series, and find it, as I found it before, a *little* greener refracting than corresponds to its place. According to a certain graphical construction which I have employed, if the places of glasses be laid down on a plane, and if the points representing two glasses be joined by a straight line, the points of that line will represent, as experiment shows, the places of glasses of intermediate composition. If, in the figure, *C*, *F* represent the places of crown and flint-glass, and downwards represent the side of green refraction, the place of water may be represented by such a point as *W*, and that of a glass such as $H108$ by a point such as *H*. Join *WH*, and produce it on the side *H*. The place of a glass made of $H108$ with the addition of a little water would be represented by a point on the line $H\overline{W}$ near *H* towards *W*, and the place of a glass such as $H108$ with water *abstracted* would be represented by a point such as *h* in *WH* produced. The glass *h* would be more determinately purple-refracting than *H*.

Purple-refracting.

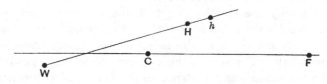

Green-refracting.

I feel pretty confident that both causes I have suggested, increase of alkali and diminution of water, would tend to purple refraction, though I should not have supposed beforehand that such small changes would have made so much difference.

P.S. I will send you back the disk, for an examination of it, now that it is polished below, might throw light on the cause of the striae it shows near the middle.

27th Feb. 1871.

I got a little sunlight to-day, and finished the indices of H109—112. I send you the results.

The 1·9 per cent. of nitre is found to have a considerable effect in H111, and little effect, indeed if anything apparently rather a reverse effect, in H105. The additional 2 per cent. of nitre in H107 and 112 seems to have had but little *additional* effect beyond the 1·9 per cent. in H111.

These results puzzled me at first, but I think I see how they may be accounted for. I suspect that it requires very sharp heating to decompose the nitre further than to a nitrite; and that while anhydrous potash tends strongly to increased dispersion and purple refraction, nitrite of potash tends slightly the other way, the increased dispersion and purple refraction due to the KO being rather more than neutralized by the diminished dispersion and the green refraction due to the NO_3. I suspect that in H105 the heating after the addition of the nitre was sufficient only to convert the nitrate into a nitrite, while in H111 ($= 107^+$) the nitrite was decomposed; and that the additional 2 per cent. of nitre introduced into H107 and 112 remained in great measure a *nitrite* of potash.

This view was I believe suggested to me by speculating on the cause of the ultramarine colour of H105. I thought at first it might possibly indicate a slight reduction of the titanic acid, but in heating a strip from the waste end of H112, which Darker sent me, on platinum wire before the blowpipe in the reducing flame I got only the amethystine tint, nor did the addition of phosphoric acid enable me to get the ultramarine. The colour of H105 made me think of the colour of the nearly spent acid in a Grove's cell. The nitrite theory seems to me very probable, though I have much less confidence in the colour of H105 being due to a nitrite.

I should be glad to know if the circumstances of the preparation accord with my supposition.

I hope to hear that you are recovered or recovering from the illness you mentioned.

P.S. Feb. 28. I may as well mention that the disk 107$^+$ contained a few crystals in stellate groups of 3 or 4 sharp crystals. These were not found in H107, but were pretty numerous in

H112. The glass *immediately* surrounding them was less refractive than the general glass, but they had no tails, whence I conclude that they were formed when the glass was too pasty to allow of currents of convection. The few crystals in the disk were no sensible detriment to the soundness of the glass.

28th Feb. 1871.

I should rather prefer the lower numbers of H104 to the higher of H111—112 for another disk. According to my measures, the irrationality would be a little over-corrected with H111, and besides, the stability of the glass has not yet been tested. It is possible there may be too much alkali in H111 for stability, though as yet the glass has shown no sign of tarnish.

I thought it very likely the bluish colour of H105 might be due to some impurity. Whether it was the colour that suggested the explanation of the difference between H105 and H111 which I gave in my letter posted this morning I don't recollect for certain. I rather think it was the other way, that having thought of the explanation I was led to connect the colour with that of the fluid in a Grove's cell. Be that as it may, I felt that there were several improbabilities in the way of supposing the *colour* due to a nitrite; but what I pick up in Gmelin as to the decomposition of nitre by heat leads me to think it not improbable that the difference between H105 and H111 as to ϖ and θ may have been due to an imperfect decomposition of nitrite of potash or magnesia in the former.

To Mr W. P. HORN.

25th May, 1872.

My dear Sir,

I am not sure whether the glass experiments you propose are intended (1) to fill up the slight desideratum still remaining in Mr Harcourt's researches, or (2) as an independent research standing in your own name. I will therefore answer on the two suppositions.

(1) As Mr A. Harcourt has undertaken to work it out, or at least to try, it would be proper to leave it in his hands *if he wishes to go on with it.* I rather imagine, however, he would be glad to be relieved of it.

The late Mr Harcourt's researches might very well be published as they stand. But if this one point could be cleared up it would be so much the better. The point is to ascertain, if possible by direct experiment, if a silicic glass containing 8 or 10 % of titanic acid would prove what I called purple-refracting, as the phosphatic glasses containing titanium did *. This would require us to make a silicic glass with titanic acid, good enough to show, when cut, the principal dark lines of the solar spectrum. I fear this would be a difficult matter on account of the pasty character of silicic glasses. I don't think it would be worth while to spend much trouble over it, but if it could be done with a moderate amount of trouble it would be worth doing. I need not say that I should be ready to examine the prism or prisms, and in drawing up a full account of Mr Harcourt's researches to acknowledge fully your completion of the work. In any case I should mention the share you had in carrying out the researches under Mr Harcourt's direction.

(2) If you prefer to work independently I shall still be ready to measure prisms for you if the number is not too great. I am assuming that Mr A. Harcourt does not wish to work the thing out. In that case I should close my account as the researches now stand, and what you might add could stand as an independent paper in your own name. As the researches are published, though as yet in abstract only, I conceive that any member of the public may take the matter up, though if the parties who first engaged in it were known to be going on with it courtesy would indicate that it should be left in their hands. But I cannot do more unless someone supplies the glasses, and if Mr A. Harcourt likes to close the research it is closed as far as the original parties are concerned, and then you as a member of the public could take it up.

P.S. May 28. Before posting the above I received a letter from Mr Edward Harcourt, who told me what he had written to you. I think there is a substantial agreement in what I had written with what he wrote to you, only I don't think that he and I had quite the same idea of what it was that you contemplated doing. It never entered my head to suppose that if we closed the research at the present point, and you were to work out in your own book with my assistance the point as to a titano-silicic glass,

* Answered in the negative, *Math. and Phys. Papers*, vol. iv. p. 357 (1875).

the results would not be published until after the full account of Mr Harcourt's researches had appeared. And on the alternative supposition as to your possible intentions, namely that you contemplated working out this point to be included in the research, that of course depended on Mr A. Harcourt's being willing to hand over the research to you.

As to the use of such glasses as (among other objects) we were aiming at, see Sir J. Herschel on Light, art. 477, in the *Encyclopædia Metropolitana*.

CORRESPONDENCE WITH LORD RAYLEIGH.

TERLING PLACE, WITHAM,
March 7 [1871].

DEAR PROF. STOKES,

May I remind you of your promise and ask you to send me any of your mathematical or physical papers that you can spare? I have the one on communication of vibration to a gas in the *Phil. Mag.* I have sent my paper on Resonance back to the R. S. The additions that I have made are not long, and so I thought it unnecessary to send it to you.

I am, Yours very truly,

JOHN W. STRUTT.

ATHENAEUM CLUB, LONDON,
23rd March, 1871.

MY DEAR MR STRUTT,

Prof. Maxwell called my attention a day or two ago to your paper on double refraction in No. 276 of the *Phil. Mag.* I have just been reading it.

In a paper of mine on some cases of fluid motion read before the Cambridge Philosophical Society and published in the *Transactions* (I should guess about 1843, but the work is not in this Club), I obtained an expression for the equivalent inertia of an incompressible fluid moving relatively to a solid (which I

7—2

supposed symmetrical with respect to 3 rectangular planes) under the form

$$M = M_1 \cos^2 \theta + M_2 \cos^2 \phi + M_3 \cos^2 \psi,$$

where θ, ϕ, ψ are the angles made with the principal axes by the direction of vibration. I saw at the time that this would lead to a theory of double refraction, differing from Fresnel's in having reciprocals of velocities in place of velocities themselves. But having calculated the velocity at 45° to the axis in Iceland spar, I found it to differ from that given by Huyghens' construction by a quantity large enough to deter me from publishing the result without a careful scrutiny of the observations of Wollaston and Malus to see whether such an error could be tolerated *. I had always a hankering after this theory, and developed it for myself much as you have done, and even investigated the form of the wave surface. After my experiments on diffraction came out Rankine published in the *Phil. Mag.* a similar theory. About four years ago I carried out the suggestion in my Report for examination by prismatic refraction on a crystal of Iceland spar, which was cut to my order so as to permit of examination in directions lying within a range including an inclination of 90° and another including an inclination of 45° to the axis. The result was perfectly decisive. The *difference of inertia theory must be rejected, and Huyghens' construction adhered to.* The difference between the results of the two theories is something like 100 times the probable errors of observation. I ought to have published the results before this †.

CAMBRIDGE,
24th May, 1872.

I have just read your paper on gratings forwarded to me by Mr Spottiswoode.

I am glad to find you are so far recovered as to be able to work at these things again.

I will take charge of the paper for the *Proceedings*.

I am not sure whether you have seen a short paper by Bridge in the *Phil. Mag.* for Oct. 1855 (Vol. x, p. 251). You do not refer to it, but it is hardly requisite you should; as it is not for the

---

* See Report on Double Refraction, 1862, *Math. and Phys. Papers*, vol. iv. p. 182.
    † A brief statement was published as above, *Roy. Soc. Proc.* June 1872 : *Math. and Phys. Papers*, vol. iv. p. 336.

idea of using photography for reducing apertures for diffraction experiments that you send in your paper, but for the actual carrying out of the idea in the case of such delicate objects, which is a very different matter.

PORT BALLANTRAE, BUSH MILLS, IRELAND,
10th August, 1872.

I was a good deal hurried in leaving home in consequence of an unexpected interruption the day before, but I did not like to go without looking out for my papers so long promised. I got most of them, but I think I have still some copies of one or two more. The principal ones, so far as I recollect, are (1) On the change of refrangibility of light, No. 1, (2) On the steady motion of incompressible fluids, (3) On some cases of fluid motion, (4) Supplement to (3), (5) On the total intensity of interfering light, (6) On attractions and on Clairaut's theorem. Of some of them I know my copies are out, (1) is in the *Phil. Trans.*, (2) (3) (4) in the *Cambridge Transactions,* (5) in the *Edinburgh Transactions,* (6) in the *Cambridge* (or *Cambridge* and *Dublin*) *Mathematical Journal.*

I was in a great hurry in leaving and had not time to write your name in the papers, nor do more than give my servant hurried directions about sending them by book-post or rail. Please to send me a line to let me know if they have arrived, that if not I may make enquiries. Should you have written already your letter will be forwarded from Cambridge.

P.S. I was forgetting to write a few words about diffraction and its bearing on the direction of the vibrations in polarized light. I too think that one of the strongest arguments is that derived from the polarization of light by reflection from excessively fine particles. I mentioned this as an argument in my paper in the *Phil. Trans.* for 1852 *. The passage can readily be found from the alphabetical index at the end of the paper. I think you asked a question about the result obtained by gratings †. You have got my original paper. My experiments were repeated by Holtzmann, who at first worked with glass-gratings and seemed

---

* The memoir on " Change of Refrangibility of Light," § 183 : *Math. and Phys. Papers,* vol. iii. p. 361, with footnote added in 1901.

† See *Math. and Phys. Papers,* vol. ii. (addition made in 1883), p. 327.

to obtain my results, but found irregularities in his experiments which induced him to have recourse to smoke-gratings. With them he obtained results which led him to the conclusion that the vibrations were *in* the plane of polarization. F. Eisenlohr investigated the dynamics of the problem, but in a way which I never could assent to. Finally Lorenz investigated the question both mathematically and experimentally. Mathematically Lorenz obtained the same formula that I had given, though I don't think his method quite so complete. Experimentally he completely verified my results. He showed that in using smoke-gratings erroneous results were obtained unless certain precautions were observed. Lorenz's work seems to have settled the question, and I had not thought of going on with the investigation. It is satisfactory that the results obtained from gratings and from fine particles are in perfect accordance, and conspire to show that the vibrations are *perpendicular* to the plane of polarization.

LENSFIELD COTTAGE, CAMBRIDGE,

24th May, 1877.

DEAR LORD RAYLEIGH,

On returning from calling to-day I found the copy of your book on Sound which you were so kind as to send me. I first dipped into it as I cut the pages, but of course a book of the kind will require study, which I have not time to give it at present on account of my lectures which however are now nearly over. Meanwhile pray accept my best thanks for your interesting book.

I was reminded by seeing a series I gave which you refer to, that I neglected to send out at the proper time my copies of a later paper on the same subject. As some of the formulae of the other are here given in a more general shape, I take the liberty of sending you a copy which possibly might be convenient to refer to.

4 CARLTON GARDENS, S.W.

June 2/77.

DEAR PROF. STOKES,

In consequence of our conversation the other evening I have been looking at your paper "On a difficulty in the theory of Sound," *Phil. Mag.* Nov. 1848. The latter half of the paper appears to me to be liable to an objection, as to which (if you have time to look at the matter) I should be glad to hear your opinion *.

By impressing a suitable velocity on all the fluid the surface of separation at A may be reduced to rest. When this is done, let the velocities and densities on the two sides be u, ρ, u', ρ'. Then by continuity

$$u\rho = u'\rho'.$$

The momentum leaving a slice including A in unit time $= \rho u \,.\, u'$, momentum entering $= \rho u^2$.

Thus $\qquad p - p' = a^2(\rho - \rho') = \rho u (u' - u).$

From these two equations

$$u = a\sqrt{\frac{\rho'}{\rho}}, \quad u' = a\sqrt{\frac{\rho}{\rho'}}.$$

This, I think, is your argument, and you infer that the motion is possible. But the energy condition imposes on u and u' a different relation, viz.

$$u'^2 - u^2 = 2a^2 \log \frac{\rho}{\rho'},$$

so that energy is lost or gained at the surface of separation A.

It would appear therefore that on the hypotheses made, no discontinuous change is possible.

I have put the matter very shortly, but I dare say what I have said will be intelligible to you.

CAMBRIDGE,

5th June, 1877.

Dear Lord Rayleigh,

Thank you for pointing out the objection to the queer kind of motion I contemplated in the paper you refer to. Sir W. Thomson pointed the same out to me many years ago, and I should have mentioned it if I had had occasion to write anything bearing on the subject, or if, without that, my paper had attracted

* In the reprint, Math. and Phys. Papers, vol. ii. 1883, the correction is made in an explanatory note on p. 55.

attention. It seemed, however, hardly worth while to write a criticism on a passage in a paper which was buried among other scientific antiquities.

P.S.   You will observe I wrote somewhat doubtfully about the possibility of the queer motion.

TERLING PLACE,
Dec. 29/79.

DEAR PROF. STOKES,

Some time since you asked me whether I could offer any suggestion as to the cause of the discrepancy between the value of the coefficient of friction calculated by you from Baily's pendulum experiments and that obtained by Maxwell and others more directly. In Rühlmann's *Handbuch der Mechanischen Wärmetheorie,* an explanation is put forward on the ground that Baily's results do not really give the whole effect of the air, since the vacuum that he used would still contain gas enough to produce a considerable effect. It would seem that the influence of the confinement, small at ordinary pressures, would become important *in vacua,* and that a residual effect might be produced nearly constant in ordinary *vacua,* and only to be got rid of by an extreme rarefaction, such as is even now beyond our powers to produce. I have not examined the matter closely, but thinking there may be something in this view I have thought it worth while to call your attention to it.

CAMBRIDGE,
2 *December,* 1880.

DEAR LORD RAYLEIGH,

I am much obliged to you for calling my attention to Rühlmann's explanation of the discrepancy between Maxwell's determination of the coefficient of viscosity and my own. What he suggests had long ago presented itself to my own mind, but I still felt great difficulty, as I will explain *.

Since it appears from the molecular theory of gases that it is what I called mu—my machine has not got Greek type—and not what I called mu', that is constant when the pressure changes, the gas must become, in regard to its own motion, more and more viscous as it gets rarer. If a sphere were swinging in free air,

* See letter to Gen. J. T. Walker, R.E., of date April 1874, in *Indian Trigonometrical Survey Memoirs,* vol. v. p. [87], reprinted *infra.*

as the viscosity increased the portion of air (estimated by volume) which we may suppose moving with the ball goes on increasing. But most of my reductions were those of Baily's experiments. Now Baily used a vacuum chamber only 3 inches diameter. Hence when the viscosity became at all considerable the motion the air would tend to take would be prevented by the walls of the chamber. As the rarefaction increases indefinitely—I call it indefinitely, though I suppose that we do not go to those extreme degrees of rarefaction that Crookes has been working with—we tend towards a state of things in which the inertia of the air may be neglected, so far at least as its own motion is concerned, and its instantaneous motion may be deemed to depend simply on that of the pendulum. Hence ultimately the effect of friction or viscosity would tend, as to its principal term at least, to fall wholly on the arc and not at all on the time. Therefore if this extreme limit were practically reached or very nearly reached, the effect of viscosity on the time at high exhaustions would be very small, as I supposed it very small, though for a totally different reason, namely, not that the viscosity itself considered in relation to the pendulum, not the gas, was small—in other words that the μ was small—as I had supposed, but that the effect of the viscosity, which at full pressure lay partly on the arc and partly on the time, came to fall on the arc alone. Therefore my coefficient ought not to be very much out.

But if the viscosity at the greatest exhaustion used, which I think ran about to an inch of mercury, was not so small as to throw the effect practically on to the arc, then no doubt my coefficient would be too small, the best coefficient, that is, got from the totality of the experiments. But what puzzles me is how to account for the wonderful consistency of the experiments when compared with theory. For they embrace forms and dimensions of the greatest diversity.

No doubt the most satisfactory way would be to examine the thing mathematically and see what it would lead to. This I have not at present done. I see how to do it for the case of a cylinder within a coaxal cylindrical case, though I could not have done it at the time when I wrote my paper on pendulums. But I have not time at present to work it out. And the case of a sphere within a cylindrical case would present enormous difficulties. The case of a sphere within a concentric spherical case is given in my paper, so that for this it would be a question of numerical reduction.

So the puzzle I mentioned to you, I think, as we were walking in St James's Park, still remains to me, namely how to account for the wonderful agreement between theory and observation if the coefficient be, as I suppose it is, very greatly too small *.

Wishing you a Happy New Year and satisfaction in your new work†, I remain, yours sincerely,

TERLING PLACE,
Jan. 7/80 [81 ?].

DEAR PROF. STOKES,

I think when I was writing to you it had escaped my mind that your calculation of μ was based upon the *time* and not upon the *arc*. It would be a good thing if you could carry out the calculation you suggest, though as you explain the removal of the difficulty does not seem very probable.

With thanks for your good wishes, yours sincerely,

4, CARLTON GARDENS, S.W.
Feb. 1/86.

Many thanks for your letter which clears up the question as far as relates to the sphere. I had in my mind rather the case of a cylinder vibrating torsionally without displacement of centre in a coaxal cylinder envelope. Here there is nothing (except the envelope) to prevent the whole fluid moving with the cylinder as a solid body without dissipation, and so it would seem that there can be no finite limit to log. dec. as the fluid rarefies independently of envelope.

On Thursday‡ you can tell me whether you agree to this.

---

* See a further discussion in a note appended to the reprint of the paper on Pendulums, *Math. and Phys. Papers*, vol. iii. (1901), pp. 137—141. On the absence of inertia effect cf. the Supplement to Crookes' memoir on viscosity of rarefied gases, *Phil. Trans.* Feb. 17, 1881; *Math. and Phys. Papers*, vol. v. pp. 100—116.

† The Cavendish Professorship of Experimental Physics at Cambridge, in succession to Maxwell.

‡ At the Royal Society; Prof. Stokes became President and Lord Rayleigh Secretary in 1885.

Feb. 25/82.

DEAR PROF. STOKES,

I have had the magnetic moment of a chance bar of steel, comparable with a pendulum rod, taken. The length was 86 centims., and the magnetic moment in c.g.s. measure 107. If we suppose all the magnetism at the ends the strength of the poles will be 1·25 c.g.s.

Hitting the bar in a vertical position did not make much difference.

I hope that this may help you to say whether a steel rod is inadmissible or not for the pendulum rod of the proposed clock. Yours very truly,

CAMBRIDGE,
9 *Dec.*, 1882.

DEAR LORD RAYLEIGH,

In reading over your paper on the dark plane *, it occurred to me, in considering the experiments you describe, that the probable cause was to be sought in a tendency in the motes to get from places of stronger to places of weaker gliding. To fix ideas, suppose fluid to be moving in planes perpendicular to the axis of x, the direction of motion being parallel to the axis of z, and w being a function of x. Then dw/dx measures the rate of gliding. Suppose this to diminish, numerically, or without regard to sign, as x increases. Then a mote would occasion less disturbance of the motion of the fluid if it were moved so as to increase its x, and more disturbance if its x were diminished. That being the case, it seems to me very likely—but I have not at present gone further into this matter—that there would be a tendency for the mote to be moved outwards, to a place, that is, where if it were it would occasion less disturbance†. Now when a fluid ascends past an obstacle, the gliding is very strong close to the surface of the obstacle. I can imagine therefore that in

---

* "On the Dark Plane which is formed over a heated wire in dusty air," *Roy. Soc. Proc.* xxxiv. 1882, pp. 414—418; reprinted in *Scientific Papers*, vol. ii. 1900, pp. 151—154, with an added note,—"It seems clear that gravitation and a movement from hot to cold, somewhat as in Crookes' radiometer, are both concerned."

† Compare generally Korteweg's theorem of minimum dissipation of energy in steady motion, *Phil. Mag.* 1883, extended by Rayleigh, *Phil. Mag.* 1893; Lamb's *Hydrodynamics* (1895), § 297.

this way a film of fluid near the surface of the obstacle might get purged of motes, which would get into a stratum a little further from the solid.

Postscript. ROYAL SOCIETY, *Dec.* 10*th.*

I called at your house yesterday on my way to the 4.30 train, and found you had gone to London and were going to Florence. I take for granted, therefore, that I may send the paper to the press at once ; any small addition which you might feel disposed to make in consequence of the remark of the referee—a remark he himself thought hardly worth making as the quantity it related to was so very minute—could be inserted in the proof.

Did you mention Florence to the physician you consulted in London, or merely mention Italy generally ? I have heard that the climate of Florence is very trying in the winter, such excessively cold winds come down from the mountains. However, in such weather I suppose you would be careful to keep indoors.

I think the purgation from motes of places of intense shearing in a gas is analogous to the purgation from sand of the ventral portions of a vibrating plate ; that is to say, that in both cases alike you must call in the doctrine of chances. I don't think that under the circumstances centrifugal force would account for it. Again, if the mote were perfectly regular and the motion of the fluid regular and (though interfered with by the mote) stable— suppose the uninterrupted motion of the fluid in strata perpendicular to Ox and moving parallel to Oz, and the mote spherical —in such a case the supposition that there might be an outward force on the mote depending on the 1st power of the velocity, is negatived by the consideration that it ought in that case to change sign with the velocity; whereas from symmetry it must remain the same.

CAMBRIDGE,
28 *December,* 1882.

DEAR LORD RAYLEIGH,

I was at first inclined to put a note to your paper, in case it should be agreeable to you that I should, to put in an explanation which I threw out *vivâ voce* at the meeting, and which commended itself to me. It was that near the surface there is a very strong shearing action when air flows past an obstacle, and that as motes interfere with the regularity of the motion they get rolled over, and wriggle about a little in this stratum of strong shearing, until by accident they get out of it, when they are merely carried with the fluid without disturbing it, and so remain there, just as sand remains on the nodal lines of a vibrating plate.

But on further looking into it it appeared to me that there were difficulties in the way of this explanation, at least in the form in which it first presented itself to me, for on making the usual suppositions as to the condition at the surface, and appealing to the formulae, it appears that there is no such diminution of shearing in passing from close to the surface to a moderate way from it.

Nevertheless I am disposed to think that the leading idea of the explanation is the right one; only the conditions at the surface are not those commonly taken. When viscosity is taken into account, we commonly assume that there is no relative motion of the solid and the fluid at the surface, that is, that the relative motion at a point P in the fluid decreases indefinitely as P approaches indefinitely to the surface, and that the gliding is regular, the shearing being finite at the surface, and changing continuously as you go inwards into the fluid.

But I am disposed now to think that even in such slow motions as those that you have been experimenting on there is a narrow film which is the seat of a great number of very little eddies, the regular motion which I mentioned in the first instance being of an unstable character in the immediate neighbourhood of the solid, and that the motes get expelled from this narrow stratum of eddies. The eddies I refer to I suppose to be due to the viscosity, and not to involve any centrifugal force that is sensible in such experiments as yours.

I have abstained from putting in a note about this, because the existence of such a narrow stratum of eddies is at present only conjectural.

Possibly this may throw some light on what has long been a mystery to me; namely, how I got such a wonderful agreement between theory and observation as to the effect of the motion of a viscous fluid on the time of vibration of a pendulum, or rather, of pendulums of a great variety of forms, and yet the coefficient of viscosity I obtained is much smaller than results from the experiments of Maxwell or of Meyer*.

Perhaps when we get to a sunnier season I may make some experiments in smoky air to see whether a smokeless film is formed where the air glides over the surface of a body vibrating as a pendulum.

I cannot think that the purgation of the air from motes in your experiments can have been due to centrifugal force. Indeed your own experiment with the rotating cylinder seems to disprove it. I think you have gone in the right direction when you say, "Indeed it would almost seem as if this kind of contact were sufficient to purify the air without the aid of centrifugal force." I should be disposed to substitute "altogether" for "almost."

You are doubtless enjoying the sunny south, though at this season I suppose that even at Florence there are cloudy days now and then, and cold winds too, which you will have to be cautious about. I heard from your servant when I went to get your address that you were much better since your arrival.

TERLING PLACE,
Nov. 15/86.

DEAR PROF. STOKES,

I want to have a talk with you some time about secondary waves in Dynamical Theory of Diffraction. I have never understood your investigation very well, as it seemed to me that the law you find is only one of a probably infinite number of laws which in a mathematical sense might be considered to be solutions. Perhaps I can explain best from Sound. The motion in front of any plane parallel to wave front may be supposed to have its origin at that plane. The motion of the plane as a whole is one of vibration perpendicular to itself, and we may analyse this as due to a number of parts in each of which a small area moves as the whole *plane* does, *while the remainder is at rest.* In this case the secondary wave is equally intense in all directions, thus differing from what you give.

My difficulty turns, I think, on your meaning when you say that you suppose the unperforated parts *merely to stop* the light.

* See *Math. and Phys. Papers*, vol. iii. (1901), added footnote, pp. 76–7, and Appendix pp. 137—141, referred to *supra*, p. 106, for explanations on this subject.

The result as regards the secondary wave appears to me to depend on *how* the stoppage is effected. What for instance would be the machinery for stopping in the case of Sound? Please do not trouble yourself to write an answer, but I should like to know your views on this point.

CAMBRIDGE,
20 *Nov.* 1886.

DEAR LORD RAYLEIGH,

I have satisfied myself that my method is quite right. It is a perfectly legitimate mode of superposition. It is merely counting your sheep as they go through the gate instead of as they stand in the field ready to go through.

The legitimacy will, I think, be seen at once if you first think of a series of independent pulses, say, to fix ideas, separated by intervals which are small compared with the breadth of a pulse, and then suppose the pulses indefinitely close and numerous, the function expressing the disturbance being arbitrary.

If the problem be regarded from my point of view, that is, taking the medium as a whole, and enquiring what, emanating from each element of the surface of a front of a wave, will replace the actual motion, you cannot substitute a force acting at the element, and regard that as the source of disturbance. For the forces required to act as sources for the motion in front and those required to act as sinks to the motion in the rear are equal in magnitude but *opposite* in sign. You may think only of the disturbance in the front, and regard it as produced by a force acting all over the surface at which the disturbance takes place, thus endowing the force with an ideal entity; but in doing so you shut your eyes to the medium in the rear, and regard the disturbance as emerging from the womb of chaos. Or you may think of such a force as acting as a sink to the motion in the rear, looking no more after what is thrown into the sink. But as I remarked the same force will not suit for both these purposes. In fact the required forces are simply the action and reaction whereby the two parts of the medium act on each other.

CAMBRIDGE,
22 *Nov.*, 1886.

DEAR LORD RAYLEIGH,

When we were talking at Burlington House I looked in my book for some words, which I thought were at the end of the paper or nearly so, which indicated that I thought the mathematical results had to be combined with common sense in the actual application to the results of experiment. I have since found them, but they are at the beginning. They will be found at p. 249 * from "The results appeared," near the top, to the end of the paragraph over the page.

I think I shall be able to convince you, should you still feel doubt of it, of the perfect legitimacy of taking stock successively instead of simultaneously.

From the time I noticed it, which was in the course of experiments on fluorescence, I always regarded the polarisation of light by finely suspended particles as one of the strongest proofs of the correctness of Fresnel's views. See *Phil. Trans.* for 1852, p. 530, Arts. 182, 183†.

P.S.   When the suspended particles were metallic, or of quasi-metallic opacity, it sometimes seemed to me that the polarisation was less complete, in a direction perpendicular to the incident light, than when they were not thus powerfully absorbing‡.

CAMBRIDGE,
25 *April*, 1887.

DEAR LORD RAYLEIGH,

With respect to the Adams Prize Subject, my own leaning would be rather to ask definitely for a discrimination between stability and instability of motion in the case of a fluid when viscosity is taken into account. I would accept, either a general investigation of the condition of stability, notwithstanding

---

* *Math. and Phys. Papers*, vol. ii.; showing that his inference as to plane of polarisation did not depend on the details of the analysis. There seems to be a misunderstanding here of the nature as well as the scope of Lord Rayleigh's criticism, which is repeated in *Encyc. Brit.* 1888, Art. "Wave Theory," *Scientific Papers*, vol. iii. p. 169, and is referred to in his Obituary Notice of Prof. Stokes, reprinted in *Math. and Phys. Papers*, vol. v., see p. 13.

† *Ibid.* vol. iii. pp. 361—363, with footnote of date 1901.

‡ For calculations relating to infinite conductivity, see J. J. Thomson, "Recent Researches" (1893), § 378 ; also more generally Rayleigh, *Phil. Mag.* 1897, *Scientific Papers*, vol. iv. pp. 317—324, and J. C. M. Garnett, *Phil. Trans.* 1904.

that it might be so complicated that it could not be worked out even in a simple case, or an actual solution in a simple case.

I am afraid that if the condition were asked for, for a perfect fluid, we might get merely a sort of *rechauffé* of what you have done, with some slight addition.

We can settle the thing finally on Thursday, when we three are to meet. I thought that I would mention my view to you beforehand that you may think of it.

3 May, 1887.

DEAR LORD RAYLEIGH,

Darwin and I are ready to agree in the enclosed. If you approve, please to sign it, and you may as well post it to the Vice-Chancellor, the Master of St John's. If however there is the slightest change which you wish to make, you had best bring it to the Royal Society, where we three should meet on Thursday.

I do not think it likely that there is a value of μ which gives a maximum of instability. I think the smaller be μ, the less will be the stability; but at the same time the longer will be the time which must elapse before the system reaches its average state of eddying motion. If T be the time elapsed since the commencement of the motion, and M the coefficient of viscosity, the state at the end of the time T may be regarded as defined by the independent variables T and M. Now this may have two perfectly different limits according as we first suppose T infinite, and then diminish M indefinitely, or first make M vanish and then increase T indefinitely*.

CAMBRIDGE,
6 May, 1887.

DEAR LORD RAYLEIGH,

I will put more definitely my views as regards stability in the motion of a fluid, whether viscous or not, taking for simplicity the case of steady, if stable, motion.

Let S be the initial state of motion, which would be continued indefinitely assuming the mathematical absence of all disturbance; suppose the fluid initially to be very slightly disturbed; let T be the time elapsed, S' the state reached by the end of the

---

* See *Math. and Phys. Papers*, vol. i. p. 311, note added 1880.

time T, M the coefficient of viscosity. Then what I believe to be true is this:—

Let T be given, however great; then the limit of S' when M vanishes is S. But instead of T being given, let M be given, however small; then the limit of S' when T becomes infinite is or may be altogether different from S. I think that in general it will be. It does not seem to me likely that, when a motion which for a sufficiently great M is stable is converted by diminution of M into one which is unstable, it will again become stable by a further diminution of M, except in the indirect sense in which we may define the motion as stable by calling it stable in case it be so when, instead of seeking the limit for M vanishing of the limit for T infinite of S, we seek the limit for T infinite of the limit for M vanishing of S. The two limits may be profoundly different, and such I believe to be actually the case.

CAMBRIDGE,

7 *March*, 1890.

DEAR LORD RAYLEIGH,

I could not help having misgivings as to the correctness of your paper read last afternoon, but the thing required more thought. I don't of course question the accuracy of your experiment, or the inference you drew from it, that at any rate with a fresh surface there was not much change produced in the superficial tension by the addition of a little soap or saponine. The point I question is, whether you were justified in saying that in a capillary film the surface tension could not be constant.

I believe on the contrary that a film *can* be supported with a constant surface tension. And I am disposed to think that this is the way in which it actually is supported. At the same time it is likely enough that the support may be modified if the richness in soap gets to be different in different parts, as is likely to be the case after a little time*.

* This theory of Willard Gibbs and Rayleigh, assigning difference of tension in a vertical film to difference of concentration of the soap in the surface layers of the film, which is only slowly equalized, is now generally accepted: see a subsequent paper in the same year by Lord Rayleigh, *Scientific Papers*, vol. iii. p. 363.

8 *March*, 1890.

I did not much believe in the necessity for a difference of surface-tension, and my first idea was that there was a motion of some kind going on, which was excessively slow in consequence of viscosity; but when I worked out the result I sent you I rather jumped to the conclusion that that was the explanation. But I see now that it won't do; for the value of d^3y/dx^3 is too large to allow y to be as small as we know it can be over a considerable extent of film. In fact, what I sent you is nothing but the ordinary equation of capillarity in a different shape from the usual.

I am still I confess sceptical about the *necessary* difference of surface-tension, and I am disposed to fall back on the former idea of a very slow motion resisted by viscosity, which you seem to have hinted at but rejected.

STORNOWAY,
Aug. 15/90.

MY DEAR PRESIDENT,

You will see that I have already escaped to one of the most distant points in the British Isles. I expect to be home again about Sept. 10. In the mean time anything marked to be forwarded will be sent on from Terling.

I expect you pay too much attention to the whips. People have run away to fish and shoot grouse in all directions.

Thanks for the references, they will be useful when I turn my attention again to Electricity, as I may do before long. At present my thoughts run on grease*. Yours very truly,

RAYLEIGH.

4, WINDSOR TERRACE, MALCHIN,
18 *Sept.*, 1890.

DEAR LORD RAYLEIGH,

. . . .

I was reluctant to dissent explicitly from Dixon's statement near the bottom of p. 56 of Part I [of the Report on Vision through a Fog].... But I think the thing is too important to pass over in silence, for the importance of the burr is vastly

* See *infra*, p. 120.

underrated if it be supposed to be produced by reflection and
refraction, and I don't think its true nature is generally recognised.
I did not myself recognise it till lately. I propose the following
change, about which I will write to Thomson :—I quote part of
the draft.

In the first place it may be remarked that the relative
efficiency of two lights, seen we will suppose through annular
lenses, for illuminating fog is very far indeed from being proportional to their efficiency for direct penetration. (For the light
by which the burr is produced is) deflected from its course
through an angle which is not by any means very small, and
therefore it cannot matter much whether the regular light from
which it originally came was concentrated by the lens within
a more or less small angle ;...

For ( ) read

The light by which the burr is produced is in the main not
that which is reflected or refracted by the globules of water
forming the fog (this light would be too widely scattered to be
useful) but diffracted, and retains an approximation to its primitive
direction even after being diffracted more than once.   It is
however

If we take the brightest parts of the light refracted and
diffracted by a globule of radius a at a distance r, the former
being calculated as for geometrical optics, which will suffice, and
the intensity of the incident light being taken as unity, we have,
neglecting the loss by reflection, which would be only 4 per cent.,

$$\text{refr. } \frac{\mu^2}{4(\mu-1)^2}\frac{a^2}{r^2}, \qquad \text{diffr. } \frac{\pi^2 a^4}{\lambda^2 r^2},$$

$$\text{ratio di : re} = \frac{4\pi^2(\mu-1)^2}{\mu^2}\frac{a^2}{\lambda^2} = 2\cdot46\,\frac{a^2}{\lambda^2}\text{ nearly.}$$

If we take a at 1/4000 inch, agreeing with a corona I once
measured, λ at 1/50000, the ratio comes 384 : 1.   The sum total
of the diffracted and of the refracted would be the same (the
reflected being still neglected), and the largeness of the ratio
indicates the relatively great concentration of the diffracted as to
direction.   And yet the assumed value of a is probably decidedly
too small for a dense wetting fog.

Unless something unforeseen prevents me, I am to leave for
Cambridge to-morrow.

P.S.   Do you approve of the insertion ?

19 *Sept.*, 1890.

I should like to omit the words in double parentheses below, in the passage where I speak of the relative importance of fog-illumination to direct penetration being greater in fog than in haze or mist. The reason is that the thing cannot well be explained without going into mathematical considerations which would be out of place in a report such as we are concerned with. The passage might I think stand provided "diffracting" were substituted for "light-scattering," which would be liable to be misunderstood; but even so, one reason among others is that in haze the particles or globules are comparatively fine, and light is dispersed by diffraction through a larger angle, and is therefore less concentrated.

Again there is reason to think that even for the same light and the same lens, the relative importance of fog-illumination and direct penetration will depend on the thickness of the fog, supposed uniform, and the consequent distance from the lighthouse at which the light could first be seen directly. In haze or mist (most of the light-scattering particles intervening between the eye and a point a little to one side of the source of light are too far from the source to receive much illumination, and) the first thing to be seen on proceeding towards the lighthouse would probably be the direct light; or if a little burr were sooner seen, the increased distance of guidance thus afforded would be of little importance. In such cases therefore direct carrying power is the chief thing to look to.

But if the fog were comparatively thick, &c.

I expected to have started for England to-day, but it is now put off to Monday.

TERLING PLACE,
Sept. 20/90.

MY DEAR PRESIDENT,

I quite agree with the suggested emendation emphasising the passage.

It is clear to me that fog and dust act as you describe. I experimented a good deal at one time on the acoustic behaviour of smoke jets. It was desirable to work with smoke as thin as could be well seen, and to this end it was important that the light leaving the smoke for the eye should make *as small an angle* as possible with its original direction.

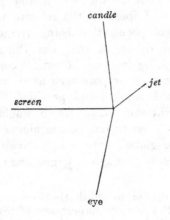

CAMBRIDGE,
4 *Oct.*, 1890.

DEAR LORD RAYLEIGH,

I have not yet heard from Sir William Thomson as to whether he accepts and has signed and sent in the report. I should think there is little doubt that he has or will, as I adopted nearly all his suggestions....

What I write about is a statement I made that of the light abstracted from the direct light by a globule of water half was reflected and refracted, and half diffracted. I did not give the demonstration, though I said it was simple. I don't know whether it (or some other demonstration) occurred to you, and whether you talked of the matter to Thomson. I may however as well mention it, and perhaps Thomson might like to see it unless you did the thing in your own way.

Suppose a wave of light (or rather a series of waves) coming from a luminous point, to fall on a globule of water of radius a. I will suppose the distance from the source to the globule, and from the globule to the eye, to be both very large compared with a. The portion of the wave which falls upon the front hemisphere, the area of which portion is πa^2, is reflected and refracted, chiefly of course refracted, and the light-energy corresponding to this area is withdrawn from that of the regular light. When the wave has just passed the globule, or say passed its

SMALL OBSTACLES DEFLECT AND SCATTER LIGHT EQUALLY 119

widest section, it is left with a hole in it of area πa^2. To get at the effect of this, imagine the quiescent hole in the wave to be the seat of two equal and opposite vibrations, one, which I will call A, being the same as that of the uninterrupted wave, the other, say B, equal and opposite.

As regards the uninterrupted portion of the wave together with A, everything will go on as if there had been no interruption at all, and the light when viewed by the eye would be brought to a point (very nearly a point) on the retina. As regards B, everything will take place as if the wave was stopped by a screen in which was pierced a circular hole of radius a. This will produce a diffraction pattern on a screen at the distance of the eye, and the total light-energy in this will be that corresponding to the size of the aperture, equal therefore to that belonging to the reflected and refracted light. This energy is supplied at the expense of the regular light. Hence the total loss to the regular light is double that due to reflection and refraction alone*.

I do not think the reflected and refracted light would ever come into account in practical cases of lighthouse warning. I think Mr Dixon was misled by the very exaggerated form in which fog-illumination was shown, as we saw it, in the experimental gallery. In that case no doubt reflection and refraction went for a great deal, as it does in the light of the sky on an overcast day. But before matters came to this condition in a practical case I think the light would be too faint to be seen.

TERLING PLACE,
Oct. 9/90.

MY DEAR PRESIDENT,

You spoke to me last year about the waves with holes in them †, and the calculation of lost energy.

I send herewith an enlarged photograph of " b " done with a Rowland grating. I believe it is the best as yet of this group.

* In other words, the disturbance scattered from an area a of a wave-front must just neutralize that scattered from the remainder β of the wave-front, for there is no light scattered from the complete wave. But when an aperture a is small enough, all the disturbance-energy from it is scattered. Thus the amount of light that is scattered from a wave-front β having a hole a in it is equal to the amount that would be deflected by an obstacle a. Cf. also Rayleigh, " On the Theory of Illumination in a Fog," *Phil. Mag.*, 1885 ; *Scientific Papers*, vol. ii. pp. 417—420.

† The subject of the last letter.

You will be interested to hear that I have found * the slight oily film on ordinary water to have a marked effect upon reflection at the polarising angle. While camphor is still spinning vigorously the difference from a clean surface is fully evident. I tried the experiment last year when first occupied with these films and got no result, being misled as I now find by a bad nicol. Afterwards when I had determined the thickness of such films I thought it was too small for an optical effect. Your query a few months ago led me to look into it again. I expect the polarization at a clean surface is absolute.

CAMBRIDGE,
8 *Oct.*, 1892.

DEAR LORD RAYLEIGH,

I did not hear Professor Rutherford's presidential address to the Biological Section of the British Association at Edinburgh, but I read it afterwards in Ireland. It is mainly devoted to colour-vision. He evidently does not agree with our report, and I for my part don't agree with his views. I wrote to him about some points, and he sent me a revised copy of his address. He said he would write later, but he had just lost a near relation.

This turned my attention again to the subject, and an idea has occurred to me relative to a peculiarity which you mentioned regarding the colour-vision of two of the Balfours, namely that though their colour matches were self-consistent, and not less accurate than those of people in general, yet they were different from the matches made by people in general who were not colour-blind. Thus the proportion of green to red in order to match daylight was in their case abnormal. You expressed the opinion that the primary sensations of colour must be somewhat different in different individuals.

I think however it might possibly be otherwise explained. It constantly happens that a colour match ceases to be a match when the two objects compared are viewed through a coloured glass. Now† before reaching the cones in the yellow spot, which

---

* See *Phil. Mag.* Nov. 1890, and Jan. 1892: *Scientific Papers*, vol. iii. pp. 396, 496.

† See letter from Maxwell, p. 20, *supra*.

there is very strong reason for believing are the percipient organs in this part of the eye, the light has to pass through the yellow colouring matter. Might not this act the part of the glass as above mentioned? I don't know whether you have tried them on colour matches made by looking a little to one side, so that the images fall outside the yellow spot; nor do I know whether matches could be made accurately enough with this part of the eye even in the case of a normal-eyed person. I do not know whether you have examined your brothers-in-law as to the facility with which they see Haidinger's brushes, or the coloured spot seen on first looking at the sky through a solution of chloride of chromium, or the dark spot similarly seen when an ammoniacal solution of a copper salt is used.

If these phenomena should indicate an abnormal excess or defect of the yellow colouring matter, might not the cause of the anomaly lie in this, and not in a difference of the three fundamental colour sensations?

WHITTINGHAME,

Oct. 15/92.

DEAR SIR G. STOKES,

I quite agree that the peculiarity of the Balfour vision might be due to looking through an absorbing medium. It would have to be one like weak permanganate, partially opaque to green, while transparent to red and yellow spectrum rays. The reason for connecting the peculiarity with the character of the primary sensations depended upon a comparison of this with the two types of dichromic vision. But the question is open.

I will try some time, when I have a nicol by me, whether they see Haidinger's brushes easily; but I am not a good person to introduce the subject as I cannot see them myself.

CAMBRIDGE,

17 *October*, 1892.

DEAR LORD RAYLEIGH,

I got your letter this morning. After breakfast I went to my lecture room, and tried some glasses to aid in showing Haidinger's brushes. They were specially distinct with a deep blue (cobalt) glass, or two such superposed. The best thickness is about that for which the band in the yellow left after absorption by the glass just about disappears. In seeing them for the first time it is important to use a uniform ground. Therefore a portion of cloud should be chosen in which no structure is visible for a couple of degrees or so around the point of sight. Ten seconds or so should be allowed for the eye, looking through the nicol, to get used to the field, and then the nicol should be rapidly turned through about 90 degrees. Before going back to the old position several seconds ought to be allowed to elapse. I don't know whether you have tried the aid of absorbing media in looking for them. It is important that the ground looked at should be perfectly uniform. If looked for on a ground of white paper, the paper ought to be out of focus, so that the ribs or texture should not be seen.

I had not chloride of chromium by me, but one can't well do better than use a cobalt blue glass of suitable depth of colour. We want especially the part of the spectrum from a little before F to a little before G. A very thick blue glass which absorbed the blue, leaving only the violet and extreme red, failed to show them.

If you fail to see them even with the assistance of a blue glass of such depth of colour as I have mentioned, I think your eyes must be rather peculiar.

I see something of the two absorption bands which Maxwell mentioned, or I should rather say bands which he mentioned and *believed* to be due to absorption by the yellow spot, but I should think Maxwell must have seen them much more strongly than I do. I had frequently seen these two (with me) ill-pronounced minima of illumination in looking at a spectrum of daylight, but till I read what Maxwell had written about them I believe I was disposed to imagine that they were weak places in the spectrum, perhaps from abundance of dark lines. I see them however, though not easily, in the spectrum of the flame of a lamp.

CAMBRIDGE,
28 Oct., 1892.

I tried again to-day about Haidinger's brushes. I think an ammoniacal solution of a copper salt answers slightly better than a blue glass. The strength should be such as just to absorb the less refrangible part of the spectrum nearly up to F. I have clear proof, from a sort of natural vivisection of the right eye which occurred perhaps 30 years later than the injury to which I attribute my peculiar perception of colour, that Maxwell was right in attributing the seat of Haidinger's brushes to absorption in the yellow spot. There are two bands of absorption, one adjacent to F on the less refrangible side, the other between F and G but nearer to the former, and in this region the strongest absorption takes place, but it continues, though not quite so strongly, from thence onwards. To see then the brushes to perfection, it is well to absorb the rays till you get near F.

If you don't see Haidinger's brushes even with the assistance I have mentioned, I think your eyes must be peculiar. For a first trial to see them it is important that the ground looked at should be perfectly uniform, therefore one should not be distracted by the fixed lines of a spectrum, and if paper be used for viewing, the fibre should not be seen.

P.S. To my eye there is an almost abrupt transition from green to blue a little before F. It is just here that Haidinger's brushes begin to appear, at least to be conspicuous.

TERLING PLACE,
Nov. 12/94.

DEAR SIR G. STOKES,

Thanks for your letters.

Have you seen Michell *, *Phil. Mag.* t. 36, p. 430 ? It will be worth your while to look at it, I think.

CAMBRIDGE,
31 *July*, 1899.

DEAR LORD RAYLEIGH,

Some years ago Thomson or Kelvin (I forget which he was then), you, and I were together at the Royal Society, and Kelvin asked me what I thought of a result you had arrived at that the appearance of bands of interference in a spectrum did not prove regularity in the light, but only high definition in the spectroscope.

---

* "On the maximum wave on water, of uniform propagation." See *Math. and Phys. Papers*, vol. v. p. 159.

If this meant what it appeared to mean I utterly disbelieved it, it seemed so manifestly untrue. Did it mean that the limit to relative retardation for which interference can experimentally be exhibited depends, as far as we have hitherto gone, on the defining power of the spectroscope? If so, that may very well be. But the words used seemed to imply more than that; they seemed to mean that the appearance of regularity was no indication of any regularity in the light itself, but was a mere creation of the spectroscope*. That I certainly could not for a moment believe. The matter got laid aside and dropped out of sight.

But I have got from the R. S. a paper to report on which obliged me to take the thing up. You refer to Gouy as having in part anticipated you, and speak of his paper with high approval. In your own paper I felt somewhat uncertain whether the apparent difference between us was more than a mere difference of enunciation. But in Gouy's paper, which you seem to endorse, I can lay my finger on what I hold to be a mathematical oversight which vitiates his conclusion.

But for fear of ambiguity I must first explain what I mean by regularity in light. The disturbance which we ordinarily consider, in applying mathematics to the undulatory theory of light, is the simple harmonic disturbance expressed by a sine or a cosine. The mathematical expression denotes a disturbance going on for ever, and if we contemplate it as consisting of plane waves travelling, then travelling from *minus* infinity to *plus* infinity. Nobody of course supposes that the actual vibrations of light can be thus mathematically regular †.

Terling Place,

Aug. 10/99.

Dear Sir G. Stokes,

I am afraid that I shall stand condemned, for I do think that "a vast succession of independent impulses following one another casually" would show interference, of course with the aid of a spectroscope. I understand you to hold that there

---

* It seems likely that there is a misunderstanding here, Prof. Stokes thinking of the interference of *the whole of* the light belonging to a bright line which the spectroscope is employed merely to isolate, whereas Lord Rayleigh's thesis relates to a narrow band of light separated from a pure spectrum, continuous or not.

† This is Prof. Stokes' copy, apparently unfinished; perhaps the letter was not sent. Cf. Larmor, *Phil. Mag.* Nov. 1905; also Rayleigh, *Phil. Mag.* Jan. 1906.

could be no interference when two identical trains of such disturbance are superposed with a relative retardation, because there would be no relation between the two disturbances which then become simultaneous. But in my view the effect of the spectroscope is to introduce such relation.

The case of a grating is the easier. The most arbitrary disturbance, *e.g.* a single impulse, is reflected from the grating with a periodicity imposed upon it, and it then becomes capable of interference. And the greater the number of lines in the grating, the "higher" is the interference possible. All of which may be illustrated acoustically.

My views are to be found in *Ency. Brit.*, Wave Theory, § 7 ; also § 14, p. 438.

I do not understand how your position can be reconciled with Fourier's theorem, according to which the most arbitrary disturbance possible can be represented as made up of systems of pendular waves.

GLENFERNESS,
Sept. 12/99.

DEAR SIR G. STOKES,

Thanks for your letters.

I had seen your Röntgen Ray lecture. Indeed I had it in my mind when I wrote a short note to *Nature* in the Spring of /98*. I shall be going home in a few days.

10, DOWNING STREET, S.W.
Feb. 12/00.

DEAR SIR G. STOKES,

My doubts † as to Godfrey arose at quite a different point from yours ; and as to the paper I can say no more than I have already.

If I understand your views, I do not agree. It seems to me that *whatever* be the disturbance incident upon a grating, the light sent off in a particular direction will have been *made* regular, and be capable of giving interference bands.

I am off to Manchester to-morrow morning, or I would have tried to see you at Burlington House.

* On the subject of the last letter; *Scientific Papers*, vol. iv. p. 353.

† See Rayleigh, "On the influence of collisions and of the motion of molecules in the line of sight upon the Constitution of a Spectrum Line," *Proc. Roy. Soc.*, July 1905, pp. 440–4.

To Prof. THOMAS ANDREWS.

LENSFIELD COTTAGE, CAMBRIDGE,
4th Dec. 1876.

MY DEAR DR ANDREWS,

I have now worked out the formulae for the determination of the change of volume of one of your cylindrical tubes (*i.e.* of the interior of it) when subject to certain pressures inside and out.

Following the notation employed in a paper of mine published in the 8th vol. of the *Cambridge Phil. Trans.*, let B be the constant of rigidity of the glass and $\frac{1}{3}A$ the constant of compressibility, so that $\frac{1}{3}A\delta$ is the pressure corresponding to a cubical compression δ or linear compression $\frac{1}{3}\delta$. Let a, b be the inner and outer radii of the tube; P, P' the inner and outer pressure, e the extension of a unit length parallel to the axis, q the radial extension to a distance r from the axis (so that a particle of the glass originally at a distance r is now at a distance $r+q$) then

$$e = \text{const.}, \quad q = cr + c'/r,$$

and if ϵ be the expansion of a unit volume of the interior [space],

$$\epsilon = 2c + 2c'/a^2 + e, \tag{1}$$

the first two terms being due to the change of section and the last to the change of length.

The following are the expressions for c, c' and e given by the theory of elasticity. Put for shortness

$$\frac{a^2 P - b^2 P'}{b^2 - a^2} = \Pi, \tag{2}$$

so that Π is the longitudinal tension referred to a unit of surface produced by the pressures P, P': then

$$c = \Pi/A, \tag{3}$$

$$c' = \frac{a^2 b^2 (P - P')}{2(b^2 - a^2) B}, \tag{4}$$

$$e = \Pi/A, \tag{5}$$

whence
$$\epsilon = \frac{3\Pi}{A} + \frac{b^2 (P - P')}{(b^2 - a^2) B}. \tag{6}$$

If g be the cubic dilatation at any point of the glass,

$$g = \frac{d\epsilon}{dr} + \frac{q}{r} + e = 2c + e = \frac{3\Pi}{A}, \tag{7}$$

which is independent of r.

If we take for the starting-point of our measurements the condition under atmospheric pressure then we should by rights take for P, P' the excesses of the pressures inside and out over one atmosphere.

If we have pressure inside only we have from (2) and (6),

$$\frac{\epsilon}{P} = \frac{3a^2}{b^2 - a^2}\frac{1}{A} + \frac{b^2}{b^2 - a^2}\frac{1}{B}, \tag{8}$$

and if outside only,

$$\frac{-\epsilon}{P'} = \frac{3b^2}{b^2 - a^2}\frac{1}{A} + \frac{b^2}{b^2 - a^2}\frac{1}{B}. \tag{9}$$

In these expressions the first terms depend on the yielding of the glass to change of volume, the second to a yielding of rigidity irrespective of change of volume. In your experiments b is *at least* as great as $8a$, so that a^2/b^2 is at most $\frac{1}{64}$. Also according to the theory adopted by Poisson and Lamé, $A = 5B$.

I believe that in general A is considerably greater than $5B$, though, according to Everett, for *flint glass* $A : B$ is not much different from 5. Hence the second term of ϵ in (8), which depends on the resistance of the glass to change of form, is more than 100 times as great as the first term which depends on the resistance of the glass to change of volume.

We see from (9) that when a pressure P is applied outside, instead of in, the part of the change of interior volume due to a yielding of the glass to change of form is the same as before, while the part depending on compression of volume of the glass is no longer insignificant unless A be practically infinite compared with B, whereas according to Everett, for flint glass (however it may be for crown) A does not much differ from $5B$. For a pressure P' equal to P applied outside only, $-\epsilon$, so far from being insignificant, *exceeds* the ϵ for a pressure P inside by

$$3P/A.$$

If μ be the coefficient of compressibility for a liquid inside, and λ the coefficient of *apparent* compressibility when the pressure is applied inside only, and λ' the coefficient of apparent dilatation when the pressure is applied outside only, we have from (8) and (9),

$$\lambda = \mu + \frac{3a^2}{b^2 - a^2}\frac{1}{A} + \frac{b^2}{b^2 - a^2}\frac{1}{B}, \tag{10}$$

$$\lambda' = \frac{3b^2}{b^2 - a^2}\frac{1}{A} + \frac{b^2}{b^2 - a^2}\frac{1}{B}. \tag{11}$$

The best way to determine μ seems to be this. Determine λ by experiment, and determine the *constant of rigidity* of the glass by the resistance to torsion of one of the hollow cylinders employed. The formula is

$$\frac{\theta}{z} = \frac{2M}{\pi(b^4 - a^4)B}, \tag{12}$$

where θ is the angle of torsion for a length z along the axis, and M the moment of torsion. For the sake of merely determining μ there would be no need to determine A, for the term in (10) in which it occurs is practically so small that no sensible error would be committed if we put $A = 5B$ or even $A = \infty$. Everett would probably be willing to determine B by his method.

If λ' were determined as well as λ the equations (10), (11), (12) would give μ, $1/A$, and $1/B$. The determination of λ' would however involve great experimental difficulties, or at least the adoption of different apparatus, and it would apparently be easier to insert a slender glass rod (not tube) of the same kind of glass, going loosely into the bore, the greater part of which it fills, and fill the interval with mercury, and compare the result with that obtained when mercury fills the whole of the bore of not necessarily the same but a similar tube.

If V be the internal volume, U the portion filled up by the glass rod, λ_g the coefficient of apparent compressibility when the pressure is inside,

$$\lambda_g = \frac{U}{V}\frac{3}{A} + \frac{V - U}{V}\mu + \frac{a^2}{b^2 - a^2}\frac{3}{A} + \frac{b^2}{b^2 - a^2}\frac{1}{B}, \tag{13}$$

and (10), (12), (13) give $1/B$, $1/A$, and μ.

Or else the difference of compressibility of mercury and glass (of the same kind as the tubes) might be determined in Oersted's manner, and then the value of $\mu - 3/A$ thus got, combined with (10) and (12), would give μ, $1/A$, $1/B$.

I think the absolute compressibilities thus determined would be more trustworthy than any that have hitherto been given. Purely hydrostatic methods like that of Oersted give only differences of compressibility. Purely elastic methods like that of Everett give theoretically $1/A$ and $1/B$, but the former goes for so little in the result that its experimental determination is precarious. Regnault's method, and that of those who have attempted to determine cubic compressibilities from the extension of wires, and to correct Oersted's compressibilities of liquids, are vitiated by the assumption that $A = 5B$. Your method of a capillary tube supplies by the aid of equation (10) the missing link. The danger is that the term $\dfrac{b^2}{b^2 - a^2}\dfrac{1}{B}$ may be very much larger than μ, so that the magnitude of λ may be mainly due to that term, in which case errors of observation whether of λ or of B would tell seriously in their influence on μ. It ought not to be so if Regnault's values of the absolute compressibility of mercury and of glass are anything like right; but I am by no means sure they may not be much too large.

In my former letter I used Green's notation. Distinguishing his A, B (the A, B of my former letter) by accents A', B', we have
$$B' = B, \quad A' = \tfrac{1}{3}(A + 4B).$$

I have verified that my formulæ reduce themselves to Lamé's when the relation $A = 5B$ is assumed.

I return the first sheet. The marked sentence on page 4 does not seem correct. If P' were $= P$ we should get from (2) and (6),
$$\epsilon = -3P/A,$$
so that there is *contraction* of volume notwithstanding that when the pressure is internal only there is expansion, as given by equation (8).

The sentence might perhaps be altered thus:—

"Nor would any useful...diameter of the bore, and any change in the capacity would be too small to demand attention."

It is true that the change, pressure inside, is due mainly to the alteration of section, not length; but it is not true that this would not be prevented by an equal pressure outside. As to the most important term in (8), the glass is simply pushed out in the manner of a cylindrical annulus of incompressible fluid; there is as much change of volume outside as in, but the *radial* displacement is of course less in the proportion of a to b.

I see I put the printed notice to authors into your paper instead of sending it to you, I now enclose it. Please say whether one was sent you with the proof.

Believe me, Yours sincerely,

(Signed) G. G. STOKES.

P.S. I have suggested "yielding" for "compression" on page 5, as the change of internal volume is almost entirely of the nature of that of an india-rubber tube subject to liquid pressure inside.

Once the absolute compressibility of one substance, such as mercury, is known, those of others are probably best got by Oersted's method.

FROM PROF. P. G. TAIT.

38 GEORGE SQUARE, EDINBURGH,
10/7/86.

MY DEAR STOKES,

Is Andrews' paper* in type yet? I ask because Mrs Andrews is desirous of getting private copies.

Can you tell me what was the process, finally determined on between you and Andrews, for finding the <i>absolute</i> compressibility of mercury? I am stopped in my pressure work by this very question or, at least, by a question which can be reduced to this. I will gladly make the experiments. Andrews told me some 13 years ago that you and he had decided on a plan †, ...

Yours truly,

P. G. TAIT.

P.S. Might I send you proofs of a paper on "Kinetic Theory of Gases"? Cayley was kind enough to look at it in MS., but of course I cannot inflict it on him now.

Sylvester spent the day with me, on his way to seek seclusion, for work, at St Andrews.

---

* Posthumous paper on compression of gases, in <i>Phil. Trans.</i>, 1886.

† <i>Supra</i>, p. 128.

38 GEORGE SQUARE, EDINBURGH,
26/12/89.

I had to get De la Rive's book out of the Library to read the passage you referred to. But it seems to me that you have misapprehended my difficulty. Long ago (1880?) you told me about gravity as the cause of the *large* separation of the electricities:—and, in my lecture of that year, I spoke of you as holding that opinion. I can find a copy for you without trouble.

What I want, and have been seeking for thirty years at least, is the *initial* cause of the electrification. I could get no trace of electrification by any process of evaporation (unless there was fizzing or sputtering, and then friction was to be suspected, as in Faraday's explanation of the Armstrong hydroelectric machine). Hence I was led to look for it as a contact effect of particles of air and water-vapour. I have not been able to prove that this is a *vera causa*, but neither have I been able to disprove it.

The student who was working the quartz spectroscope has got an appointment in Madras, and I have been too busy of late to make any experiments myself. But I hope to resume work with it next month. I had several pieces of spar cut, according to your plans, by Hilger:—and I have at last got an arrangement which seems to leave nothing to be desired. The two images are exactly in one plane, and polarized in *rigorously* perpendicular planes. Their only defect is that they are somewhat small:—but if I enlarge them I get results of obliquity which render measurements uncertain.

30/12/89.

There are two points in which your proposed experiment differs quite from any that I have tried:—viz. the rise of an admixture of light gas to pull away the negatively electrified air upwards, while gravity pulls the positive drops down; and the introduction of the previous filtering. I send you a copy of my 1880 lecture. At p. 27 you will find what I said of you. And at pp. 26, 30 you will find, respectively, the theory I sought to establish and the nature of the experiments I made. The last sentence of the marked passage on p. 26 was, of course, written before we knew of Aitken's experiments.

Prof. Michie Smith told me, a few weeks ago, much the same thing as that you say of Clement Ley:—but he put it in the form that a "thunder-head" (the top of an ascending column of hot moist air) does not become active till it has risen *through* a cold stratum of cirrus or, rather, cirro-stratus cloud. I saw at once that I had often made the observation myself, without thinking what it implied:—because whenever I see anything like *this* I know that thunder is pretty sure to follow. I have often escaped a wetting on St Andrew's Links by attending to this, while those who disregarded it got deluged.

I think it is Van d. Mensbrugghe who ascribed the separation of the electricities to capillary forces:—but, if that be so, how does the *air* get one of them? That was the difficulty which led me to ascribe the separation to *contact*-difference between air and vapour. Your experiment will not decide this question.

P.S. Look at my article in the *Phil. Mag.* for January, soon due.

2/1/90.

You will find Helmholtz' view of the genesis of a thunder-cloud (an observation like that of Mr Ley) in the *Berichte* of the *Berlin Physical Society*, 22/10/86. A short version of it is in *Nature*, xxxv. p. 24.

When you have leisure, I should like much to know what you think is the cause of the WHITENESS of snow, as seen on a dark night far from any terrestrial source of light and when the only illumination is from clouds which seem (at least) not white nor neutral grey, but INDIGO. Is snow a fluorescent substance? I have thought of this every winter for many years: but I have never had a really satisfactory trial with a spectroscope. When observations have been possible, I have always been unable to get into a part of the country where they could properly be made. But on two or three occasions, when I was obliged to be out long before sun-rise, but after the gas-lights were out, the contrast of tint between the snow and the cloud-hemisphere has much impressed me.

Have you seen E. Wiedemann's modification of the 2nd Law of Thermodynamics, made to include Fluorescence and Phosphorescence?

ON OBSERVATIONS OF WAVES AND SWELLS AT SEA.

To Sir EDWARD SABINE, Pres. R.S.

BLACKROCK, BUSH MILLS,
22 *Sept.*, 1870.

MY DEAR SIR EDWARD,

Will you be so good as to give me Mr Melsens's address and his initials? I am thinking of writing to him to give more explicit directions about observing the periodic time of rollers. My attention was attracted to the subject by a heavy ground swell which we had to-day along with fine weather and otherwise smooth sea. I observed a few series at intervals, No. 1 when I went down with Mary and the children to look at the waves, Nos. 2 and 3 perhaps 20 m. or half-an-hour later, Nos. 4, 5, 6 after we returned to the house. I got the following results, n being the number of intervals (or $n+1$ the No. of waves) in the series, and t the mean (or rather the most probable) interval in seconds.

Series No....	1	2	3	4	5	6
$n=$	12	8	9	4	2	10
$t=$	17·08	15·17	15·37	14·40	14·50	13·93

Taking the means of those series observed in quick succession we have

$$\text{No. 1} \qquad t = 17\text{·}08$$
$$\text{2 and 3} \qquad 15\text{·}27$$
$$\text{4, 5, 6} \qquad 14\text{·}28.$$

This seems to show a progressive diminution of periodic time a result which I had not anticipated, but which is easily explained. The ground-swells are doubtless travellers from a distant region of the ocean which has been disturbed by a heavy gale, and as in deep water the velocity of propagation varies as the square root of the wave-length, the formula being $v = \sqrt{\dfrac{g\lambda}{2\pi}}$, the longer waves excited in the disturbed region outrun the shorter, and come first to our shores.

If t be the periodic time, λ the wave-length *in deep water*, v the velocity of propagation also *in deep water*, $\lambda = \dfrac{gt^2}{2\pi}$, $v = \dfrac{gt}{2\pi}$; when the waves are propagated into shallow water the velocity and wave-length decrease according to a known formula.

For the periodic times

	$t =$ 17	15	14 sec.
I find	$v =$ 87·10	76·86	71·73 feet per sec.
	= 59·38	52·40	49·82 miles per hour
	$\lambda =$ 1482	1152	1005 feet
	= 247	192	167·5 fathoms.

Taking the velocity at 50 miles an hour for round numbers, if the age of the waves was 24 hours the seat of disturbance would be 1200 miles off. A gale might well exist at such a distance contemporaneously with fine weather in the neighbourhood.

The longest waves in open ocean mentioned by the late Captain Stanley were of 40 or 50 fathoms. Waves of the enormous length of 200 fathoms, of a height measured by inches rather than feet, and of a form such that a section of the surface is the curve of sines, would at sea pass well-nigh or absolutely unperceived. It is only when they are propagated into the shallow water round our coasts—and for waves of such length, water of 100 fathoms must be deemed shallow—that they attract attention. The same amount of *vis viva* per wave having to be kept up in spite of the restricted depth and diminished length of disturbance, *i.e.* of water disturbed, the height of the waves would be increased very greatly when the water becomes as shallow as 10 fathoms, even if the form of the surface remained such that a section was the curve of sines. But besides this the elevated portions become narrow and steep, and the troughs comparatively flat, so that the height of the former must be still further increased.

The waves proved more regular than I had anticipated; thus the first series gave the intervals, in seconds, 19, 17, 16, 18, 17, 18, $\frac{3}{2}4$, 14, 18, 18, 18.

I can't help thinking the observation of ground-swells would be useful in connexion with the discussion of the logs of ships. The only element which I see my way to observing quantitatively with considerable precision, at least the only element worth observing quantitatively, is the periodic time, and that can be done with no more apparatus than a common watch provided with a seconds' hand. This could easily be done by the coast-guard man at 4 or 5 stations along the Atlantic coast of Scotland and Ireland.

We go to Armagh to-morrow for perhaps 3 weeks.

Although I began my letter by merely asking for Mr Melsens's address I have let my pen run on, on a subject which I was led to take a renewed interest in yesterday.

To Mr MELSENS, St Helena.

OBSERVATORY, ARMAGH, IRELAND,
29th Sept., 1870.

Dear Sir,

Some considerable time ago Gen. Sir E. Sabine showed me a paper of yours relative to the rollers observed at St Helena. We have not far to go for an explanation. We know that a heavy gale, still more a hurricane, out at sea, would produce a great disturbance in the ocean, heavy waves in fact. This disturbance would not be confined to the region in which it was excited, but would be propagated outwards into what would otherwise be smooth water, or else, it may be, into water in which smaller waves already exist, and the two would go on independently of each other. The waves would thus reach a region which the storm had not yet come to, or perhaps never would come to at all. In the region originally disturbed the motion is rather complicated, but it may be regarded as made up of different series of waves superposed, the waves of each series being regular, but the wave-length, and with it the periodic time, differing from one series to another. Now the velocity of propagation of waves in deep water varies with the periodic time, the velocity of propagation (v), periodic time (t), and wave-length (λ), being thus connected:

$$v = \frac{gt}{2\pi}, \qquad \lambda = \frac{gt^2}{2\pi}, \qquad \therefore v = \sqrt{\frac{g\lambda}{2\pi}},$$

g being gravity, and π the ratio of the circumference of a circle to the diameter.

Hence the long waves would outrun the shorter, and be the first to arrive at a distant shore. The shorter waves too in a long course would be much reduced by friction, which would hardly affect the long waves. After diverging to a considerable distance from the disturbed region, the long waves would be much reduced in height, and to the crew of a vessel far out at sea would perhaps be hardly sensible. But on arriving in the comparatively shallow water which surrounds an island (and for such long waves water which would ordinarily be regarded as pretty deep must be deemed shallow) the waves mount up, and at last, it may be, become very formidable.

The velocity of propagation and the length of the waves (or distance from crest to crest) both diminish as the waves are propagated into shallow water, while as I have said the height increases. But there is one element which remains unchanged during the life of a series of waves, namely, the periodic time. This element is easily observed, and not only helps to identify (or the reverse) series of waves observed at different stations (such as St Helena and Ascension), but also makes known the velocity of propagation in deep water, furnishing thereby a relation between the distance the waves have travelled from the region of disturbance and the time which has elapsed since the disturbance took place. And my chief object in writing is to give some hints as to the best mode of observing the periodic time.

My attention was attracted to the subject during a recent sojourn on the north coast of Ireland by noticing a heavy surf which came in one day though the weather was fine and calm.

I found I got better results by observing when the waves

reached a particular rock than by observing when the ridges of the waves seen sideways, as they entered a bay, were in a line with a particular object on shore. The observation requires no more apparatus than a common watch with a seconds' hand. A single person can observe very well; but it is more satisfactory when two work in concert, as one can then keep his eye on the rock while the other keeps his eye on the watch, when the time draws near when the wave is about to reach the rock. In that case the second notes and writes down the time when the first calls out "now."

Waves usually come in sets followed by a comparative lull during which observation might be uncertain. Suppose we observe the times when a set of say 10 or 12 consecutive waves reach the rock. Of course, if we can observe more so much the better, provided it can be done with certainty. But say we are able to observe 10 or 12. By dividing the interval of time between the first and last by the number of wave-intervals, we shall get a very good value of the mean interval. But in this way all the observations except the first and last would be thrown away, and it is obvious that a better result would be obtained by utilizing all the observations and assigning to them their proper weights respectively. The rule for this is simple enough. Let n be the number of wave-intervals in the set, so that $n + 1$ is the number of waves observed. Taking the observed times, write the first under the last, the second under the last. but one, and so on, and take the differences. Multiply these differences by n, $n - 2$, $n - 4$, &c., and take the sum of the products. Divide this sum by $\frac{1}{6} n (n + 1) (n + 2)$, and you will get the best value of the mean interval deduced from the whole set.

If n be even the multipliers are all even, and it saves trouble to take the halves of the multipliers, namely, $\frac{1}{2}n$, $\frac{1}{2}n - 1$, $\frac{1}{2}n - 2$, &c., or, beginning at the other end, the natural numbers 1, 2, 3..., and to divide by $\frac{1}{12} n (n + 1) (n + 2)$.

Should the time of one of the waves be uncertain from any accident, the mean of the times of the waves next before and next after may be taken as its time, and the rule carried out.

I take for example the first series I observed. The seconds observed were 49, 8, 25, 41, 59, 16, 34 [51], 8, 22, 40, 58, 16, one wave being missed or uncertain, and its time 51 being supplied from $\frac{1}{2} (34 + 68) = 51$. Supplying the minutes reduced to seconds,

which it was not necessary to write down, and proceeding according
to the rule, we have

49	68	85	101	119	136	154	171	188	202	220	238	256
							136	119	101	85	68	49
							35	69	101	135	170	207
							1	2	3	4	5	6
							35	138	303	540	850	1242

sum $= 3108$,

$$\tfrac{1}{12}n(n+1)(n+2) = \frac{12 \times 13 \times 14}{12} = 13 \times 14 = 182, \quad t = \frac{3108}{182} = 17 \cdot 08.$$

If the first 12 waves only had been observed, the process would
have been

49	68	85	101	119	136	154	171	188	202	220	238
						136	119	101	85	68	49
						18	52	87	117	152	189
						1	3	5	7	9	11
						18	156	435	819	1368	2079

$$\text{sum} = 4875, \quad \frac{n(n+1)(n+2)}{6} = \frac{11 \cdot 12 \cdot 13}{8} = 286,$$

$$t = \frac{4875}{286} = 17 \cdot 03.$$

As a test of the trustworthiness of the result, I will take the odd
and even waves separately, and determine the periodic time
independently from each group, doubling the denominator, as
otherwise we should get the time of a double interval.

49	85	119	154	188	220	256
				119	85	49
				69	135	207
				1	2	3
				69	270	621	sum $= 960.$

$$\text{Denominator} = 2 \cdot \frac{6 \cdot 7 \cdot 8}{12} = 7 \times 8; \quad \frac{960}{7 \times 8} = \frac{120}{7} = 17 \cdot 14.$$

68	101	136	171	202	238
			136	101	68
			35	101	170
			1	3	5
			35	303	850	sum $= 1188.$

$$2 \times \frac{5 \cdot 6 \cdot 7}{6} = 70, \quad \frac{1188}{70} = 16 \cdot 97.$$

These two last, which are totally independent of each other
except as to the influence of the single wave supplied, not directly

observed, differ from one another only by 0·17, about the $\frac{1}{6}$th of a second or the $\frac{1}{100}$th part of the whole interval.   The first and last waves *taken alone* would have given $\dfrac{256-49}{12} = \dfrac{207}{12} = 17\cdot25$, differing little from the other determinations, but probably not quite so near the truth, as depending on the observation of two waves only, the number of waves intervening being merely counted.

The observation mentioned was taken about 11 a.m.   Somewhat later in the day the periodic time had sunk to 15·2 s., and next day, when there were still swells, though not deserving the name of surf, it had sunk to 14·2 s.   In consequence of this change, it would be desirable to take the periodic time 2 or 3 times a day while the rollers last, and each time on two or three series so as to make sure of the result.   The slight change in periodic time is quite in accordance with what I said about the longer waves outrunning the shorter.

For the object of determining the velocity of propagation of the waves in deep water, an approximate determination of the velocity would suffice, as there is no object in knowing the velocity with great precision.   The chief use of a more accurate determination is to assist in determining the probable identity (or the reverse), as to origin, of rollers observed at distant stations, such as St Helena and Ascension.   For it does not seem likely that in the rollers produced by different storms the periodic time would agree, except occasionally by chance.   In some heavy waves accompanying a storm felt at the place, on the north coast of Ireland, I found the periodic time was only 7 or 8 seconds instead of from 14 to 17.

The numerical calculation involved in the application of the rule I have mentioned is not very serious, unless the series comprises a large number of waves.   In that case if the application of the regular rule be thought too irksome, it would be well, instead of taking the first and last only of the series, to take the mean of a few at the beginning and a few at the end.   To go back to the first example, I will take the mean of three at the beginning and three at the end.   The mean of 49, 68 and 85 is 67·33 and that of 220, 238 and 256 is 238·00.   The difference 238·00 − 67·33 or 170·67 corresponds to 12 − 1 − 1 or 10 intervals, and we have in this way $t = \frac{1}{10} \times 170\cdot67 = 17\cdot07$ agreeing almost exactly with the

result got by the regular rule. The mean of 2 at each end would
have given

$$\text{mean of 49 and 68} = \ 58\cdot5$$
$$\text{,,} \quad 238 \text{ and } 256 = 247\cdot0$$
$$\overline{188\cdot5}$$

corresponding to 11 intervals, giving $\quad t = \dfrac{188\cdot5}{11} = 17\cdot14.$

Suppose we wished to take 3 at each end, but deemed the last
wave but two uncertain; we might then take 3 at the beginning
and 2 at the end,

$$\text{mean of 49, 68, 85} = \ 67\cdot33$$
$$\text{mean of 238 \& 256} = 247\cdot00$$
$$\overline{179\cdot67}$$

corresponding to $10\frac{1}{2}$ intervals, giving $\quad t = \dfrac{179\cdot67}{10\cdot5} = 17\cdot11.$

I have multiplied examples perhaps to prolixity, as I wished
to put you in full possession of the method of procedure, and any
explanation which might be found requisite would take a long
time to travel by post.

A word or two may be useful as to the mode of observation.
I found it better to observe when the waves reached a particular
rock than when their ridges, seen edgeways as they entered a small
bay, were in a line with an object on shore. A rock should be
selected, if possible, in sufficiently deep water for the waves not to
break before they reach it. If the form of the coast permit of the
observer's stationing himself so as to get a side view of the waves
as they approach the rock, so much the better, as the observer can
then see their ridges as they approach the rock, and is better
prepared to note the exact moment when they reach it. The
calculation prescribed gives the time in watch-seconds. If the
watch used has a gaining or losing rate of sufficient magnitude
to be sensible in the observation, the calculated wave-interval
should be reduced to true seconds.

In case you should have occasion to write, I may as well
mention that my normal address is Lensfield Cottage, Cambridge.
I am writing to get your address from Sir E. Sabine.

To Captain A. H. TOYNBEE, Marine Superintendent,
Meteorological Office.

Observatory, Armagh,
5 *September*, 1878.

My dear Sir,

I am much obliged by the forwarding of the extract from
Captain Watson's log. It reached me during the bustle of the
[British] Association week at Dublin, and I left the *study* of it
till I should be uninterrupted. It is very interesting, and points
I think unmistakably to a violent gale, in all probability a cyclone,
which passed somewhere between Bermuda and Cape Hatteras.
The vortex passed the ship, or rather the ship passed it, about
5.30 on the 5th, but by that time the cyclone had in good measure
died out. The very heavy swell which was felt at 5.30 p.m. on
the 4th was produced a good way to the S.S.W. of the then place
of the ship, at an earlier time by perhaps 2 or 3 days and at a
considerable distance to the S.S.W.

I have been laying down the places of the ship, and you may
expect to hear from me again at greater length in a day or two.

It is curious to see that captains seem to have so little idea of
the propagation of waves, excited in a stormy region, into a region
where, as regards the wind, it is a comparative calm. As to the
production of such swells by currents, I hold it to be quite out of
the question.

I am, dear sir, yours faithfully,

G. G. STOKES.

Extract from Meteorological Log (4289) of S.S. "Algeria."
Captain WILLIAM WATSON.

Sunday, 3rd March, 1878.

5 to 6 p.m. (Lat. 43° 33′ N., Long. 50° 44′ W.). A confused sea
S.W., W.N.W., and N.; at intervals of 2½ minutes there was a swell

running in well-defined ridges from W.N.W., having a period of 6·7 seconds and about 12 feet high; ship's course S. 84° W. 12 knots; depth of water 35 fathoms on Bank of Newfoundland.

This swell first appeared low, each succeeding undulation being higher, up to the fifth or seventh, when it gradually subsided and became lost in the other seas. Each series of from 12 to 15 undulations occupied from 80 to 95 seconds; and they followed each other with remarkable regularity.

Monday, 4th March.

From 2 a.m. (Lat. 43° 22′ N., Long. 52° 43′ W.) to 8 a.m. (Lat. 43° 9′ N., Long. 54° 9′ W.), wind S. with hard squalls, an occasional one going up to force 9; at 9 a.m. the squalls less severe and with longer intervals between them ; from 9.30 to 10 a.m. the anemometer gave 38 miles per hour as the velocity of wind, but the force varied between 6 and 8; from 11 to noon velocity by anemometer 28·7 miles per hour, but the force was sometimes as low as 4 with gusts going up to force 7. From noon to 2 p.m. wind steady at force 6, velocity by anemometer 31 miles per hour, from 2 to 4 p.m. wind increasing with hard squalls. Anemometer showed 76 miles as distance travelled by wind in that time. At 7 p.m. wind moderating and rain passing away, the clouds breaking in the west; at 8 p.m. wind S.S.W. 5; Midt. wind S.W. 4 with passing showers of rain.

The sea kept smooth until 8 a.m. (Lat. 43° 9′ N., Long. 54° 9′ W.) at which time it was coming from S.S.W. about 10 feet high, but by 10.30 a.m. it had risen to 20 feet, still running true from S.S.W.; from 10.30 to 10.45 a.m. 136 wave-crests passed under the ship; course of ship S. 80° W. 10 knots, depth over 1000 fms.; 1.30 p.m. sea rising and the period had increased to 8·5 sec. At 4 p.m. some of the wave-crests were so high that, when standing on the bridge in the middle line of the ship with the eye 30 feet above the water-line, as they approached they rose some feet above a line joining the eye and horizon and the same as they passed to leeward. At this time I put the ship's head more to the sea. At 5.30 p.m. waves at their highest and having a period 13·4 sec. and at least forty feet high. Course of ship S. 40° W. 5 knots; at this time a sea was now and then seen coming from the W.S.W. When the ship was nearly end on to the sea and in the trough of it, the crests of the advancing and receding waves were seen high above the bow and stern.

As there was comparatively little wind the S.S.W., sea had to some extent the character of a swell, and only when the W.S.W. came along did the water break.   The ship rode easily and did not take any water aboard till towards midnight, when the Wly. sea had gained some height and force and I had put the ship on her course S. 80° W.   While heading the S.S.W. sea I watched the ship closely and counted how often she rose and fell to it; after doing so for 45 minutes and taking an average I made her rise and fall 8·9 times in 123 seconds.   At midnight the S.S.W. sea falling, but the Wly. sea running fast with an occasional heavy break.

Tuesday, 5th March.

Up to 6 a.m. (Lat. 41° 57′ N., Long. 58° 8′ W.) wind quite moderate and backing to the Sd.   7 a.m. breeze freshening, with a wild-looking sky.   Wly. sea going down.   8.15 a.m. wind south 7. Cum. having a ragged and torn appearance, flying rapidly from W.S.W., low cirs. from W., heavy cums. gathering in the W. and rising fast.   8.45 a.m. wind flew into N.W. 9, a heavy but short rain shower—just before the wind shifted the clouds overhead, and they were very low, were whirled about with an apparent motion from right to left $\circlearrowleft$ as could be seen against the upper clouds; the moment the wind came from the N.W. they were all driven to leeward and nothing but low scud was seen flying over the sky, now covered with high cirs., while in the N.W. was a low bank of nim. dense and black as ink; the squall lasted an hour, and by noon had moderated to force 2.

At 8 a.m. (Lat. 41° 53′ N., Long. 58° 36′ W.) the S.S.W. sea was still high and coming along in ridges as straight as a ruled line, but the period had increased to 15 sec. and the height was 36 feet measured by ascending the rigging until the eye, the crest of the

coming wave and the horizon were in one line. It was a fine sight
to see those masses of water rolling along in rather a calm sort of
way, the ship one moment on the crest, and in a very few seconds
after, away down in the trough with a hill of water on each side;
the Wly. sea appeared so insignificant as hardly to be worth notice,
though it now and then made us aware of its presence by sending
a shower of spray over the bridge. At noon (Lat. 41° 45′ N.,
Long. 59° 32′ W.) the S.S.Wly. sea was running with a height of
25 feet and a period of 15 sec. Course of ship S. 79° W. 12 knots.
At 6 p.m. it had fallen considerably and was coming from south
with a period of 11·6 sec. and a height of 17 feet; by midnight it
had almost passed away.

Where did this sea come from? We had no wind to speak of,
the atmospheric pressure was not lower than 29·45, yet I was
obliged to bring a large and powerful steamer head to the sea, not
wishing to run any risk of doing damage. I have on more than
one occasion found the sea, for a given force of Sly. wind, higher in
this locality than in any other part of the Atlantic.

Last November, when in about the same position, I had to take
in the square sails and bring the ship's hd. to the sea. Yet this
sea got up all at once, ran heavily for four hours, and then fell as
fast as it had risen. Whether the currents, the Gulf and Labrador
streams, which are not very far apart in the position of the
"Algeria" on the 4th and 5th of March, have anything to do with
this I can't say, but from my own experience I have come to the
conclusion that the highest Sly. seas on the Atlantic are met with
here.

OBSERVATORY, ARMAGH,
12 *September*, 1878.

MY DEAR SIR,

I have been rather going about of late or I would have
written to you before this at greater length.

Before I got your letter containing the heights of the barometer, I had plotted the places of the ship, and drawn my conclusions. I send you a copy of the plotting.

I have represented swells by parallel pencil lines in the direction
of the ridges. Perpendicular to these I have drawn pencil lines
in the direction from which the swell came.

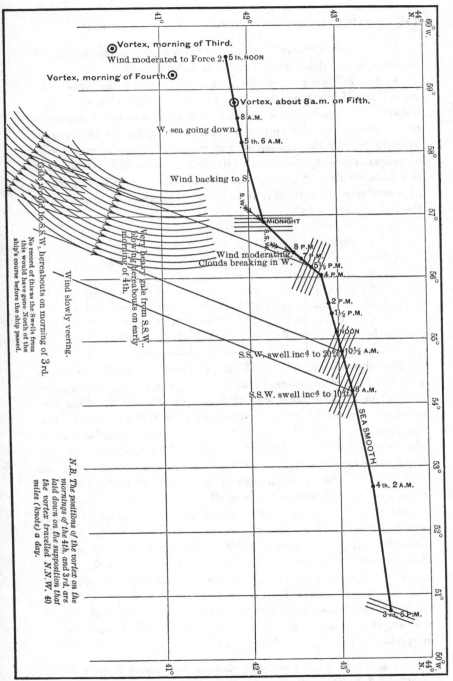

Storm-Diagram constructed from meteorological log of S.S. "Algeria,"
Captain W. WATSON, March 3—5, 1878.

I have also corrected the apparent periods for the motion of the ship by the data furnished by Captain Watson.

In the following table the first column gives the date; the second the ship's speed in knots per hour; the third the angle between the ship's course and the direction from which the swell came; the fourth the apparent period; the fifth the true period as corrected for the motion of the vessel; the sixth the velocity of propagation—V.W.—of an individual wave-crest; the seventh the distance run—D.R.—[by the waves] in a day, in knots, taken, in accordance with theory, at half the velocity of propagation of the individual waves*.

Date	V. S.		A. P.	T. P.	V. W.	D. R.
3rd 6.0 p.m.	12	28°·5	6·7	9·25	27·9	335
4th 10.5 a.m.	10	57°·5	6·6	8·06	24·4	297
4th 1.5 p.m.	10	57°·5	8·5	10·01	30·8	362
4th 5.5 p.m.	5	17°·5	13·4	14·83	44·7	537
Later	5	17°·5	13·8	15·23	45·9	556

The observation at Lat. 43° 33′ N., Long. 50° 44′ W. mentions a *confused* sea from different quarters, together with a very regular swell from W.N.W., this was at 6 p.m. on the third. This disturbance appears to have subsided, for it is said afterwards " the sea kept smooth until 8 a.m." on the fourth—which was at a place about 140 knots from the former. The roughness on the third appears accordingly to belong to a distinct disturbance from that of which the very heavy swell afterwards mentioned formed a part. The *regularity* of the swell on the third indicates that it had travelled a considerable way from the place of its birth, which must have been somewhere along a line drawn to W.N.W. from the ship's position on the 3rd at 6 p.m. I shall accordingly dismiss without further notice the disturbance on the third. As neither height nor period of the swell at all came up to what was afterwards met with, the disturbance which produced the W.N.W. swell was evidently far less severe, though of course, for aught we can tell, it *may* have belonged to very severe disturbance, of which it formed an outlying part.

---

* In accordance with Sir G. Stokes' theory of wave-groups, *Math. and Phys. Papers*, vol. v. p. 362.

I come now to the grand disturbance, beginning with the observation of 2 a.m. on the fourth.

The S.S.W. swell which was encountered beginning with 8 a.m. on the fourth indicates a heavy disturbance somewhere in a direction drawn S.S.W. from the then place of the ship, a disturbance in which the wind was from S.S.W., and which was supported and increased by the wind for some considerable time.

If this disturbance were cyclonic in character, as is probable, and is fully confirmed by the subsequent observations, the ship being considerably to the E.N.E. of the part of the ocean where the disturbance was produced, the wind at the place of the ship might be expected to be more to the south, say south, which is just what was observed.

Later on the swell increased both in height and in period, up to 5.30 p.m., when it was at its maximum, and was of very unusual magnitude.

This magnificent swell indicated a severe and long-continued action of a S.S.W. wind on a region lying S.S.W. of the place of the ship at that time. The increase of the swell observed from 4 a.m. to 5.30 p.m. I take to be due, not so much to an alteration of conditions with the time, as to the ship's change of place, her westerly course bringing her into a region where she got into wave-shot of a more severe part of the cyclone.

The ship's W.S.W. course took her gradually out of the region of wave-shot of the portion of the cyclone where the wind was S.S.W. and within wave-shot of the region where the wind was westerly. The progress of the cyclone, which in this portion of the ocean would be travelling in a direction N.N.W. or thereabouts, would contribute to the same result. This accounts for the "westerly sea running fast," which was experienced about midnight.

The wind which had been S., and had veered to S.W., about 4 a.m., backed to S., and became more severe. The ship was now nearing the vortex. At 7 a.m. the westerly sea was noted as going down. The heaviest westerly seas would now lie south and also east of the ship. For a heavy sea requires a wind which is lasting as well as strong, I mean lasting as regards the same disturbance, *i.e.* agitating the same portion of the ocean, or better still following the disturbance at about the same rate as the disturbance itself progresses. Hence considering the portion of

10—2

a cyclone where the wind blows west or nearly so, though near the vortex the wind may be stronger than a little further out, still in the latter region the water would be longer under the influence of the westerly gale, and therefore a heavier sea would be raised than nearer the vortex, in spite of the westerly wind being not quite so severe. Also if we mark the region where the wind is westerly, the westerly seas will of course increase from the west to the east of that region, and will travel further to the east, as a westward swell, into a region where the wind is no longer westerly. The same of course, *mutatis mutandis*, applies to the seas excited by any other portion of the wind-system.

The change of the ship's place therefore accounts for the subsidence of the westerly swell.

The ship was now very near the vortex; but as the account given of the wind by no means seems equal to the production of such a severe S.S.Wly. swell as had been encountered, it is probable that between the night of the third and morning of the fourth,—at which times the heavy swells experienced at 5.30 p.m. on the fourth may be deemed to have been under the influence of the powerful S.S.W. wind which gave them birth,—and the early morning of the fifth, the violence of the cyclone had a good deal subsided.

The sudden shift of wind, the whirling motion of the clouds, and the immediate change in the character of the clouds observed when the wind had shifted, show that the vortex was passed, at a very short distance, about half-past-eight on the morning of the fifth.

This is confirmed by the indications of the barometer, which had been steadily going down till about that hour, when it began to rise.

The described shift of wind shows that the ship had passed a *little* to the *south* of the vortex.

What is said of the motions of the clouds about the vortex accords very well with Mr Clement Ley's conclusion that the line of vortex is not vertical, but slopes upwards in the direction of the progress of the cyclonic disturbance.

On second thoughts I send you my drawing, such as it is, rather than taking the trouble to copy it. I should be glad to have it again, but of course if you think it worth while to take a copy of it, you can get one of the staff to make it, or rather to make a better plotting of the course of the ship.

I think this example shows how much may be learned by combining good observation of swells with the ordinary observations of other elements*.

16 *September*, 1878.

MY DEAR SIR,

I did not suppose that the cyclonic disturbance, as I called it, indicated by Captain Watson's observations, was cyclonic in more than a rough sense. I did not consider it a regular cyclone.

I felt all along that the westerly swell was what seemed to fit worst with the supposition of a regular cyclone. About midnight on the 4th the vortex, as I supposed it, would lie about W.S.W. of the ship. If the vortex were there, and the wind blew in exact circles round it, the wind ought to be S.S.E., and the swell, if any, might be expected to be more from the south, say S. or S.S.W. For though there would be, on the supposition, an easterly element in the wind at the place, there would not be, as you have suggested, an easterly swell. For an easterly swell would be excited in a region where the wind was easterly, or approximately so, and would increase from the eastern to the western edge of that region, and then would travel onwards in a westerly direction as an easterly swell. But the ship is, by hypothesis, already to the east of the region where the wind is

---

* Further considerations on the forms of ocean rollers, and on their necessarily irrotational character, are included in the Appendices added in 1880 to the reprint of the Memoir of 1847, " On the Theory of Oscillatory Waves," *Math. and Phys. Papers*, vol. i., cf. pp. 224, 228, 320.

It may be of interest to mention that for such rollers as those described at the foot of p. 143, having a period of 15 secs. and an amplitude $\frac{1}{2} h$ of 18 feet, the velocity v is 76 feet per sec. and the length λ is 1140 feet; and the velocity of leeway due to the surface current required by the irrotational motion (*Math. and Phys. Papers*, vol. i. p. 208), being $\left(\dfrac{\pi h}{\lambda}\right)^2 v\ \dfrac{1 - e^{-4\pi\delta/\lambda}}{4\pi\delta/\lambda}$ for a ship of depth δ, amounts to about $\frac{2}{3}$ of a foot per sec.

Recent navigation statistics give the average period of N. Atlantic waves as 12 secs.; the longest waves hitherto recorded were 2800 feet long and had a period of 23 secs. On the generation of waves by wind, see Kelvin, *Phil. Mag.* 1871, and two papers by v. Helmholtz, *Sitz. Berlin Akad.* 1889–90. At the end of the latter, reprinted in *Wied. Ann.* xli. pp. 641–62, Helmholtz's observations on sea-waves at Cape Antibes are discussed.

easterly. She lies west of the region where the wind is westerly,
and so far would be favourably situated for being exposed to a
westerly swell; but the difficulty is that she lies too much to the
north of it. However the wind at midnight on the 4th was S.W.,
not S.E.; and if we make a corresponding change in the direction
of the swell it brings it nearly right for west. To account for
this, continuing the hypothesis of a regular cyclone, we should
have merely to place the vortex some 50 miles or so further north
than I supposed. I was perhaps rather led away by the assemblage of phenomena described as occurring between 8.15 and
8.45 a.m. on the 5th into supposing that the vortex was then close
to the ship, but a little north of it, whereas it may have been
further north than I supposed.

In my first drawing I supposed that the ship was put back
on her course at 6 a.m. on the 5th, having overlooked the statement
that at midnight the ship *had been* put back, though at what
hour was not stated; and though I marked the line as crossed
out it was perhaps copied instead of the line put in its place,
which from the dead reckoning proves to have been nearly right.
The placing of the ship too far to the north in the first instance,
in case that were adopted in the copy, increases the difficulty of
accounting for the swell from the west*.

I find a difficulty in accounting for the westerly swell on the
supposition of a furrow of the form of a ridge running nearly
N. and S., on the east of which the wind is S.W., and on the
west N.W., with a rapid veering through west as the furrow
travels eastwards, on the ground that a swell from the west seems
to indicate the prevalence of a westerly wind for some considerable
time over the same region of water or rather of disturbance, as
I explained in my last. The wind as Captain Watson describes
it "*flew*" from S. to N.W. Where he was, the interval during
which it was west was only one of some minutes; half-an-hour
at the outside. This would not account for the raising of such a
sea from the west, even though it did not approach in magnitude
to the grand swells from the S.S.W. experienced earlier. The
only question in my mind is, would a N.W. wind exciting a N.W.
sea, followed over the same region of disturbance, after a brief
interval during which a westerly wind prevailed, by a S.W. wind,
produce a sea from the west as a sort of resultant? I confess it

* This was corrected in the chart, p. 145.

seems to me that, as the principal winds on this hypothesis are at right angles to each other, I should rather expect a " cross, confused sea."

What I have written is far too crude for a paper before a learned Society. Its chief use is to indicate directions in which to pursue further inquiry.

What I feel most confidence about is the existence, somewhere about the region where I have drawn the arcs of circles in pencil, and at the date of March 3 or thereabouts, of a severe gale of some continuance from the S.S.W., or thereabouts. I shall feel curious to see the records of the log of any other ship which may happen to have been about at that place at that time. I don't know whether an ordinary route of ships lies in that region. A S.S.W. swell falls in just as well with the supposition of a ridge such as you have mentioned as with that of a more nearly cyclonic disturbance. Indeed I felt all along that if the disturbance were cyclonic, it could be only rudely so, the S.S.W. swell so far exceeded in magnitude that from any other direction. I could think of only one cause why a S.S.W. swell should be the strongest in a perfectly regular cyclone, and that was that as the cyclone would be advancing in a N.N.W. direction, or thereabouts, the progress of the cyclone in the region where the wind was S.S.W. would cause it to follow the disturbance it was exciting. But the progress of the cyclone would be so slow, compared with the rate of travelling of long swells, that this could not apparently make much difference in favour of the S.S.W. swells.

18 *September*, 1878.

Swells give some indication of the minimum violence of the wind which produces them, in this way. It seems plain that wind cannot increase an already existing swell unless it is travelling faster than the swell is propagated. Now the rate of travelling is known when the periodic time is known, and thus we get a minimum limit to the rate of the wind which, though possibly not blowing so strong the whole time the swell was being raised, at any rate gave it its finishing intensity.

A swell of 15 seconds' period would travel 52·4 English miles per hour in deep water, and the velocity of the producing wind must have been a good deal greater than this. I do not say that the *mean* velocity over any considerable time must have been greater than this, but *at least* there must have been frequent gusts considerably exceeding 52 miles per hour in velocity.

Captain Watson's results with the anemometer did not give a mean velocity nearly as great as 52 miles an hour, but then he mentions gusts of much greater than the mean velocity. Still, as far as I can judge from the description, I should not suppose that the wind he afterwards encountered was at all adequate to raise such a swell, and therefore it is that I suppose that the violence of the cyclonic disturbance had a good deal subsided by the time its most severe part reached him.

P.S.   If Captain Watson is in town, I should rather like him to see my remarks on his account of the great swell.

19 *September*, 1878.

The difficulty I feel about the hypothesis that the westerly swell encountered by Captain Watson was produced by a westerly wind which, though of brief duration at a particular spot of the ocean, followed the swell it was exciting is that the velocity of propagation of a swell of even as low as 7·5 seconds' periods is so great that, if the westward velocity of propagation of a barometric depression be anything like as small as the velocity of progress of the vortex of a cyclone, the swell would so enormously outrun the transfer of the wind that little would be gained by the latter in keeping the same disturbance for any length of time under the influence of the exciting wind.

Thus the vortex of a cyclone may travel perhaps 50 miles a day, and I suppose—but you will know—that the rate of propagation of a furrow of barometric depression would be something like this, say even 100 miles a day. Now a swell of 7·5 seconds' period would, as regards the rate of the individual crests, travel about 26·2 miles an hour, and as regards the transfer of the disturbance half that. This would make 317 miles a day.

Doubtless in the earlier stages of the excitement, when the period was still considerably short of 7·5 seconds, the travelling of the swells would be slower; but still, unless the westward progress of the furrow were considerably over 100 miles a day, the transfer of the furrow would not much prolong the time during which the disturbance would be under the influence of the disturbing cause.

If captains were aware that the observation of swells at sea supplies to some little extent the want of weather telegraphy, they would take more interest in them. But so long as they attribute them to currents, &c., &c., and not to their real cause, namely the gales with which they (the captains) are so nearly concerned, they will care less about them.

Doubtless we want a good deal more experience to make them available for weather forecasts at sea; but we must first get and discuss the observations in connexion with the other information contained in the log of the ship in which they were recorded *and of other ships* that happened to be in the neighbourhood at the time.

ARCHDEACONRY, AUGHNACLOY,
7th Sept., 1878.

Will you kindly send me, to the address Observatory, Armagh, the heights of the barometer from Captain Watson's log No. 4289 from 1878 March 3, 6 p.m. to March 5, noon? The uncorrected heights will suffice. I merely wish to see whether they bear out my notion of a cyclone, within wave-shot of which the ship came, and passed near the vortex about 8½ a.m. on the 5th, by which time however the cyclone had in good measure spent its fury.

I believe such observations of swells will afford very valuable information, though not of a kind which can be easily expressed in numbers.

FROM W. FROUDE, F.R.S., DIRECTOR OF THE ADMIRALTY EXPERIMENTAL STATION, TORQUAY.

CHELSTON CROSS, TORQUAY,
17 *Jan.*, 1873.

MY DEAR SIR,

I am very much obliged to you for the copy you have sent me of your Memorandum on the Measurement of Waves*.

* Drawn up for the Meteorological Council.

In particular I am glad, for my own information, to have a copy of the formula for the speed of waves in shallow water.

At present it interests me because I have lately been trying some experiments on the rolling of two ships in a considerable swell outside Plymouth, when I was struck with the fact that the period of the waves was evidently a good deal longer than that due to their apparent length: which however (the length) I was not able to observe with much exactness, because it was necessary to keep the two ships broadside to the waves.

The results of the experiments (which were repeated on three different days with somewhat different states of sea) include a continuous record, on paper, extending over more than an hour, of the rolling of each ship, the angles being expressed simultaneously in two forms, (1) the angle between the ship's mast and the vertical or the "absolute" angle, (2) the angle between the ship's mast and the normal to the effective wave slope...which I suppose is nearly that of the "surface of equal pressure" passing through the ship's centre of buoyancy: a correct time scale being also automatically marked on the paper as it travelled.

In one of the ships, the "absolute angle" was obtained only by an observer who on a raised observatory perpetually sighted the horizon with a rocking bar, the motions of which were communicated to the apparatus below. In the other ship the absolute angles were obtained, not only by the observer as just described, but also by an automatic apparatus which answered extremely well, the basis of which was a heavy fly-wheel (200 lbs.) suspended by a hardened steel axis on hard steel rocking segments (virtually parts of friction rollers of large diameter), and which was altogether so delicate that when its centre of gravity was so near the point of suspension that its "period" was 75 seconds, if set oscillating when the frame was steady it would continue to oscillate for 20 minutes, if started at 90° from the position of rest. Thus if the wheel was placed at rest, the frame (or the ship carrying the frame) might rock about under it freely without putting it in motion, so long as the ship's period was sufficiently short of 70''; in fact it was only 7·75''. The wheel recorded the ship's rolling, by a method which is easy to imagine and which I need not describe, for the description is already overlong. All I want to be understood is, that there was produced continuously in each ship a trustworthy and exact record of the "absolute" and the "relative" angles, simultaneously, and the difference between the two records is of course the wave angle.

On a separate piece of paper I give a sketch showing the nature of the two curves, and of the resulting wave curve. If you cared to see it I would send a tracing of some of the actual diagrams produced by the apparatus, but the resulting wave curve in the actual diagrams is very complex (much more complex than shown in the sketch), evidently consisting of several superposed series of waves. Still it is resolvable occasionally into waves of tolerably regular period, though of rather irregular profile, and if the length due to the period can be deduced from the known depth of water, the wave diagram could be interpreted in linear dimensions: as it stands the horizontal scale is purely a time scale.

At present the diagrams are under examination specially with a view to the determination of the degree to which the actual motions of the ships correspond with those which theory prescribes to them: namely, that a given "relative angle" should impress on the ship a righting or inclining moment of approximately the same magnitude as the ship would experience if inclined to the same angle in deep water, and that the accelerating or retarding moment which the ship experiences is the moment due to the inclination and that due to the resistance of various sorts which the ship experiences while performing her oscillations. The resistance is experimentally determined by taking a diagram, with the same apparatus, of the rate at which her oscillation becomes extinguished in still water, after it has been artificially instituted and she is then left to herself.

By careful graphic integration performed on the diagram of "relative angles," by help of these data, the absolute angle comes out with an almost astonishing degree of exactness—oscillations of as much as 16° off the vertical, when thus worked out, scarcely in any case differ above $\frac{1}{4}$ of a degree from the recorded oscillation.

I am very glad you laid so much stress on the determination of "wave-periods." For the purposes of naval architecture it is of course a point of enormous importance: and considering how obvious this is in reference to a ship's rolling, it is astonishing that it should have been left till recent times to pay attention to it. I am hoping to be able to construct the self-recording apparatus which I just now described, in rather smaller compass, both as to dimension and cost, than in my first attempt: and if so I hope the Admiralty will have such apparatus fitted systematically in all their more important ships. The observations at present taken in the course of service are practically worthless and the apparatus would give results of real value; and a record of the wave-periods would be one of the results.

In the memorandum on the method of observing waves which I wrote for the Controller, and to which you refer, I fear I hardly enough emphasised the necessity of observing periods. I had to write it in great haste, and at the moment the experiments on which I was engaged led me to wish especially for some actual information as to the height of waves in connection with their length; since a knowledge of this, as involving a measure of their steepness, is essential to the determination of the ultimate angle to which a given ship will roll, either in synchronising waves or in waves which in any given degree depart from synchronising with the ship's natural period of rolling....And accurate information on the subject seemed to me to be even more entirely non-existent than information respecting "period."

But obviously it is easier to observe period than to measure length; and in deep water, as you point out, the period may be confidently relied on as determining the length.

The concluding paragraph of your memorandum relating to the observation of waves, the fruit of a distant storm, reminds me to mention a fact or circumstance, bearing on the duration of the passage of such waves from one place to another, which is not generally known; for though very probably it is not new to you, it is just possible that you may not have noticed it, and it is certainly important.

Primâ facie, the speed of such waves would determine the duration of their passage over a given distance.

But this is not really so: because the foremost waves are perpetually dying out, as they invade the undisturbed water, and are undergoing metempsychosis in the ranks behind them.

It is not easy to describe exactly what I mean, so as to convey a full appreciation of it to you if you do not happen to have noticed it: perhaps I can best explain the condition, by referring to the train of waves which follows the wheels of a paddlewheel steamship.

If we watch these, say abreast of the stern of the ship, we see that they exactly keep pace with her: and *primâ facie* one would expect that, consequently, if the wheels were annihilated and the ship were kept up to her speed by some extraneous force, the waves, having the same velocity, would also keep pace with her: but on the contrary, though they would still keep pace, as regards speed, the foremost wave would be rapidly becoming smaller and smaller, and judging by what appears to happen elsewhere I believe that very soon the foremost wave would have died out altogether; and though a train of fully formed waves is coming on behind with the same speed, each of these in turn would become less as it approaches the front rank and would in turn vanish—while at the tail end of the train a converse action is taking place, so that the energy of the series is all the while maintained.

In my long experimental tank or canal here, I have frequent opportunity of noticing this in the propagation of artificially generated waves. I have not indeed yet investigated it quantitatively *, because my hands are full: but at a later date when experiments on the oscillation of models will be the work in hand, I shall have to establish regular appliances for the generation of waves, and the investigation to which I refer will be comparatively easy.

I must apologise for the unconscionable length of this letter, but as it needs no answer, I trust you will not feel under any obligation to read it through. I fold it up with this notice outwards so that you may be on your guard at starting.

Believe me, Yours very truly,
W. FROUDE.

CHELSTON CROSS, TORQUAY,
4 July, 1876.

MY DEAR SIR,

I am shocked that I have left so many days, unacknowledged, your kind and interesting letter of the 30th. I have been extra busy getting ready for experimental trial of her resistance

---

* The explanation of the retarded propagation of wave-groups was given by Prof. Stokes in a Smith's Prize Examination Paper, Feb. 1876: *loc. cit.* p. 146. The attention of Lord Rayleigh had been drawn to the phenomenon by Mr Froude some time earlier, and the same explanation had occurred to him: see Rayleigh, " On Progressive Waves," *Proc. Lond. Math. Soc.* vol. ix. (1877), p. 21, reprinted as an Appendix in *Theory of Sound*, vol. i.

a large model (5 ft. diameter) of one of the Russian round ships, and I am now only scrawling a few lines in haste to save the post lest my continued silence should seem neglectful. The information you refer to will be most valuable and interesting. I was trying some experiments with waves in our tank the other day, the results of which in reference to this particular matter had made me intent to write to you even if I had not heard from you. I have also come in contact with and partly investigated the curious fact that the frictional resistance of water is an inverse function of the temperature, and I wanted to consult you as to the bearing of my experiments. I have already sent the details to Sir W. Thomson; it is about 4 per cent. between 70° and 40°, and then a much more rapid increase as the temperature decreases to 32, this after allowing for density, and counting force in terms of head of the fluid of the same density.

I shall trouble you with a communication shortly.

I am ever yours truly,

W. FROUDE.

18″ to 23″ was the period of the rollers I observed in Torbay, of which I wrote to you: they lasted 10 or 12 hours.

MEMORANDUM FROM PROF. W. H. MILLER TO G. G. STOKES.

"(Copy of part of a letter from Airy.)

I am glad that Mr Stokes has made something of that unmanageable integral* and I should like to know whether his method will assist to the following determinations:

1. To express approximately in a function of n, the value of m at the nth disappearance of light.

2. To express approximately the intensity of light in terms of m, when m is very large but not infinite—as for instance by a series proceeding by negative powers of m."

---

* G. B. Airy, "On the Intensity of Light in the Neighbourhood of a Caustic," *Camb. Phil. Trans.* vol. vi. In the paper m is proportional to distance from the caustic, reckoned positive towards the illuminated side. For later developments of the Caustic problem, cf. letters to Mr C. V. Boys, *infra*; also Larmor, *Proc. Camb. Phil. Soc.* vol. vi. Oct. 1888, and vol. vii. Jan. 1891.

CORRESPONDENCE WITH SIR G. B. AIRY.

PEMBROKE COLLEGE, CAMBRIDGE.

May 12th, 1848.

DEAR SIR,

I write to reply to your enquiries, communicated to me by Prof. Miller, respecting my approximate expression of your definite integral*.

I have found the following developement for your integral, where ϖ is the required integral, and $x = \left(\dfrac{3\pi}{2}\right)^{\frac{2}{3}} \cdot \dfrac{m}{3} = \cdot 9369\, m$ nearly.

$$\varpi = \left(\frac{2}{3}\right)^{\frac{1}{3}} \pi^{\frac{1}{6}} \cos\left(\frac{2}{3} x^{\frac{3}{2}} - \frac{\pi}{4}\right) \cdot x^{-\frac{1}{4}} \left\{ 1 - \frac{1 \cdot 5 \cdot 7 \cdot 11}{1 \cdot 2 \cdot (48)^2} x^{-\frac{6}{2}} + \dots \right\}$$

$$+ \left(\frac{2}{3}\right)^{\frac{1}{3}} \pi^{\frac{1}{6}} \sin\left(\frac{2}{3} x^{\frac{3}{2}} - \frac{\pi}{4}\right) \cdot x^{-\frac{1}{4}} \left\{ \frac{1 \cdot 5}{1 \cdot 48} x^{-\frac{3}{2}} \right.$$

$$\left. - \frac{1 \cdot 5 \cdot 7 \cdot 11 \cdot 13 \cdot 17}{1 \cdot 2 \cdot 3 \cdot (48)^3} x^{-\frac{9}{2}} + \dots \right\}. \qquad \text{(A)}$$

These series ultimately diverge hypergeometrically for any value of m or x however great. Nevertheless, when m is large, or even moderately large, the leading terms converge with great rapidity, and two or three terms give a very close approximation to the result.

The series shows (assuming that the convergent part is really an approximation to the required integral) that ϖ vanishes for values of $\frac{2}{3} x^{\frac{3}{2}} - \frac{\pi}{4}$ a little greater than $\frac{\pi}{2}, \frac{3\pi}{2}, \frac{5\pi}{2} \dots$ Putting $\frac{2}{3} x^{\frac{3}{2}} = \theta$, supposing θ to correspond to a root of the equation $\varpi = 0$, putting $\theta = (i - \frac{1}{4})\pi + \delta$, where i is an integer and δ is small, and extending to the third order, there results

$$\frac{\theta}{\pi} = i - \frac{1}{4} + \frac{5}{18\,(4i - 1)\,\pi^2} - \frac{13785}{18^3\,(4i - 1)^3\,\pi^4}$$

$$= i - \frac{1}{4} + \frac{\cdot 0281448}{4i - 1} - \frac{\cdot 0235614}{(4i - 1)^3} \qquad \text{(B)}$$

---

* The integral is $\displaystyle\int_0^\infty \cos \frac{1}{2}\pi\,(w^3 - mw)\,dw$; cf. preceding page. See G. G. Stokes, "On the Numerical Calculation of a class of Definite Integrals and Infinite Series," *Camb. Phil. Trans.* vol. ix., read Mar. 11, 1850, *Math. and Phys. Papers*, vol. ii. pp. 329—357; and subsequent memoirs.

and then m is known from

$$m = 3\left(\frac{\theta}{\pi}\right)^{\frac{2}{3}}.$$

The series (A) fails altogether for small values of x or m so that the formula (B) is not competent to decide whether the vanishing point corresponding to any value of i corresponds to the ith or to the $(i+1)$th, $(i+2)$th... dark band. But your calculation shows that there is no dark band till $m = 2\cdot48$ nearly, which corresponds to $i = 1$ in (B); so that i is the integer which marks the order of the band.

I have calculated the first 15 vanishing points from the formula (B). The calculation did not take $1\frac{1}{2}$ hour. I subjoin a few results that you may judge of the convergency of the series.

For negative values of m I find, writing $-m'$, $-x'$ for m, x, so that m', x' are positive,

$$\varpi = \left(\frac{3\pi}{2}\right)^{\frac{1}{6}} (36x)^{-\frac{1}{4}} e^{-\frac{2}{3}x'^{\frac{3}{2}}}\left\{1 - \frac{1\cdot5}{1\cdot48}x'^{-\frac{3}{2}} + \frac{1\cdot5\cdot7\cdot11}{1\cdot2\cdot48^2}x'^{-\frac{6}{2}}\ldots\ldots\right\}.$$

The leading term in the expression for the intensity is

$$2\cdot3^{-\frac{1}{2}}\cdot m^{-\frac{1}{2}}\cdot\cos^2\left\{\left(\overline{\frac{m}{3}}\Big|^{\frac{3}{2}} - \frac{1}{4}\right)\pi\right\},$$

for m positive,

$$(12m)^{-\frac{1}{2}}e^{-2\pi\left(\frac{m}{3}\right)^{\frac{3}{2}}},$$

for m negative.

I have not as yet assigned a limit to the error committed in stopping at any term, but I hope to be able to do so. I am however at present occupied with other investigations.

I have applied a similar method to the integral which you have tabulated up to $e = 10$ in the *Phil. Mag.* for January 1841. I am nearly sure the method will apply to the integral which occurs in the case of a circular hole in front of the object-glass of a telescope. Indeed I believe that it will apply to a variety of integrals of that nature, which are interesting from their occurring in physical problems.

Believe me, dear Sir, Yours very truly,

G. G. STOKES.

G. B. Airy, Esq., Astronomer Royal.

(*From memory*)

Aug. 3, '48.

P.S. To turn to quite a different subject may I ask if you have considered the difficulty which Prof. Challis has raised with reference to plane waves of sound, a difficulty which turns on the interpretation of a first integral of the equation which determines ϕ? It was some time before I could satisfy myself as to the interpretation*. I had some thoughts of sending my views on this one point alone to the *Phil. Mag.* without entering into the discussion, but should not do so if I knew there was anything coming from you.

PEMBROKE COLL. CAMB.
April 30*th*, 1849.

MY DEAR SIR,

I write at present chiefly to explain what led me to adopt the reasoning which I communicated to you respecting $F(\lambda/r)$.

From the general tone of your preceding letter, in which you spoke of the interference as independent of λ, without using any qualification, and from your reference to a geometrical caustic, which I probably misunderstood, I was led to suppose that you considered the existence of a sort of interference *sui generis* depending on special dynamical conditions which may obtain in the case of very minute globules, and which is characterized analytically by leading to expressions independent of λ. It was against such a notion that my argument was directed. It is quite natural that on the one hand you should think of interference independent of λ when you meant merely practically independent, and that on the other hand I, taking that independence as an analytical characteristic of a class of interferences depending (according to my idea of your meaning) on special dynamical causes, should have applied the argument I made use of, and which had just occurred to me, to disprove the existence of such a class of interferences.

* Viz. that the wave ultimately becomes a bore. G. G. Stokes, "On a difficulty in the theory of sound," *Phil. Mag.* Nov. 1848; *Math. and Phys. Papers*, vol. ii. 1880, pp. 51—55, with added note in partial correction.

With respect now to the calculation of the interference which takes place in the case of very minute globules, we must, I think, in that case have direct recourse to the dynamical equations of motion. I feel very strongly convinced that the ether must be treated as an incompressible elastic solid. The equations of motion of such a solid are very difficult to manage. The present case of interference is, however, not directly a question of polarization, and there is little doubt that the behaviour would in the main be the same as for sound.

With this simplification, namely the using the equations of motion of an elastic fluid, or of two such fluids, one forming the small sphere and the other the surrounding media, instead of the equations of motion of elastic solids, I feel almost certain that the problem will admit of ready solution if r be very small compared with λ. I feel little doubt that the analytical result can be obtained independently of the latter restriction, but I cannot well perceive what difficulties the reduction to numbers, or even the discussion of the result, may in that case present*.

I hope to be soon able to attack the problem, and I shall communicate to you the results of my investigation. Meanwhile I feel almost certain that if r be extremely small compared with λ, we shall at a considerable distance from the centre of the sphere have a disturbance in which the component of vibration varies, along a given radius vector, inversely as the distance, and at a given distance varies as the cosine of the angle between the radius vector and the direction of original propagation, the phase being the same in different directions at the same distance from the centre.

I am, dear Sir, Yours very truly,

G. G. STOKES.

G. B. Airy, Esq.

---

* See calculations by Lord Rayleigh, "On the Acoustic Shadow of a Sphere," *Phil. Trans.* 1904. On the simpler case see Prof. Stokes' memoir of Nov. 1849, "On the Dynamical Theory of Diffraction."

PEMBROKE COLL. CAMB.
Oct. 9th, 1849.

DEAR SIR,

...Now that I see clearly your real meaning I proceed to answer your question as far as I am able.

With respect to the distribution of matter within the Earth, this much is certain, that an infinite number of different modes of distribution are possible, while it is equally certain that not any mode of distribution arbitrarily assigned is admissible. I am not able to give an analytical formula (containing suppose one or more arbitrary functions with or without integral signs), which shall express in the most general way the distribution which is admissible. Such a problem has never been attempted in any case, so far as I am aware. A very simple consideration will, however, serve to show that other distributions besides those in spheroidal strata are admissible.

Conceive the mass distributed in one admissible manner. Let $d\tau$ be an element of volume adjacent to the internal point P, ρ the density at P. Conceive the mass $\rho d\tau$, or any portion of it, removed from the element $d\tau$ and distributed uniformly over a spherical shell having P for centre, or over any finite or infinite number of such shells, the largest of all the shells lying wholly within the Earth. Let all or any number of the elements of mass be treated in the same way. Since the attraction on an external particle remains unaltered, the new distribution will be admissible equally with the first; yet it is evident that if the first distribution were in spheroidal strata the second need not be.

By this redistribution of elements of mass in spherical strata, and the converse, we may pass in a very general way from one possible distribution to another: but whether this mode of passage is *the most general possible*, so that of two admissible distributions either may be derived from the other by redistribution in spherical shells, repeated, it may be, indefinitely, I am quite unable to say*.

I am, dear Sir, Yours very truly,

G. G. STOKES.

---

* Prof. Stokes recurred to this subject later, in a series of letters to Lord Kelvin. Prof. Stokes' memoirs on Clairaut's theorem belong to this year.

11—2

CAMBRIDGE.

14*th April*, 1871.

I don't think there is any difference between us except as to what we suppose Sir Wm Thomson's object to have been, which you take to have been increased accuracy and I take to have been increased convenience.

I am well aware that the determination of absolute time by lunar distances is subject to far greater error than that of local time by altitudes, inasmuch as the angular velocity of the body the place of which determines the time would in the one case be that of the moon in her orbit and in the other that of the Earth round its axis, and consequently that it is only in long voyages that such a method would be resorted to *now*, whatever it may have been when chronometer making was comparatively in its infancy. Yours very truly,

G. G. STOKES.

NOTE APPENDED TO SIR G. B. AIRY'S REMARKS ON SIR W. THOMSON'S PAPER ON SUMNER'S METHOD, *Roy. Soc. Proc.* Ap. 27, 1871*.

[From a general recollection of a conversation I had with Sir Wm Thomson before the presentation of his paper, I do not imagine his object to have been what the Astronomer Royal here describes, but partly the saving of trouble in numerical calculation, partly the exhibition, for each separate observation of altitude at a noted chronometer time, of *precisely what that observation gives*, *neither more nor less*, which introduces at the same time certain facilities for the determination of a ship's place by a combination of two observations. Of course the place so determined is liable to an error east or west corresponding to the unknown error of the chronometer; and doubtless under ordinary circumstances this forms the principal error to which the determination of a ship's place is liable. This remains precisely as it did before; and it is hard to suppose that the mere substitution of a graphical for a purely numerical process could lead a navigator to forget that he is dependent upon his chronometer, though perhaps the general tone of Sir Wm Thomson's paper might render an explicit warning desirable such as that which Mr Airy supplies.

G. G. STOKES.]

* In a further note on the use of Sumner's method, *Roy. Soc. Proc.* June 15, 1871, Sir W. Thomson undertook to avoid any possible risk by including an explicit warning on the effect of chronometer-error.

September 11, 1871.

Perhaps, from want of separation of the two parts of one subject, I may have conveyed a wrong opinion of my ideas on Sumner's method taken up by Thomson.

1. This is Sumner's method. Observe the Sun's altitude for time. Assume a latitude, or two if you like (but one is sufficient), and therewith compute ship-time and longitude. Upon your chart, draw a short line, in azimuth perpendicular to the Sun's azimuth, of such a length that its meridian projection equals the uncertainty of latitude (say 2' or 3'). That line will give you something wonderful, although the uncertainty of longitude from the chronometer-unsteadiness amounts to 20' or 30'. So says Sumner; and Thomson—as appears to me—backs him.

2. As to the calculation of the spherical triangle, if Sir W. Thomson can put this by tables into a convenient form he will be doing a great service to navigation. But this has nothing to do with Sumner's *specialité*; it applies to every case of determination of time at ship from altitude.

I am, my dear Sir, Yours very truly,

G. B. AIRY.

March 27*th*, 1872.

My dear Sir,

I have sent to Mr White my paper on the 26-day period of magnetism. I propose it for the *Proceedings*: I will bring diagrams to illustrate it. It is curious that in 1870, Hornstein's[*] year, the law in question is well marked at Greenwich, and there is not a trace of it in other years.

Mr Glaisher has confronted 100 years' sun-spots with corresponding temperatures, apparently destroying the supposed relation. I have urged him to get the comparison into shape.

Do you think it would be practicable so to modify the time of publication and the spirit of the *Proceedings* that they could

[*] Hornstein announced a terrestrial magnetic inequality having the period of the Sun's rotation; referred to by Maxwell, *Electricity and Magnetism*. Its existence is now considered to be doubtful.

be available for the insertion of irregular physical science ?   For instance, the great Aurora of Feb. 4, and the unparalleled magnetic disturbance which accompanied it.   The French have opportunity of this in the *Comptes Rendus*.   In England we have only *Nature*. I think we want a weekly publication, not interrupted by the Society's holidays, and not strictly confined to the presentations at its meetings.   I am, my dear Sir, Yours most truly,

G. B. AIRY.

LENSFIELD COTTAGE, CAMBRIDGE.
16*th March*, 1872.

As to the optical action of magnetism, experiment proves that there is asymmetry somewhere : neither of us places it in the light : *you* place it in the medium, *I* place it in the force. In one sense we both place it in the medium, namely, that the immediate action is that of an asymmetric medium on light ; but why is the medium asymmetric ?   *You* say asymmetry pre-existed, and is brought into evidence by the action of a symmetric force ; *I* say symmetry pre-existed, or [may] have pre-existed, and asymmetry is produced by the action of an asymmetric force.   Why do I say this ?

This effect, like all other effects of magnetic force, is absolutely the same whether the magnetic force be produced by a permanent magnet, or by a galvanic current, or by a temporary magnet, or by a mixture of a galvanic current and temporary magnet as in an electro-magnet.   It would be perfectly unreasonable therefore to attribute the effect to utterly different causes in these different cases—to a pre-existing asymmetry of the medium in one case, to something independent of any pre-existing asymmetry of the medium in the other.   Now in the galvanic case we have in the environment of the medium through which the light passes (and which we may suppose to be water or glass) a plain and palpable seat of asymmetry, and therefore it would be a *gratuitous* multiplication of the causes of natural phenomena, opposed to Newton's first *regula philosophandi*, to assume, absolutely without other evidence, the existence, in the medium in its natural state, of an asymmetry of the kind required.

You write as if the effect were capable of exhibition only in exceptional media; I think from what Faraday wrote, the *non*-exhibition was quite the exception, if we omit doubly-refracting substances, in which it would seem certain *a priori* that we should not get the effect, except *perhaps* when they were examined exactly in the direction of an optic axis. And yet the optical part of Faraday's work on the subject was rather rough. He used lamp-light polarized by reflection (therefore light by no means intense) and a Nicol's prism. With sun-light and a Jellett's analyzer, I should expect to find the effect in every solution, and uncrystallized solid, and crystal of the cubic system large and perfect enough for examination.

To this I make one exception. It has since been found that some solutions rotate the opposite way to what Faraday found. If the solvent rotate the usual way, then for one particular strength of solution the rotation would vanish, in passing from positive to negative. And however refined the mode of observation, there may be a few cases of power of rotation so feeble as to elude observation. But even Faraday's list suffices to show that non-rotation is the exception rather than the rule.

If asymmetry of arrangement, such as we suppose in your hypothesis, pre-exists in water, glass, etc., how is it that we find absolutely no trace of the phenomena shown by syrup of sugar, etc.? It is true that the optical phenomena of the action on polarized light of syrup of sugar on the one hand, and glass under the influence of magnetism on the other, are very different as regards the relation to difference of directions; but the kind of asymmetry that is required to pre-exist in glass or water in your view, to render explicable the production of the known phenomena by the action of a force assumed not to be asymmetric, is precisely what we require to explain the known phenomena of syrup of sugar in its natural state, *i.e.* not acted on by magnetic force.

As to the success of Poisson's view in explaining the action of a magnet on a magnet or on soft iron, that remains as it ever did. The law of action of an Ampère elementary *current* (as it is called: but I wish to purify the word from any necessary idea of motion) on another such current is identical with the law of action of a Poisson magnetic element on another such element, and therefore one view will explain the action of a magnet on a magnet just as well as the other. One is the mathematical equivalent of the other, and we are perfectly at liberty to work on Poisson's hypothesis as we please; just as in statics we may imagine the mass of a rigid body collected at its centre of gravity, though we know that physically it is not so collected. Doubtless if we knew nothing relating to magnetism but the action of a magnet on a magnet, Poisson's view would be the simpler of the two; but when we take in the discoveries of Oersted and Faraday we cannot adhere to Poisson's theory (considered as a physical theory, and not merely a mode of representing the action of a magnet on a magnet) without a gratuitous multiplication of the causes of phenomena.

CAMBRIDGE.
4th July 1872.

MY DEAR SIR GEORGE,

I don't say that the Faraday experiment of the optical effect of magnetizing a medium contains a mathematical demonstration that Poisson's theory of magnetism, considered as a physical theory and not as a mere convenient and correct mode of picturing to the mind the action of magnetic forces, is false; but I do say that it renders it so extravagantly and outrageously improbable that I cannot conceive how anyone after fully considering the conditions can maintain it.

You admit that a suitable galvanic coil and a permanent magnet are mutually replaceable as to their external actions.

LENSFIELD COTTAGE, CAMBRIDGE.
3rd November 1875.

MY DEAR SIR GEORGE,

I read to-day in the *Phil. Mag.* your remarks on Sir Wm. Thomson's paper*. There is not a word, I think, that you have there written that Thomson would object to, except that he would change your "perhaps" into "certainly" K_4 *is* to be determined by the conditions at the boundaries. And what he

---

* Papers in *Phil. Mag.* "On Laplace's Theory of the Tides," elucidating Laplace's work against Airy's previous destructive criticism. For references to subsequent improvements and simplifications see Prof. Lamb's *Hydrodynamics*.

has done is to *prove* (which Laplace omitted to do, and the omission of which forms an important lacuna in what he has left) that the particular value of K_4 assigned by Laplace's process, and that alone, satisfies the condition of non-discontinuity at the equator, a condition which must be satisfied if the whole earth be covered with water.

In fact, what Thomson has done is to *prove* (what Laplace left unproved, perhaps did not understand) that the process given by Laplace has for effect to pick out from the infinite number of values of K_4 which all permit of satisfying *the general* equation, which general equation applies to all boundaries by parallels of latitude symmetrical N. and S., that *particular* value which belongs to the *particular* and *evidently determinate* problem belonging to the case when there is no boundary at all.

I have not Thomson's first paper by me, and K_4 have written from memory, but I think that is the notation.

I say "evidently determinate problem" because the free oscillations have for period one or other of what are practically the roots of a transcendental equation which will differ in general from the given period of the forced oscillation.

LENSFIELD COTTAGE, CAMBRIDGE.
16th November 1875.

MY DEAR SIR GEORGE,

It sometimes happens that...difficult problems are elucidated by the consideration of more simple ones which involve the same general principles. With respect to the tidal question, will you allow me to invite your consideration of a problem so simple that we can turn it about and view it from all sides, and yet which contains some of the leading general principles of the other?

Suppose then we had to consider the vibrations, of the nature of those of sound, of the air within a very slender circular tube, the air being subject to a periodic disturbing force. The tube is supposed to be so narrow that any element of it may be treated as straight, and the disturbance to be very small. I will suppose the disturbance symmetrical right and left of the radius vector passing through a particular point O which I will take for origin. Let x

be measured from O along the tube, the length being 2π, and let ξ be the small displacement along x, and a the velocity of sound. Then, if there were no disturbing force, the partial differential equation of motion would be

$$\frac{d^2\xi}{dt^2} = a^2 \frac{d^2\xi}{dx^2},$$

but if there be a disturbing force X along x, X must be added to the right-hand side. Let X (which must change sign with x in order that the equation may be symmetrical with regard to O, and which is by hypothesis periodic as regards t) be of the particular form $c \sin x \sin nt$. Then the equation of motion will be

$$\frac{d^2\xi}{dt^2} = a^2 \frac{d^2\xi}{dx^2} + c \sin x \sin nt \tag{1}.$$

The integral of this equation might of course be written down; but as the equation is designed to illustrate one which cannot be integrated, I will ignore the possibility of integrating it. Corresponding to the disturbing force there will be a periodic disturbance whose period is $2\pi/n$ and which will involve $\sin nt$ or $\cos nt$ or both. As the equation (1) involves $\sin nt$ only, I will take $\xi = u \sin nt$, u being a function of x without t.

Substituting in the general equation, we have

$$a^2 \frac{d^2u}{dx^2} + n^2u + c \sin x = 0 \tag{2}$$

the integral of which equation, subject to the condition of symmetry right and left of O, is

$$u = \frac{c \sin x}{a^2 - n^2} + A \sin \frac{nx}{a} \tag{3}$$

so that

$$\xi = \frac{c \sin x \sin nt}{a^2 - n^2} + A \sin \frac{nx}{a} \sin nt \tag{4}$$

A being an arbitrary constant.

Observe, that though the last term of (4) would satisfy our equation if there were no disturbing force (*i.e.* the general equation and the condition of symmetry) this term does *not* represent an initial disturbance which may be superadded to that kept up by the disturbing force. For a free periodic disturbance, recurring at each point of the tube at the expiration of its period, must have for period the period of the fundamental note of the tube or of one of its harmonics, *i.e.* n must have one of a series of *definite* values; whereas by hypothesis n is a *given* constant not determined by anything related to the tube. The problem is, in fact, from its very nature *essentially determinate*, and A must be properly determined.

Now ξ must = 0, not only for x = 0, but also for x = π. We get then from (3) or (4)

$$A \sin n\pi/a = 0,$$

and therefore n being a given constant so that the equation cannot be satisfied by putting

$$\sin n\pi/a = 0,$$

we must have

$$A = 0 \qquad (5)$$

and therefore

$$\xi = \frac{c \sin x \sin nt}{a^2 - n^2} \qquad (6)$$

is the definite solution of the definite problem proposed.

To what then does the arbitrariness of A in (4) correspond? Why (4) satisfies the general equation and part, *but only a part*, of the necessary equation of condition. Up to (4) inclusive, the tube might just as well have been blocked at x = − α, and at x = + α instead of returning into itself. In which case we should have the further equation of condition

$$\xi = 0 \text{ when } x = \pm \alpha,$$

giving

$$A = -\frac{c \sin \alpha}{(a^2 - n^2) \sin n\alpha/a},$$

and therefore

$$\xi = \frac{c}{a^2 - n^2} \left\{ \sin x - \frac{\sin \alpha \sin nx/a}{\sin n\alpha/a} \right\} \sin nt.$$

When α = π we fall back on the former solution applying to a tube returning to itself, which, on account of the symmetry with respect to 0, is as good as blocking at x = ± π.

Now the simple sound problem I have discussed is analogous in its leading features to the more complicated tidal problem about which Laplace and you and Thomson have written. The assumption that the disturbance must depend on the cosine of double the hour angle, as we know the disturbing force does, is the analogue (if I may use the word) of the assumption in the tube problem that the disturbance depends on the sine or cosine of nt; it therefore binds down the disturbance to have for its period half a day, and thereby sets aside the free periodic disturbances, which have periods determined by the roots of a certain transcendental equation—periods which are the analogues of the periods of the fundamental note of the tube and its harmonics. (2) is the analogue of the tidal equation numbered (2) in Thomson's first paper; (3) with its one arbitrary constant A is the analogue of Thomson's (3) with K_2 given by (4), (i.e. $K_2 = H$), K_4 arbitrary and the other K's determined in terms of $K_2 (= H)$, and K_4 by (5). The constant K_4, up to this point arbitrary, is the analogue of A which *so far* is arbitrary in the sound problem, though K_4 like A must be determined before the solution is complete. The equation $K_4/K_2 =$ Laplace's continued fraction is the analogue of $A = 0$, and just as $A = 0$ and that value alone reduces (3) which in general contains *two* circular functions in x to a special form which contains but *one*, so, as Thomson has proved, Laplace's value of K_4 and that alone reduces the integral with one arbitrary constant K_4 from a form in which the differential coefficient of the height with respect to the latitude is finite at the equator to one in which the same differential coefficient $= 0$ at the equator; and at the same time reduces the series from a form in which the *coefficients* of the powers of x^2 taken by themselves form a series of only neutral convergence to one in which they converge with great rapidity, at least in the higher orders. It is conceivable that if we knew less than we do about differential equations we might be able to get the special integral of simpler form $u = \dfrac{c \sin x}{a^2 - n^2}$ of (2) though unable to get the more general integral (3), which special integral applies as we have seen to what may be regarded as the case of greatest dignity when there is no barrier at all. So it is conceivable that Laplace's value of K_4, though derived only from a process derived from the general equation, may have for effect to pick out a special solution of special form which applies to the case of greatest dignity, that of an ocean without a barrier. And Thomson has proved that it *has* this effect.

Equation (3) with A arbitrary is the analogue of your equation with K_4 (Thomson's notation) arbitrary. And just as A may be determined to suit a tube blocked at $x = \pm \alpha$, so K_4 may be determined to suit a sea extending from one pole to a parallel of latitude. Only as Thomson has shown, the parallel must be on the same side of the equator as the pole. Your method would *apparently* apply to the case of a barrier at the equator itself (and therefore to the case of no barrier at all) but it practically fails because the series by which H and K_4 are multiplied are of only neutral convergence. If it did not fail it would necessarily give the same numerical value for $K_4 : H$ as is got from Laplace's process.

P.S. Permit me to make a slight correction at the top of the last page of my letter of the 3rd inst. I wrote this hastily (as I felt at the time but it did not affect the general question) for the *general equation* applies whether the boundaries of parallels of latitude are symmetrical N. and S. or not, though of course the particular process employed by Laplace and you, involving the use of a series according to even powers only of the sine of the N. P. D., indeed I might say......, could only apply when the motion is symmetrical with respect to the equator.

PLAYFORD, NEAR IPSWICH.
January 3, 1876.

MY DEAR SIR,

......Now *first* I have to recall that my objection to Laplace's solution was, to the solution of his equation as it is by him exhibited. And this objection stands; and no greater condemnation can be given to Laplace than your suggestions of vast lacunæ which must be filled to give any meaning to his process.

Secondly, the paper which you lately sent me appears to relate mainly to questions of longitude in an equatoreal or parallel canal (the Laplace question relating to latitude). This is worthy of all attention. Now the following appears to me to be a point worthy of consideration. The question of a circumfluent ring, whether of water or of air, divides itself into two questions, *videlicet*: A. Is there a barrier? B. Is there no barrier, but perfect ring? On these :—

A.   The question of a canal with barrier is fully treated in my
Tides and Waves, the solution applying (*mutatis mutandis*) to
air or water, and requires no farther notice here.

B.   The forced wave exists; and there may be an infinity of
free waves.   And these resolve themselves into two classes:
B_1 whose period is commensurable with that of the forced waves
(and which will cause strange but periodical alterations of magnitude), and B_2 whose period is incommensurable with that of the
forced waves (and which will cause disturbances something similar,
for a limited duration, to those of B_1, but never continuing their
similarity to B_1).

What a mercy it is that friction prevents us from revelling
very deeply in all these complexities.

And thus end my tidal speculations for the present.

Can you answer me the following question :—

In my *Tides and Waves*, written, I think, in 1837, I produced
the theories of "Free Waves" and "Forced Waves;" then sub-
jectively new and original to me.   Do you know any place in
which they had been exhibited before that time ?

I am, my dear Sir, Yours very truly,

G. B. AIRY.

LENSFIELD COTTAGE, CAMBRIDGE.
4th Jan. 1876.

MY DEAR SIR GEORGE,

I am sorry that my bad handwriting should have caused
you so much trouble.   I rather envy you and others like you the
legibility of your writing.

As to the little problem to which I invited your attention,
I did not mean my canal to be either meridianal or parallel to the
equator.   In fact the canal had no relation to the Earth at all, but
related to a purely pneumatic question, only bearing on the tides
at all in so far as both problems (the little sound problem and the
tidal problem discussed by Laplace and by you) involved the same
general principles as to the application of partial differential
equations.   And I thought that the easy sound problem, which
we can turn about and view from all sides, would be useful in

throwing light on the more difficult and complicated tidal problem,
where we meet with a differential equation that cannot be integrated
in finite terms.

With respect to your question as to what was known about
waves when you wrote your treatise, I cannot answer better than
by referring you to my report prepared for the British Association
1846 of which I had a copy.

I have given the dates of memoirs read before Societies.
I have not mentioned the date of publication of your article
in the *Encyclopaedia Metropolitana*. I hardly should have thought
it was quite so early as 1837. I recollect Dr Whewell's speaking
to me about it, I think when it first came out, and he was not
Master till 1841, and I did not know him till some time (?) after
that. Besides I think I should naturally have referred to the
various investigations which I mentioned in chronological order.

It is curious that Poisson and Cauchy seem not to have
condescended to discuss the motion represented by a single
element of their own formulae. Thus Cauchy (whose treatise
I happen to have taken out of the library) finds (p. 62, eq. 54
with a slight change of notation)

$$\phi=\int_0^\infty e^{my}\cos mx\cos\sqrt{gm}\,t\,\phi(m)dm+\int_0^\infty e^{my}\sin mx\cos\sqrt{gm}\,t\,\phi_1(m)dm$$

$$+\int_0^\infty e^{my}\cos mx\sin\sqrt{gm}\,t\,\psi(m)dm+\int_0^\infty e^{my}\sin mx\sin\sqrt{gm}\,t\,\psi_1(m)dm$$

for the function which gives the most general small motion in two
dimensions produced from rest in a deep liquid, and later on (in
the notes) he gives the expression in the case in which the depth
is finite and uniform. The differential coefficients of ϕ give the
pressure and velocity. This expression represents the most general
motion as the result of the superposition of an infinite number of
"simple harmonic" disturbances corresponding to different values
of m (i.e. to different wave-lengths), I mean *will* so represent it
when we replace the products

$$\frac{\cos}{\sin}(mx)\frac{\cos}{\sin}\sqrt{gm}\,t$$

by sums and differences, the disturbances being propagated in
the positive and negative directions with the velocity $\sqrt{\frac{g\lambda}{2\pi}}$.
But he did not stop to discuss the nature of the motion in the

case of one of these "simple harmonic" disturbances, but put out his force in the difficult problem of determining the result of a single splash.

I am, dear Sir George, Yours sincerely,

G. G. STOKES.

LENSFIELD COTTAGE, CAMBRIDGE.
5th Jan. 1876.

MY DEAR SIR GEORGE,

I have taken Poisson's memoir on waves out of the library (*Mémoires de l'Académie*, Tom. 1, p. 1). In the preface he gives an account of prior researches. Laplace appears to have been the first to deduce from rigorous theory anything relating to waves. In a paper published in the memoirs of the Academy for 1776 he considered the case of small motion in two dimensions in which the surface of a fluid at rest is deformed to a trochoidal curve, and then left to itself. I have not been able to find this paper, but I think I once saw it. I think it refers to deep water only, and answers to what you would call a stationary oscillation, which of course may be split into two representing waves travelling in the positive and negative directions respectively. This was about 10 years before Lagrange gave the theory of waves propagated in water of small depth, in fact long waves.

For small motion in two dimensions (he considers three dimensions also), Poisson gives the following as the complete integral of the equations (p. 84)

$$\phi = \Sigma B \left(e^{a\,(h-z)} + e^{-a\,(h-z)}\right) \cos\left(ax + a'\right) \sin ct$$
$$+ \Sigma B' \left(e^{a\,(h-z)} + e^{-a\,(h-z)}\right) \cos\left(ax + a'\right) \cos ct$$

where B, B', a, a' are contrary constants, and

$$c^2 = ga\,\frac{e^{ah} - e^{-ah}}{e^{ah} + e^{-ah}}.$$

Poisson and Cauchy don't seem to have condescended to discuss the character of the motion represented by a single element of their integrals or sums. They set themselves to the more difficult, and in many respects less interesting, task of determining the motion which would result from a given initial disturbance.

I am not aware that anyone before you showed so simply, from first principles, the equality of the horizontal motion of the particles in a vertical line in the case of long waves, or carried the investigation beyond the first order of small quantities, or considered the case of forced waves except in so far as it is virtually involved in the problem of the tides, or examined in detail the motion represented by one element of Poisson's integral. It was indeed considered in detail by Gerstner ("Wellenlehre*") but on a theory in some respects radically erroneous, though in certain other features it was right, and accordingly in some features agrees very nicely with experiment while in other respects it is quite at fault. As the theory is not rigorous I don't count it.

I have lately perceived a result of theory which I believe is new—that the velocity of propagation of roughness on water is, if the water be deep, only half the velocity of propagation of the individual waves. This is of importance in connecting records of long swells which may be found in ships' logs with records like those of Ascension or St Helena†.

Yours sincerely,

G. G. STOKES.

October 8, 1877.

The behaviour of those rings in the experiments of which you sent account to me is very curious:—I do not see my way through it.

When I was engaged on similar subjects, I used a prism with flat sides, and when need required I cemented to the lower side a plano-convex lens. If you are engaged on diamond experiments you will find some such course prudent, for in one or two touches of the diamond the surface of the glass is destroyed:—the loss of a plano-convex lens is unimportant, but the injury of a prism is serious.

I am, my dear Sir, Yours very truly,

G. B. AIRY.

* Viz. expounded in "Wellenlehre auf Experimente gegründet," von den Brüdern E. H. und W. Weber, Leipzig, 1825. The theory is examined in *Math. and Phys. Papers*, vol. i. pp. 219—229, Appendix added in 1880.

† Prof. Stokes elaborated directions for the utilisation of this subject in ocean meteorology: see *supra*, p. 133.

CAMBRIDGE, 11*th December* 1877.

DEAR SIR GEORGE AIRY,

I was very sorry that you were prevented from being present last Thursday when your paper was read, and also for the cause of your absence. I hope that with a little care the hurt to your knee will be of no consequence.

The "seiches*" clearly point to some oscillation of a limited portion of the sea forming a sort of basin or channel, and two questions arise:

(1) Supposing the water to have a certain swing-time, how are the oscillations produced?

(2) What is the channel, and what the nature of the oscillation?

As to (1) the case differs from that of a wind instrument, in which a disturbance once produced tends to perpetuate itself so long as the blast lasts, in a manner which is not I think difficult to explain. In the present case I can see no such reproductive element in the disturbance supposing the disturbing cause to be a uniform wind, and I am therefore led to look to a variable wind or barometric pressure.

Now the automatic registrations of the height of the barometer show frequently in unsettled stormy weather fluctuations of a

---

* These letters were referred to Prof. G. Chrystal, who has recently published important memoirs on seiches ("On the Hydrodynamical Theory of Seiches," *Trans. R. S. Edin.* xli. 1905; also "Calculation of the Periods and Nodes of Lochs Earn and Treig, from the Bathymetric Data of the Scottish Lake Survey," with E. Maclagan-Wedderburn, *loc. cit.*), and is now conducting an experimental investigation on the Scottish lakes. He permits the following quotations from his reply:

"The Stokes letters are, to me at least, intensely interesting. Every word is to the point. He takes exactly the view of the matter at which I had arrived, and which I have been discussing with Shaw for some time back. The only point of difference is that, in view of the experience of the intervening 28 years, we look with more expectation towards the barometer and less to the wind.

"It is curious how completely he anticipates all the points I have been thinking over. There is, for instance, the question of the degree of isolation of an estuary, which is necessary in order that Seiches may arise. ...The phenomenon is beautifully shewn incidentally in a diagram in Baird's pamphlet on the Indian tides....

"Of course you should publish the letters. Remember that they were written very soon after Forel had begun his work on Léman, when the subject of Seiches was in its infancy. It is wonderful to see the keen intellect of Stokes cut right to the marrow of the question....

"If all the apparatus we have accumulated will only work together at some favourable time, we may get at Loch Earn confirmation of Stokes' Theory."

moderate period, such as a fraction of an hour, or a few minutes, and such fluctuations would probably not be synchronous over places as far apart as the two ends of a basin large enough to give oscillations of the water of a period as great as 15 or 20 minutes.

Again, we know that in a storm the wind is very frequently gusty*, and accordingly the lateral force of the wind on the water would be far from uniform.

Now if at any time these barometric changes or the alternations of stronger or less strong wind were rudely periodic, with a period approximating to the natural swing-time of the water in a basin such as supposed, the disturbing force acting for some time would be competent to produce sensible oscillations, even though the disturbance of the water produced by the action of the disturbing force for a single period only might not be sensible.

(2) The basin might be either a land-locked portion of the sea, of such size and depth as to furnish the observed periodic time, or as you suggest a deep basin partially isolated by banks on which the water is comparatively shallow. The latter seems to me the more probable, because on looking at the map we cannot pick out merely by the coast-line a basin even rudely isolated; and if any part be deemed a basin which is in free communication with the Mediterranean at large, oscillations cannot go on accumulating in it under the action of a roughly periodic force, because almost as fast as they were produced they would be lost, to the basin, by being propagated into the sea at large. There appears therefore a better prospect of obtaining a roughly isolated basin by means of banks, over which the water is comparatively shallow.

In either case there would be subsidence (in the basin) of the oscillations, supposed produced and not kept up, by communication

---

* "The limnogram gets 'gusty' too; that is to say, it acquires an embroidery of oscillations of comparatively short duration, more or less irregular in configuration and sequence. These may be due—(i) to seiches of short period; (ii) to trains of progressive surface waves caused by wind gusts or squalls; (iii) to wave groups, *i.e.* interference of the first order between trains of progressive waves; (iv) to interference effects of the second order between such trains. See Börgen, *Ueber den Zusammenhang zwischen der Windgeschwindigkeit und den Dimensionen der Meereswellen, nebst einer Erklärung für das Auftreten von Wellen von langer Periode an freigelegenen Küstenpunkten* in the *Annalen der Hydrographie und Maritimen Meteorologie*, Jan. 1890. Observations on Loch Earn with the statoscope have conclusively shewn that (ii) is one *vera causa*. It is now probable that the secondary oscillations shewn by tidal diagrams are not all due to the same cause; some fall under (i), others under (ii), (iii) or (iv), or some other category."

to the sea at large; but if the isolation be pretty fair, the percentage of energy lost by communication to the sea at large in a single oscillation will be but small, and so if the disturbing force be roughly periodic and of the proper period, an oscillation of sensible amount may be produced.

To determine exactly the periodic times corresponding to the different possible stationary undulations in a basin of given shape, is a problem which in general we cannot manage. It can easily be solved for a rectangular basin whether the motion be in two dimensions or not.

Probably it could be solved for a hemispherical basin, but I have not tried. I should think it would not be difficult for a semi-cylindrical basin, with the axis of the cylinder horizontal, and bounded at the ends by planes perpendicular to the axis of the cylinder, the motion being in planes perpendicular to the axis.

Perhaps it might be done for such a basin though the motion were in three dimensions. I recollect a long time ago working it out for a cylindrical basin with axis vertical. I did not publish the result because I found that Ostrogradsky* had already got a paper on the subject.

A propos to the fortnightly tide I made a rude calculation to see how long it would take to fill up the Mediterranean, regarded as a vessel filled through a small orifice (the Strait of Gibraltar) from a reservoir (the Atlantic) in which the water stood at first 4 inches higher than in the vessel. I took roughly from a map the area of the Mediterranean not including the Black Sea, and made a rough guess from such data as I could readily lay my hands on of the section of the Strait at the narrowest point. By narrowest I mean smallest in area, which is not probably where the coasts come closest, as the water shallows towards the Atlantic. With rude data I made the filling time something like three weeks. I took no account of friction, but friction would of course increase the time of filling. I should think the fortnightly tide in the Atlantic must be distinctly, perhaps somewhat considerably, greater than in the Mediterranean. I remain, Yours sincerely

* *Mém. Sav. Ét.* iii. 1832, not much to the point.   G. C.

LENSFIELD COTTAGE, CAMBRIDGE.
12th Dec. 1877.

DEAR SIR GEORGE AIRY,

I forgot to mention in my letter last night that the "seiches" at Toulon have a period of I think 15 m.* In all probability, in fact almost certainly, the basin of oscillation is here different. The same basin would give a theoretically infinite series of periodic times, the period decreasing indefinitely in the higher numbers; but practically it would be only the simplest, or possibly two or three of the more simple modes of oscillation, that we need attend to.

I spoke of (i.e. wrote of) the necessity of an approximate isolation of the basin of oscillation in order that the oscillations, supposed to be excited in any way, should not quickly die away by communication to the water of the great basin formed by the Mediterranean at large. (I neglect here the small outlets at Gibraltar and Gallipoli, and regard the Mediterranean as a closed basin.) Theoretically one of the possible "harmonic" oscillations of the great basin, of a very high order, would belong to a disturbance (or stationary undulation) in which the chief disturbance would be in the local basin and its neighbourhood, and this disturbance, friction apart, would go on for ever. But to produce such an oscillation would require a continuance and regularity in the disturbing force which we cannot expect in a force depending on meteorological conditions; so that practically I don't think we shall err by treating the shores, except those in the neighbourhood of the local basin, as being at an infinite distance, and regarding the outward-travelling undulations as never to return by reflection, but lost to the local basin.

By means of tide-gauges temporarily placed for a short time at a variety of places on the coast of the Mediterranean and its islands, and by classifying the "seiches" observed, if any, by their periodic time, it might be possible to learn much as to the localities of basins of oscillation.   I remain, Yours sincerely

* 15 m. is about the uninodal period of Loch Earn.   G. C.

17th December 1877.

My dear Sir,

I write mainly to acknowledge your letters of 11th and 12th. The theory of sea-waves in ill-shaped basins is so impossible, and that in the best well-shaped basins so difficult, that I am driven, *nolens volens*, to atmospheric analogies.

Looking to the movement in an open-ended organ-tube, I do not require *periodic* forces of wind, &c.; and I am not fearful that oscillations will be rapidly destroyed by the communication of the basin, at one end, with an open ocean otherwise tranquil. And looking to the effect of the lateral holes of a flute, I should expect the effect of laterally diverging channels to be that of altering the time of oscillation (this, however, does not appear to apply to the present instance).

All this is very vague, I fear unavoidably so.

Can you refer me to the authority for the Toulon seiches of 15 m.? They have a bay of their own, or a basin of their own, of which I have no accurate knowledge.

I have written to the Academy at St Petersburg, asking for a self-registering tide-gauge on some island in the Baltic.

I inclose some carefully-made copies of barometer-registers. There seems to be no uniformity in the petty disturbances, either from hour to hour or from day to day. I believe that you may trust these records for time and for duration of oscillations, but scarcely for magnitude; because the movement, impressed on the plate which gives (photographically) our record, is *mechanical*, by a float on the mercury and an intervening wheel-on-axle. For perfectly reliable results in these small movements, nothing but a beam of light passing over the mercurial column can be trusted.

I am, my dear Sir, Yours most truly,

G. B. AIRY.

ATHENAEUM CLUB, PALL MALL, S.W.
19*th* Dec. 1877.

DEAR SIR GEORGE AIRY,

I return to Cambridge on Friday, and will take an early opportunity of measuring the telescope models.

I hope also to write shortly pointing out what seems to me to be the reproductive element in the disturbance of the air in the case of a wind instrument, and which I fail to see in the oscillations in a basin. I am much obliged by the copy of the barometric registration. I hope to point out why I conceive the disturbance need not be more than very rudely periodic.

I am writing to Sir Wm Thomson [to ask] where (if anywhere) the results as to the "seiches" at Toulon are published.

I remain, Yours sincerely

CAMBRIDGE.
21*st* Dec., 1877.

DEAR SIR GEORGE AIRY,

As to approximating to the time of oscillation in a natural basin by taking the basin as of some geometrical form, since the fit must be a *bad one at best**, it seems hardly worth while going beyond the simple case of a rectangular basin which is to be made to fit the natural basin as best may be. If a, b be the lengths of the sides, and h the depth, the series of periodic times τ corresponding to the different possible oscillations are given by the formula

$$\left(\frac{2\pi}{\tau}\right)^2 = mg \frac{e^{mh} - e^{-mh}}{e^{mh} + e^{-mh}} \quad \text{when} \quad m^2 = \frac{i^2\pi^2}{a^2} + \frac{j^2\pi^2}{b^2}$$

i, j being any two integers either of which may be zero.

This however would be a rude process, sufficient to give a general idea of the magnitude of the time τ in a natural basin, but not more, on account of the rudeness of the representation. But if we know the form it would not be very troublesome to

---

* Not so very bad in some cases. G. C. [Cf. his two memoirs, *loc. cit.*]

construct a model of it, suppose with sculptor's clay (Mr Froude uses paraffin), which after drying could be painted, or covered with coal-tar, and then filled with water and thrown into oscillation. The times in the natural basin and in the model would be as the square roots of homologous lines. However we have not got soundings enough, except near shore, to give more than a rough representation of a sea basin.

I heard this afternoon from Sir Wm. Thomson. He thinks the period is more nearly 20 or 25 minutes than 15, as I had supposed*. He says they (the seiches) are to be found not merely in the Mediterranean but at Brest and Portland, and he believes everywhere where a sufficiently delicate tide-gauge is kept. He refers me for further information to Mr Roberts at the Nautical Almanac Office, who is working at tidal reductions for him and Captain Evans, and has instructions to investigate them as far as may be from all the tide-curves he has. As far as I can find out, little or nothing seems to have been published about them, except by M. Forel as regards the Swiss Lakes.

I must explain what I meant by the reproductive element in the disturbance of the air in a wind instrument. I have mentioned the thing in my lectures, but I have not come across it in any book. Let us take for example a pitch-tube giving the fundamental note. How is the oscillation kept up?

The figure represents rudely the pipe in section, the blast passing through a channel very narrow in one direction, and

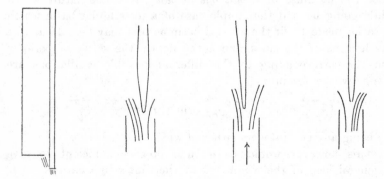

impinging on a sharp edge. If there were no oscillation the blast parted by the sharp edge and the filaments of air would pass

* 2′ 7″ is about the uninodal period of Loch Tay.   G. C.

partly on one side and partly on the other and remain so parted. Suppose now an accidental cause to push the blast a little inwards. This would produce a wave of condensation; which after reflection at the top would reach the orifice at the end of a half period. There the outrush would tend to throw the filaments of the blast outwards, and the upward-travelling wave of rarefaction due to the mere reflection of the wave of condensation at the open end would be reinforced by the wave of rarefaction due to the outward deflection of the filaments of the blast, and so on. Hence we can see how when the blast attains a certain strength an accidental disturbance tends to perpetuate itself.

Now I fail to see any such element of perpetuation in the oscillation of water in a basin, supposing the wind perfectly uniform, and that leads me to look to irregularities of wind or possibly barometric pressure as the exciting cause.

In order to produce a comparatively large disturbance, the disturbing force need not be more than rudely periodic of *about* the requisite period. We may compare the oscillations to those of a pendulum in a slightly resisting medium, the resistance varying as the velocity, for which the equation of motion is of the form

$$\frac{d^2x}{dt^2} + 2q\frac{dx}{dt} + (n^2 + q^2)x = f(t)$$

the integral of which, supposing the motion to go on indefinitely, is

$$x = \frac{1}{n}\int_{\infty}^{t} e^{-q(t-t')}\sin n\,(t-t')\,f(t')\,dt',$$

the form of which points out what I said. To take a particular case, supposing the disturbing force expressed by a simple circular function with a period $2\pi/n$, different but not greatly different from that $2\pi/n$ of the natural oscillation, then the integral would contain a term with $q^2 + (n - n')^2$ in its denominator, and if q be pretty small and n' moderately nearly equal to n the term in the disturbance will be much multiplied. When a disturbing element exhibits fluctuations, the range of the period of which contains the period which is required for greatest effect, the chances are that in the long run we shall get several successive disturbances working in the same direction.

I hope to-morrow to send you measures of the telescope models. I remain, Yours sincerely

May 6, 1879.

MY DEAR SIR,

Many thanks for your capital letter of April 3 regarding the harmonics of diurnal temperature-records*. It really goes very far to show that inferences as to the general laws are very fairly mastered by harmonics. Long life to No. 3.

There are some of Mr Ellis's results which naturally cannot be reduced by such a process—as the law when a N. wind follows a W. wind—&c., &c.

I am, my dear Sir, Yours most truly,

G. B. AIRY.

ROYAL OBSERVATORY, GREENWICH, S.E.
24 *July*, 1879.

MY DEAR SIR,

Returning from absence and troubled not only by accumulations but also by business-callers and progressive business, I have little time to remark on your letter of July 11, except that I think you have well established the sufficiency of harmonic terms to the third order to represent the meteorological results which you had taken in hand.

I am, my dear Sir, Yours very truly,

G. B. AIRY.

ROYAL OBSERVATORY, GREENWICH,
3 *March*, 1881.

MY DEAR SIR,

I received your letter of December 31 in the country; I have kept it in sight, and have referred to it from time to time; but business which seemed to demand my attention every moment has prevented my answering it. I fear that what I have to say now will not be very complete.

* See extracts from records of the Meteorological Council relating to harmonic analysis, *infra*.

The objects, to which the correspondence (inclosed in your letter) applies, appear to be these, or principally these:—(1) The nearly continuous photographic register of solar spots: (2) The nearly continuous register of the sun's radiant heat.

1. I suppose that the uses of the solar spot register are two: (A) for the changes from year to year, (B) the changes from day to day.

Now on (A) we have a great deal of information, beginning from Wolff, and going on regularly at Greenwich and I suppose at other places. On what they show, the best evidence is that in Mr Ellis's paper on the comparison of spot observations with magnetic observations. The exhibition (incidental to that comparison) of the numerical record of sunspots shows, in short periods extreme irregularity, and in long periods an approximation to order—I cannot call it a law—of a rude kind, which cannot be stated without leaving ample licence for departure from anything like law. The question (as regards the matter before us) applying to this is, Is there reason for believing that we should approach nearer to order or law by multiplying the observations on the proposed scale? I do not think there is. Is it worth the trouble to keep up the present system? I think it is. It gives us, though in a very rude shape, a cosmical recurrence of much interest*; and it gives us the first grounds for thoughts on causal connexion that might in time suggest distinct reasons for extensive observations in India or elsewhere.

In (B) are such phenomena as the rise of spots and faculae, the diminution of spots &c., the bridging across of spots, the changes that go on in their spectroscopic phenomena, the definition of the solar regions which they most affect, &c., &c. The record of these, at Greenwich for instance, is very imperfect. But it is not contemptible. There are to be found, on comparison of consecutive photographs, a great number of instances of every result of changes from day to day that can be suggested. I think that until these have been well *studied*, with the view of ascertaining how observations can be extended advantageously at home, and on what points (*very distinctly indicated*) observations at other stations are *wanted*, we should be wrong in undertaking a series, though in itself better, at a distant station†.

* Cf. Schuster, *Roy. Soc. Proc.* 1905.

† See extracts from Reports of Committee on Solar Physics, *Math. and Phys. Papers*, vol. v. pp. 125—137.

2. The record of solar radiant heat, as a broad natural fact, unconnected (at the present time) with theory, appears to me well deserving of attention. Its bearing on almost everything in human life is evidently important. And its bearing on possible cosmical theories may be valuable, though I would not act on anything so vague, except in conjunction with more distinct reasons. But there appears to be at present a difficulty as to the mode of measuring. It does not appear, I think, that we have a satisfactory actinometer*. Our thoughts ought first to be directed to that instrumental arrangement.

In the mean time there is another mode of measuring the radiant heat received by the earth—that indicated by the deep-sunk earth thermometers. The evidence of their efficiency, as distinguishing the heat received in different years, and in different months, will be found in the "Reduction of 20 years' photographic, &c., and 27 years' observation of earth-thermometers, &c. at Greenwich," especially Plates IX. and X. I took the liberty some months ago of suggesting this to the Committee on Solar Physics; and I take the present opportunity of repeating my confidence in its value.

I am, my dear Sir, yours very truly,

G. B. AIRY.

LENSFIELD COTTAGE, CAMBRIDGE,
12 April, 1881.

MY DEAR SIR,

Some considerable time ago you were so good as to address a letter to the Committee for Solar Physics in which you advocated observations with thermometers sunk in the earth, as the best means, if I understood you aright, of studying a possible variation in the amount of solar radiation.

It was my intention to have written to you at the time on this subject, but in some way the matter passed over.

You have more recently been so good as to address another letter to us in which you refer to your former letter containing the above-mentioned suggestion. There has not been a meeting

---

* See *Math. and Phys. Papers*, vol. v. pp. 137—140; also correspondence on the design and performance of B. Stewart's and other actinometers, *infra*.

of the Committee since the receipt of the letter last mentioned, but I think it well that I should lay my views on the subject before you.

What you have published in the "Reduction of Greenwich Meteorological Observations" shows how well the mean temperature of the air from year to year tallies with the indications of the sunk thermometers, allowance being of course made for the known laws of progress of waves of heat downwards, according to which, for a given period of fluctuation, there is a retardation of epoch proportional to the depth, and a diminution of amplitude of which the logarithm varies proportionally to the depth.

The inequalities of short periods being practically eliminated by thermometers sunk to a moderate depth, the observation of such thermometers affords a mode of ascertaining the mean annual temperature of the place, involving so little labour that it can be carried on without the staff of a regular observatory.

It is true that the situation of the Royal Observatory appears to be somewhat unusually favourable to observations with sunk thermometers. Situated as it is on a hill with a gravelly soil, there is little danger of the readings of the thermometers being vitiated by the flow of water underground coming from some unknown, and possibly distant, place.

There would, however, probably be little difficulty in obtaining suitable sites for sunk thermometers elsewhere, so that I do not think much of this objection.

But I do not think that any direct conclusion can be drawn as to a possible variation of solar radiation from the observation of sunk thermometers. The temperature of the air depends no doubt *ultimately* on solar radiation. But more immediately it is greatly influenced by the presence or absence of cloud, by rainfall and evaporation, by the direction and intensity of the great currents of the atmosphere which in our latitudes vary in so irregular a manner. A summer warmer than usual in England may synchronise with a summer colder than usual in Italy. The weather is, so to speak, a very complicated integral of a system of differential equations; all the more complicated in consequence of the incessant conversion of stable into unstable equilibrium by the heating of the lower strata of the atmosphere.

So complicated is the assemblage of the conditions, that I do not see that we have much chance at present of getting further in a theoretical direction, in seeking to connect the weather with solar radiation, than we are carried by falling back on the general principle that,—if a system be subject to a periodic disturbing force, a disturbance of the same period will be produced, which if the observations are continued for a sufficient length of time will show itself in mean results, the time being sufficient to cause the practical disappearance of casual irregularities. And even if we had got the periodic disturbance, we could only infer from it the length of the period of the inequality in the disturbing cause, not the epoch. The difference in time between the maxima of the disturbance and the maxima of the disturbing cause would be liable to vary from one place to another, though not from one time to another if the disturbance were regular. And the disturbance, if we may judge by the visible alterations of the sun's disk, does not seem to have more than a somewhat rough periodicity.

But there is one mode of observation which seems to me most hopeful; I mean that of endeavouring to get direct measures of the solar radiation by actinometers. By going to a place several thousand feet above the level of the sea, and otherwise favourably situated as to climate, we may hope to get rid of irregular obstructions of the sun's rays, and then suitably constructed actinometers may give trustworthy indications of the sun's radiation. We are still in the dark as to what proportionate part, of the total radiation, any variation there may be in the amount of radiation is likely to be represented by; perhaps before long we may know something more about it*.

I am, dear Sir, yours very truly,

G. G. STOKES.

* Observations on this subject are now being resumed by the International Committee on Solar Physics, with the aid of a standard balanced actinometer of Ångström's design, in which the radiation absorbed by one platinum strip is compensated by the heat developed from a known electric current in a similar one.

PLAYFORD, IPSWICH,
October 11, 1883.

MY DEAR SIR,

Some days ago I received the copy of your Biographical Memoir (*Roy. Soc. Proc.*) of General Sabine sent by you. I am very much obliged to you. I will offer to you some notes, which at any rate will show that I take interest in your work.

The early history of Sabine's work is new to me. The first thing which I knew was, his return from the elder Ross's Northern Expedition. After this I have nothing distinct down to his return from the Pendulum Expedition and the publication of his account of it.

The scheme of a wide-spread system of meteorological observations was, so far as I know, entirely Sabine's own; and so I suppose was the magnetic scheme; although in the first instance, before the diurnal changes attracted so much attention, continued magnetic observation did not seem necessary. But the discovery, in Gauss's School, of the simultaneity—to a great extent—of magnetic phenomena in different places, excited us all; my first official act at Greenwich had been to urge the establishment of a Magnetic Observatory, and now I took part in Gauss's term days. Personally, I have observed through the night, and have seen the rising sun glimmering on these eye observations. I began to appreciate the importance of the phenomena, and I felt the importance of continuous record. So at a Meeting of the British Association—was it at Hereford?—I moved, and carried, "That the Treasury be requested to offer a premium for the arrangement of a Self-recording Apparatus." The Treasury refused. I then personally made a similar application to the Admiralty, who had a fund for scientific experiments. The Admiralty assented. The result of this was the establishment at Greenwich of the apparatus arranged by Mr Brooke, which, in its general features, has been adopted ever since that time.

This history is little known. The whole of the papers concerning it were carefully bound by me in an orderly form, and are preserved in the Record Room of the Royal Observatory. I wish that I could induce you to consult them.

In a short time Colonel Sabine sent two officers—were you one?

—to look at the Greenwich apparatus, with the view, I believe, of mounting one at Kew.

I had been warned that Col. Sabine wished to keep magnetism entirely in his own hands. I am aware of some instances which seem to support this. But no sign of this was shown for a long time; till after a partial failure of some instrument-maker's work at Greenwich he made a public attack on me. The failure had been discovered by me, not by Col. Sabine, and measures for correcting it were under consideration. There are printed papers on this subject which I could probably procure. This disturbed our harmony, but it has not prevented me from observing Sabine's efforts and expressing my admiration of them when they came under my notice.

Among these was the grand collection—in *Philosophical Transactions*—of the results of magnetic observations all over the earth. On some occasion, after the exhibition of the last sheet, I expressed my feeling that it was a noble termination of a noble work. I was told—I forget by whom—that this expression reached Lady Sabine's ear, and delighted her much. I am very happy to think so.

I conclude with saying that, in my opinion, Sabine has left a strong mark on cosmical science.........

I am, my dear Sir, yours faithfully,

G. B. AIRY.

ADAPTATION OF TELESCOPES TO PHOTOGRAPHY.

To SIR W. H. M. CHRISTIE, K.C.B., F.R.S., ASTRONOMER ROYAL.

GLASTHULE LODGE, KINGSTOWN, IRELAND,
August 16, 1886.

A happy, thought occurred to me to-night as to a mode of adapting the Greenwich Refractor that is to be put to photography. I hope it may solve the difficulty. It is to make the crown-glass lens of the objective extremely nearly, but not quite, equi-convex, and to mount it in a cell which could be put either way in, the difference of radii of the two surfaces being made to accord with a calculated difference.

For vision, the flatter surface would be turned outwards, and the lenses placed nearly in contact. For photography, the lens would be reversed, and its distance from the flint-glass would be greater than before by a known quantity which would reduce the chromatic correcting power of the flint to what is required for photography.

The separation alone would make the spherical aberration positive, and the reversion alone would make it negative, and if the small difference of radii has the right value these two will compensate each other. Thus the spherical aberration will remain at zero while the chromatic compensation is altered from that suited for vision to that suited for photography.

The focal length of the objective will of course be shortened a little by the separation, but not I think to an inconvenient degree.

August 17, 1886.

When I wrote to you last night, I had not made any calculations as to the disparity of radii in the two surfaces of the crown-glass required for carrying out my suggestion: I merely assumed that it would be very small, knowing how very sensitive spherical aberration is to an alteration of protuberance, without alteration of focal length, in the crown, and supposing that the effect of separation on the spherical aberration must be much smaller. I find however that the disparity of radii—at present I have only made an extremely rough calculation—must be far larger than I supposed, the proportion of curvatures of the two surfaces being something like 6 to 5, or 5 to 4. This brings out a form about midway between that which Mr Grubb employs and the Munich form.

The only disadvantage however that I see in this form—apart from the sacrifice of some facilities in the construction and testing—is that it might introduce a coma into very decidedly oblique pencils, when the more curved face is turned outwards for photographic work. The absence of coma is nearly secured in the Fraunhofer and Herschel forms, which are nearly the same; and in the crown equi-convex form, which is what Mr Grubb uses, or very nearly so, the coma is not sensible, at least with the small field you require for eye observations with large instruments. But for photography you want a larger field, and moreover the coma goes on increasing as you make the crown more protuberant in front without altering the flint.

LENSFIELD COTTAGE, CAMBRIDGE,
14 July, 1887.

I have not made any special investigation of the behaviour of an object-glass in which the flint is placed in front of the crown. I have got a series of little lenses, crown and flint, which I ordered for the purpose of trying various forms. A feature I noticed with these was that when the flint was put first, and I was near the form of no aberration, the appearance on a screen —I refer to a direct centrical pencil from a luminous point at a good distance—indicated that a section of the pencil by a plane through the axis presented three cusps, like...[see end]. I concluded that as the form changed continuously, passing through that of no aberration, the central cusp pointing one way became imaginary, and the side cusps pointing the other way coalesced into one. This is analogous to the alteration in the form of a curve the ordinate of which presents the feature, maximum-minimum-maximum; and then by an alteration of a parameter of the curve the side maxima squeeze out the intervening minimum, which thenceforth becomes imaginary, and take the place of the three, maximum-minimum-maximum, forming a single maximum. And as in the case of the curve the critical form of the parameter gives a maximum where the variation of the ordinate is specially slow, varying ultimately as the fourth power instead of the square of the increment of the abscissa, so in the case of the cusps it would seem that the critical form in which the three cusps coalesced into one, which is of course the form of no aberration, or rather one of the infinite series of forms of no aberration, might give a specially good concentration of light when the circumstances differed a little from that of a direct pencil. I know I tried three oblique pencils as well, but I do not recollect the result. Nor do I recollect whether the feature I have mentioned ran through the whole range of my experimental lenses, or only through a part of the range. It would be easy to try again, but the sun is not now shining, and I have some rather pressing work on hand.

One would say from theoretical reasons that the same thing must have been true when the crown was placed first, and the direction of the cusp was reversed in passing through a form of no aberration. The side cusps were not however in that case experimentally observable: I conclude that they were situated outside the limits of the pencil which the lens transmitted, until we got so near the form of no aberration that the observation of them was impracticable.

Supposing the pencil not limited by the finite aperture of the lenses, the forms of the section by a plane passing through the axis for the two cases would I imagine be like this,

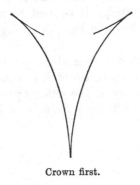

Crown first.

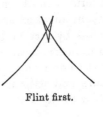

Flint first.

14 *July*, 1887.

Perhaps I may as well mention, as I was thinking of doing when I wrote this morning, what I have found to be a simple and convenient way of finding the general construction of an object-glass which you cannot or do not wish to dissect. It is to reflect sunlight,—or use direct sunlight as may be convenient,—from a condensed beam by means of a lens, such as a watchmaker's lens, passing it through a deep blue glass, and receive the focus in the body of the objective, which should be in a room at least partially darkened. The path of the rays is marked by the phosphorescence of the glass, which is pretty sure to be of different colour in the different lenses, as the quality of glass would be different. By making the focus travel across the objective, by moving either the objective or the condensing lens, you can see by the lengthening or shortening of the path within the glass which lens is of the convex and which of the concave kind. The blue glass defends the eye from the glare of the bright light, and enables you to see the phosphorescence.

9 *August*, 1888.

I think the choice lies between symmetrical images and images formed by excentrical pencils, rendered sufficiently small and round by the use of a properly placed and properly sized stop.

If we go in for symmetry and full aperture, I think the crown is best placed first. The curvatures taken very rudely would be something like $+4$, $+6$, -6, $+1$. If we go in for excentrical pencils, for the sake of a larger field, I think the flint is best placed first. In this case a stop is required. It seemed to me as if a form with the first surface nearly flat was best, as giving a better concentration and flatter field.

Unless by further trials we can satisfy ourselves about getting a good form for excentric pencils, I am disposed to think it would be safest to go in for symmetrical images, and if the defects prove curvature and astigmatism are too great at 2° from the axis, to be content with say 1·5°.

EXTRACTS FROM THE MINUTES AND REPORTS OF THE METEOROLOGICAL COUNCIL.

[From the *Minutes of the Meteorological Council*, July 16, 1878.]

ON THE USES OF HARMONIC ANALYSIS.

Submitted—The following memorandum by Professor Stokes :

Meteorological elements may be considered and discussed from two points of view. We may either contemplate the progress of the changes at any particular time, as, for example, in tracing the history of a particular storm, or in endeavouring to make out general laws connecting the changes of the various elements, such as Buys Ballot's Law ; or we may seek to deduce from large masses of observations regular fluctuations which underlie the total fluctuations presented as the immediate result of observation, of which last a more or less considerable part have no immediate relation to the time, but in contemplating the regular periodic fluctuations are to be regarded as casual.

For the first object great accuracy is not required ; what we want is to get a general comprehensive view ; and this is afforded in a very available form by the published diagrams. Moreover, for this object means are useless, or very nearly so.

It is for the second object that the publication of hourly results is chiefly, if not exclusively, demanded. Besides the more obvious regular changes which have long been known, investigators may wish to examine whether there may not be other regular periodic changes which may be discovered by discussing a great number of observations.

Suppose now the results for each day were subjected to harmonic analysis, and the numbers read off on the cylinders of the machine alone published, we should have five or seven numbers (according as the analysis was carried to the second or third order) to publish, instead of 24. And the question arises— would the results be equally valuable for the purposes of these investigators ?

Let x be the time on the scale of a day to one period, $f(x)$ some function of x, the element analysed. The five cylinders of the existing machine would give the values of the five integrals,

$$\int f(x)\,dx, \quad \int f(x)\cos x\,dx, \quad \int f(x)\sin x\,dx,$$

$$\int f(x)\cos 2x\,dx, \quad \int f(x)\sin 2x\,dx,$$

to which would be added, in the case of a machine with seven elements, the integrals

$$\int f(x)\cos 3x\,dx, \quad \int f(x)\sin 3x\,dx;$$

and the values of these integrals would be registered from

$$x = 0 \text{ to } x = 1, 2, 3, 4, \&c. \text{ periods.}$$

We may say then that we know the values of the indefinite integrals taken from 0 to x, though it is true that this is only for values of x that are multiples of a period.

Suppose now that $f(x)$ is subject to a residual periodic inequality, the period of which is either long, or not greatly differing from 24, or 12, or 8 hours. Such a term would give rise to a term in one of the integrals of far larger amount, and of long period. The integrals would, therefore, lend themselves singularly well to the discovery of such residual inequalities, and to the determination of their coefficients*.

For example, suppose the mean value of $f(x)$ is subject to a seasonal inequality, this will give a fluctuation of long period and comparatively large amount in the registrations of the $\int f(x)\,dx$ cylinder. This, or rather its excess over the product of x by the mean rate of increase, might, if desired, be subjected itself to harmonic analysis by the machine, taking a year for the period instead of a day, and using those excesses for the function to be treated instead of $f(x)$.

A seasonal inequality in a diurnal or a semi-diurnal variation would show itself in a corresponding fluctuation in the integrals

$$\int f(x)\cos x\,dx, \quad \int f(x)\sin x\,dx$$

[* Cf. Schuster, "On the Periodogram," *Roy. Soc. Proc.*, Dec. 1905.]

in the former case, and

$$\int f(x)\cos 2x\,dx, \quad \int f(x)\sin 2x\,dx$$

in the latter, which would be treated just in the same way.

A lunar semi-diurnal inequality, such as a barometric fluctuation due to an atmospheric tide, would show itself by a fluctuation in the readings of the

$$\int f(x)\cos 2x\,dx \quad \text{and} \quad \int f(x)\sin 2x\,dx$$

cylinders. Having a period of about a fortnight, the period being known, the coefficient could be got as usual, either numerically or by the machine.

There is one conceivable case in which hourly means would give something that harmonic results would not give, or rather would give only an approximation to.

It is conceivable that, in the mean of a sufficient number of observations to eliminate casual fluctuations, the curve of hourly means might present a somewhat capricious outline. Such a curve could not be exhibited with very close accuracy by periodic series without going to a considerable number of terms. It could not, therefore, be treated, except as a more or less rough approximation, by the use of periodic series at all, and, therefore, not by the harmonic analyser. If, however, the failure to represent the mean results obtained from the actual observations arises *only* from those mean results differing from the mean results of an infinite number of observations, then there is no real failure; the harmonic analysis *does* represent, so far as casual fluctuations permit, the mean result, in so far as there is any use in taking mean results at all. As regards the study of non-periodic fluctuations mean results may be looked on as worthless. Such results, if studied at all, must be studied individually.

It is not proposed to put an abrupt stop to the publication of hourly means in continuation of what has been begun and would be looked for at the hands of the Council, but it does not seem desirable to engage at present in any additional extensive work of the kind.

[From the *Report of the Meteorological Council*, 1879–80, pp. 32—42.]

DESCRIPTION OF THE CARD SUPPORTER FOR SUNSHINE RECORDERS.

Reprinted in *Math. and Phys. Papers*, vol. v. pp. 54—68.

[From the *Report of the Meteorological Council* for 1878-9.]

In the last Report it was stated that the Council contemplated the employment of the Harmonic Analyser, invented by Sir William Thomson, in order to facilitate the reduction and discussion of the photographic records of the seven observatories. Having been satisfied with the performance of a working model lent to them by Sir William Thomson, the Council have ordered the construction of an harmonic analyser containing seven cylinders, one for the mean and three pairs for the first three pairs of terms of the expansion, to be executed by a skilful engineer in a manner suitable for permanent use.

The form taken by the reductions and the mode of publication will probably be modified when the harmonic analyser comes to be regularly employed; but until the instrument is actually in use any final determination on these points would be premature.

[From the *Report of the Meteorological Council*, 1879-80.]

[By Prof. Stokes.]

I. Memorandum as to the Employment of the Harmonic Analyser in the Meteorological Office.

It will facilitate the explanation of what is to follow if we take a particular example; but it is to be understood that the remarks to be made are of general application.   Let us confine ourselves, therefore, in the first instance to the consideration of the daily fluctuation of atmospheric temperature at a particular place.

The main object of the harmonic analysis is the determination and representation of periodic inequalities,—in the example chosen, the diurnal inequality of temperature.   For this we should subject the records extending over a considerable time, such as a year, to the analysis, and thereby determine the constants in the expression.   But it is obvious that if nothing of the record were preserved for publication except this final result, the information communicated would be extremely meagre.   Not only would stated fluctuations of the variation, such as the seasonal inequality, be passed over, but it would be impossible for a meteorologist seeking after some hitherto unknown inequality (such as an inequality with a period of 24 days or thereabouts, which Professor Balfour Stewart thinks there is evidence of) to make any use of the result. Again, though the harmonic analysis is useless for following the history of the weather in all its details, it is not without use as presenting a succinct representation of the leading features; but this would not be presented if nothing but the final result were given.

The temperature for a single day is not a strictly periodic function. Even if we leave out of consideration those rapid and apparently casual fluctuations that appear to depend on alternations of cloud and sunshine, or on passing showers, the end does not fit on to the beginning. If, treating it as periodic, we were to repeat it day after day, we should be dealing with a function which, graphically represented, would exhibit a succession of curves all alike, but presenting breaches of continuity where the end of one curve joined on to the beginning of the next. At the place of junction there would in general be a sudden change in the ordinate, another in the direction of the tangent, another in the radius of curvature, and so on. Nevertheless, in anything of settled weather the discontinuity would be much smaller than the changes due to the normal diurnal change, especially if the record for the 24 hours began at midnight, or at some other hour when the sun is below the horizon.

The temperature for the day, regarded as a function of the time, may conveniently be divided into two parts, which we may call the *progressive part* and the *residual part* or *residue*, where the progressive part is a simple algebraic function of t (the time), linear, quadratic, cubic,...according to the order we please to go to,—the constants being so chosen as to give to the residue, or the residue and its first derivative, or the residue and its first two derivatives,...the same values at the beginning and end of the day. The function will thus be cleared as far as may be of the effect of progressive change, and the residue will represent, more nearly than the function itself, the normal diurnal fluctuation.

Let the time-scale be so chosen that a day is represented by a period, or 2π, and let $f(t)$ be the temperature. The development of $f(t)$ in harmonic series will be, going to three orders after the mean,

$$A_0 + A_1 \cos t + A_2 \cos 2t + A_3 \cos 3t$$
$$+ B_1 \sin t + B_2 \sin 2t + B_3 \sin 3t,$$

where
$$A_0 = \frac{1}{2\pi} \int_0^{2\pi} f(t)\,dt,$$

$$A_i = \frac{1}{\pi} \int_0^{2\pi} f(t) \cos it\,dt, \qquad B_i = \frac{1}{\pi} \int_0^{2\pi} f(t) \sin it\,dt,$$

i being 1, 2, or 3, as the case may be.

The several integrals will be given by the machine. To get the development of the residue we must deduct the development of the progressive part. According to the theory of these series* this will be got from the above by integrating by parts, and retaining only the terms which are free from the integral sign, provided we assign, as we may, the mean term wholly to the residue. Denoting by Δ the increment of $f(t)$ or its derivatives in passing from $t = 0$ to $t = 2\pi$, it will accordingly be

$$a_1 \cos t + a_2 \cos 2t + a_3 \cos 3t + b_1 \sin t + b_2 \sin 2t + b_3 \sin 3t,$$

where
$$a_i = \frac{1}{\pi i^2} \Delta f'(t) - \frac{1}{\pi i^4} \Delta f'''(t) + \ldots$$

$$b_i = -\frac{1}{\pi i} \Delta f(t) + \frac{1}{\pi i^3} \Delta f''(t) - \ldots.$$

In correcting the constants A_0, A_1, B_1, ... so as to get the expansion of the residue, it does not seem desirable to go beyond the function $f(t)$ itself. This comes to taking as the residue a function which would be represented graphically by the excess of the actual ordinate over that of a straight line drawn parallel to the line joining the extremities of the curve, and passing through the middle point in the axis of abscissæ. To this degree of accuracy the development of the residue would accordingly be, writing Δf for $\Delta f(t)$,

$$A_0 + A_1 \cos t + A_2 \cos 2t + A_3 \cos 3t$$

$$+ \left(B_1 + \frac{1}{\pi} \Delta f\right) \sin t + \left(B_2 + \frac{1}{2\pi} \Delta f\right) \sin 2t + \left(B_3 + \frac{1}{3\pi} \Delta f\right) \sin 3t.$$

The constants A_0, A_1, B_1, &c. are given by the increments of the numbers on the cylinders of the machine, and the corrections are similarly given by the increments of the function itself, supposed tabulated for the commencement of each day.

[* " On the critical values of Sums of Periodic Series," *Math. and Phys. Papers*, vol. i. p. 236.]

If, instead of applying the analysis to a single day, we apply it to n consecutive days, the increments of the numbers on the cylinders, and also the correction for reducing the development of the function to that of the residue, will each have to be divided by n, Δ now denoting the increment of the function in passing from the beginning of the first to the end of the last day in the group. Supposing the days on the average much alike, the increments of the numbers on the cylinders will increase nearly in proportion to n, and the quotient will tend to a limit in which irregular changes from day to day disappear. The increment of the function $f(t)$, however, instead of increasing nearly as n, will fluctuate in a casual manner, and when it is divided by n, the quotient will tend to vanish as n is taken greater. It is only therefore when days are considered individually, or grouped in some manner other than that of natural sequence, that the correction for reducing the actual function to the residue need be taken into account.

The seasonal inequality of mean diurnal temperature, or of diurnal fluctuations of temperature, may conveniently be got by subjecting the coefficients, obtained as above, again to harmonic analysis, the period being now a year. For this it would not be necessary to take the increments of the numbers on the cylinders for each day. It would be amply sufficient to take the days in groups of 5, 10, or even more.

If the numbers on the mean cylinder and on the pair of cylinders of the first order, or what perhaps would be better, the increments of those numbers, as likewise the initial temperatures, were published for each day, and the numbers on the cylinders of higher orders for somewhat wide intervals, the tables would pack into small compass, and would nevertheless present a tolerably complete and partially digested record of the march of temperature throughout a year; and if curves were drawn representing, one the mean temperature, and the other the coefficient of the resultant of the pair of terms of the first order, the whole could be represented to the eye, in its leading features, in a very small space.

The details of the mode of treatment and publication of records to which the harmonic analysis is applied can hardly be fixed till some substantial experience has been gained in the regular use of the instrument, as distinguished from trials having for object to test whether the instrument was in thoroughly good working order. The above is given as indicating the general direction which the changes in the mode of reducing and publishing the records seem likely to take in consequence of the introduction of the instrument. The same general principles which apply to the discussion of temperatures will apply of course to the other elements, but each element must be considered by itself as to the features which it may be most desirable to bring into prominence.

II.   Suggested First Application of the Harmonic Analyser and Rules for the Use of the same.

I think the first application had best be to temperature for some year, say, in the first instance, to the records of the dry bulb thermometer; but the same remarks will apply to the wet bulb, which perhaps might be taken up next.

The civil reckoning of time should be followed each day, commencing at midnight. As the sheets commence at 10 a.m. and go on for two days, the record of every second day will be divided, part being on one sheet and part on another. The readings on the cylinders need not be attended to when the sheets are changed, further than to see that they are not accidentally disturbed in the process of changing.

The cylinders, being set to zero for the midnight with which the work is commenced, should at first all be read for every midnight, and the readings registered. The mean cylinder and (in the case of the thermometer) the pair of the first order should continue to be read for each day, but, for the others, readings at intervals of several days will probably be sufficient, as may be determined by subsequent orders.

Should a record exhibit a small or moderate gap, which can be filled in by the eye with tolerable certainty, it should be so filled in pencil before the sheet is put on.

Should the gap be more serious, so that in the judgment of operator it would be better to reject the day altogether and interpolate, the process should be as follows:

Suppose that the record for the 7th of a month is rejected, those for the 6th and 8th being good. Let A, B, C, D be the hour readings concerned, as they would have been if the records of the three days had been perfect, these readings being those of any one of the cylinders. The readings are those for the midnights of the 5th, 6th, 7th, and 8th, so that $B - A$ gives the increment for the 6th, and so in the other cases. Then A and B are read on the cylinders, but as the record for the 7th is missing we start with the number B in commencing the paying off of the record for the 8th, and at the end of the day arrive at a provisional reading, say P. Then in default of the record for the 7th we may take the increment for that day of the number on the cylinder as the mean of the increments for the 6th and 8th. This gives $B + \frac{1}{2}(P - A)$, $P + \frac{1}{2}(P - A)$ as the numbers which are to be taken for C, D, respectively. These numbers are to be entered in the register, and the cylinder set by hand to the last of them before paying out the record for the 9th. If the gap should extend over two days, the records should be supplied by a process founded on similar principles.

If the readings of the records of the subordinate cylinders for every day should have been discontinued, the cylinders should be read for the purpose of making the interpolation even though the reading might not have been otherwise demanded.

The preservation of the records on the cylinders not only supplies information as to the element analysed, which would be lost if nothing but the final results were retained, but also affords the means of re-examining any step of the process where an error might be suspected.

A blank column should be left in the register for the subsequent entry of the midnight values of the element analysed.

In what precedes the applicability of the harmonic method, with a limited number of terms in the expansion, is assumed. The mean hourly value of any meteorological element is of course a periodic function of the time of day. We know that any periodic function may be expressed by a harmonic series, and the only question is what number of terms it may be practically necessary to employ. The publication of the "Reduction of Greenwich Meteorological Observations" seemed to afford a good opportunity of testing this question in what appeared likely to be a specially crucial case. The result of the examination will be seen from the following letters to the Astronomer Royal :

LENSFIELD COTTAGE, CAMBRIDGE,
April 30, 1879.

DEAR SIR GEORGE AIRY,

I was greatly struck with the enormous amount of work of which the outcome is embodied in the "Reduction of Greenwich Meteorological Observations" which I lately received, and with the very definite character of the results arrived at. My admiration was not unmixed with some anxiety, lest after all the Thomson's harmonic analyser, which we have ordered for the Meteorological Office, should be capable only of giving results which by comparison must be regarded as of the "cheap and nasty" class.

I thought, therefore, that I would test its applicability in a severe case; I mean of course the applicability of the principle, the machine being supposed to be properly constructed, and to work satisfactorily.

I chose the observations of temperature rather than those of the barometer, because the manifest discontinuity introduced by sunrise and sunset seemed to render temperature a more severe test than height of the barometer. As I explained in a former letter, the greater the number of observations of which the mean is taken, the greater the superiority, if any, of the method of hourly means over the harmonic method is likely to prove.

These motives led me to select for trial the last column of Table 50, page 38. I did my work pretty carefully, but I have not revised it, so that it is possible that there may be an arithmetical error, but I do not think it at all likely that the general conclusions would be affected. I went as far as the sixth order. I had contemplated going further, but it did not seem necessary.

I determined the coefficients in the usual way. I may remark in passing that this does not give them correctly in the high orders. Each is, by rights, given by an integral, which is only approximately represented by a sum, and in the high orders the difference would be very sensible. I had some thoughts of working them out correctly, but the high orders proved to be so small as to render this unnecessary.

Denoting by t an angle proportional to the time measured from midnight, and going through its period in 24 hours, I found

$$\text{temperature} = 49\cdot69 - 4\cdot280 \cos t \quad - 2\cdot722 \sin t$$
$$+ 0\cdot883 \cos 2t \quad + 0\cdot563 \sin 2t$$
$$+ 0\cdot063 \cos 3t \quad + 0\cdot119 \sin 3t$$
$$- 0\cdot043 \cos 4t \quad - 0\cdot014 \sin 4t$$
$$- 0\cdot001 \cos 5t \quad - 0\cdot026 \sin 5t$$
$$- 0\cdot003 \cos 6t \quad - 0\cdot008 \sin 6t$$

For the coefficients in the different orders when the two terms are combined into one we get

5·072	1·047	0·135	0·045	0·026	0·009
Orders 1	2	3	4	5	6

It appears then that the first three orders give the result within about 1 per cent. of the whole diurnal variation. The error would be only a few hundredths of a degree.

I made an attempt to estimate the errors of the published numbers, by which I mean the difference between the published numbers which represent what I called, in a former letter, the mean of fact, and what I there called the mean of law; in other words, what would be the mean of an infinite series.

For this I chose two months, January and July, and two hours, midnight and noon. In Tables 38 and 44 I took the differences between the numbers in the midnight and noon columns and the means at the feet of the columns, and regarding those as errors I took their mean. The means were 2·475 and 2·500 for January; 1·610 and 1·710 for July; mean 2·074. As Table 50 gives the means for 240 months, I divided the result by the square root of 240, giving 0·134. This, then, we may take as a sort of estimate of the residual errors in the last column of Table 50 arising from the finiteness of the series, great though it was. It is about equal to the greatest value of the terms of the Order 3 in the periodic series.

I noticed, however, in working these figures that, on the whole, midnight and noon went together, so that the "errors" that I have spoken of arose more from variations of temperature from day to day, or week to week, than from variations within the same day. I therefore took the differences between the midnight or noon column and the last column but one, which gives the monthly mean of the mean temperature of the day. I meaned each vertical column, and took the differences between the mean and the individual numbers, and regarded these as the errors. The means came 0·28 and 0·33 for January; 0·78 and 1·20 for July; mean 0·65; same divided by the square root of 240, 0·042. This is almost exactly the coefficient of the fourth order.

It appears that the harmonic method has come very well out of its trial in a case purposely selected as being likely to be the most severe. It appears, too, that we need not trouble ourselves to go beyond the third order. The third order itself hardly emerges from casual fluctuations.

Of course the harmonic method would not *by itself alone* give such non-periodic results as those represented in Plate 4, but the modification required to introduce these is very simple.

May 10, 1879.

I have made some further calculations in pursuance of the subject mentioned in my letter of April 30, the results of which you might perhaps like to know.

From the numerical trials described in that letter, I was led to infer that the uncertainty of the diurnal inequality of temperature expressed by the differences between the numbers in the last column of Table 50 and the number at the foot of the column was something like 0·04, and accordingly of about the same magnitude as the term of the fourth order in the harmonic reduction. The uncertainty I contemplate as arising from the uneliminated residue of casual fluctuations, whereby the actual temperature at any observation differs from the mean really belonging to that day and hour.

If we can only regard the differences between the means in the last column of Table 50 and the ideal means for an infinite series as casual, it is clear that the harmonic reduction when carried to the third order will represent the result as nearly as the latter represents the ideal mean that we are seeking for. But it occurred to me that the fluctuations indicated in my former letter might be fluctuations arising rather from the coefficient than from the law of diurnal variation, in which case, though the absolute mean of fact for any hour might deviate from the ideal mean to an extent comparable with 0·04, the law of variation might possibly be a good deal more correctly given. In this case the harmonic reduction, as compared with the other, would show a minute inferiority, unless at least it were carried to orders beyond the third; though the error is so small that it is hard to imagine that any scientific conclusion that could be drawn from the observations would be interfered with. Still, I was desirous of ascertaining whether even a minute error of the harmonic reduction taken to the third order, large enough to emerge from casual fluctuations, could be discovered.

It occurred to me that this point could be decided by taking separate years or groups of years instead of the whole series of 20 years, expressing the means for the separate years or groups in a harmonic series, and comparing the coefficients with those got from the whole 20 years series. If the fluctuations indicated in my former letter depended mainly on fluctuations in the coefficient of diurnal variation, the law being but little departed from, the changes in the coefficients of the several terms in the series should bear some sort of proportion in magnitude to the coefficients themselves, so that the leading terms should be those chiefly affected, a variation like one per cent. being insensible in the small terms; whereas if the fluctuations affected the law as much as the mere amount of diurnal variation, no such subordination of the higher to the lower orders in regard to the changes of the coefficients was to be expected.

The published volume does not contain a table for air temperature analogous to Table 19, or better, to Table 20 for the barometer. Of course it could be formed from the published numbers by meaning Tables 38 to 49, as if they had been superposed and the means taken of the numbers that lay one over the other. But this would involve more labour than I was disposed to go through. So I had recourse to a plan which practically, I think, suffices for determining the question, and which permits of utilising an existing table.

I took Table 51 and made three groups of the months, the months of each group being equidistant. I meaned the numbers in each group or rather added them, reserving the division to the end. In this way the seasonal inequality would be pretty completely eliminated, and the results would not much differ in character from those got from dividing the 20 years into three groups, and seeing how far each group agreed with its fellows. Instead of working with the actual numbers of the table, I preferred taking the difference between the number for each group,

14

and the mean of the three. Subjecting these differences to harmonic reduction, I obtained

$$+ 0.006 \cos t \qquad + 0.022 \sin t$$
$$+ 0.019 \cos t \qquad - 0.009 \sin t$$
$$- 0.025 \cos t \qquad - 0.015 \sin t$$

$$+ 0.014 \cos 2t \qquad - 0.004 \sin 2t$$
$$- 0.026 \cos 2t \qquad + 0.025 \sin 2t$$
$$+ 0.011 \cos 2t \qquad - 0.021 \sin 2t$$

$$- 0.000 \cos 3t \qquad - 0.007 \sin 3t$$
$$+ 0.006 \cos 3t \qquad - 0.004 \sin 3t$$
$$- 0.006 \cos 3t \qquad + 0.011 \sin 3t$$

$$- 0.008 \cos 4t \qquad + 0.008 \sin 4t$$
$$+ 0.016 \cos 4t \qquad + 0.030 \sin 4t$$
$$+ 0.007 \cos 4t \qquad - 0.042 \sin 4t$$

I have written the numbers according to orders, and the groups in the several orders on consecutive lines, so that the development of the first group, for instance, is made up of the first lines in the several orders. For the coefficients of the different orders when the two terms of the same order are combined into one I get

Order	1	2	3	4	
	0.023	0.015	0.007	0.011	*G*. 1
Coefficient	0.021	0.036	0.007	0.034	*G*. 2
	0.029	0.024	0.013	0.043	*G*. 3

There is no such subordination of magnitude of the coefficients in the higher orders to those in the lower as must have taken place on the supposition that the deviations from the ideal mean arose chiefly from deviations from the mean coefficient, not from the law, of diurnal variation. The inference is that the small deviations in question, amounting to a few units in the second place of decimals, can only be treated as casual; and as they are comparable with the coefficient of the fourth order given in my former letter, it follows that the harmonic series taken to the third order will represent the results of observation within their limits of error. The small coefficient of the third order itself can only be deemed to be determined from the observations, subject to an error amounting to 30 or 40 per cent. of the whole.

I noticed in taking the difference of the group numbers from the mean of the three that the differences ran very much in series of *plus* and *minus*. They exhibited considerable regularity. I rather expected from this that the harmonic series would have been most significant in the leading terms, but such proved not to be the case. However, the regularity of the fluctuations indicates that the uneliminated errors are not such as would be got by the throwing of dice for the different hours independently of each other, but that the fluctuations from which they arise are such as extend their influence over several hours at a time. This character, even had their amount been larger, would have prevented them from interfering much with the smoothness of the mean curve obtained, so that smoothness of curve is not by itself alone a proof that we have got very near the ideal mean.

I should mention that my Group 1 is made up of January, April, July, and October, and 2, 3 are similarly formed beginning with February and March respectively.

ADDITIONAL MEMORANDUM ON HARMONIC REDUCTIONS.

In the " Reduction of Greenwich Meteorological Observations " there is no table of mean hourly temperatures given in which the months are combined but the years kept distinct, though the materials for forming such a table are published, as the hourly means are given for the separate years and months.

Sir George Airy has subsequently had these means taken, and has been so kind as to furnish me with a copy of the results. I have now subjected these to harmonic reduction, proceeding as far as the fourth order. The first term, giving the mean, was not calculated, as it is given in the last column but one of Table 52 on page 39 of the Greenwich Reductions. The values of the eight remaining coefficients in the series,

$$A_0 + A_1 \cos T + B_1 \sin T + \ldots + A_4 \cos 4T + B_4 \sin 4T,$$

are given in the following table:

COEFFICIENTS in the HARMONIC REDUCTION of the MEAN HOURLY TEMPERATURES in the individual years 1849—1868.

Year	A_1	B_1	A_2	B_2	A_3	B_3	A_4	B_4
1849	$-4\cdot105$	$-2\cdot454$	$+0\cdot822$	$+0\cdot498$	$+\cdot134$	$+\cdot138$	$-\cdot068$	$-\cdot013$
1850	$-4\cdot316$	$-2\cdot753$	$+0\cdot817$	$+0\cdot524$	$+\cdot101$	$+\cdot136$	$-\cdot060$	$-\cdot038$
1851	$-4\cdot287$	$-2\cdot728$	$+0\cdot907$	$+0\cdot512$	$+\cdot064$	$+\cdot117$	$-\cdot053$	$-\cdot005$
1852	$-4\cdot367$	$-2\cdot717$	$+0\cdot936$	$+0\cdot643$	$+\cdot046$	$+\cdot106$	$-\cdot060$	$-\cdot049$
1853	$-3\cdot822$	$-2\cdot225$	$+0\cdot753$	$+0\cdot494$	$+\cdot101$	$+\cdot125$	$-\cdot067$	$+\cdot040$
1854	$-4\cdot693$	$-3\cdot194$	$+1\cdot049$	$+0\cdot670$	$+\cdot034$	$+\cdot143$	$-\cdot078$	$-\cdot058$
1855	$-4\cdot281$	$-2\cdot698$	$+0\cdot918$	$+0\cdot597$	$+\cdot022$	$+\cdot195$	$-\cdot029$	$-\cdot036$
1856	$-4\cdot024$	$-2\cdot633$	$+0\cdot963$	$+0\cdot501$	$+\cdot065$	$+\cdot096$	$-\cdot065$	$-\cdot014$
1857	$-4\cdot624$	$-2\cdot882$	$+1\cdot004$	$+0\cdot589$	$+\cdot073$	$+\cdot167$	$-\cdot088$	$+\cdot003$
1858	$-4\cdot636$	$-2\cdot017$	$+0\cdot993$	$+0\cdot571$	$+\cdot006$	$+\cdot102$	$-\cdot045$	$-\cdot002$
1859	$-4\cdot452$	$-2\cdot793$	$+0\cdot891$	$+0\cdot680$	$+\cdot051$	$+\cdot102$	$-\cdot030$	$-\cdot024$
1860	$-3\cdot720$	$-2\cdot324$	$+0\cdot794$	$+0\cdot567$	$+\cdot045$	$+\cdot079$	$-\cdot027$	$+\cdot003$
1861	$-4\cdot256$	$-2\cdot261$	$+0\cdot820$	$+0\cdot621$	$+\cdot054$	$+\cdot057$	$-\cdot038$	$+\cdot013$
1862	$-3\cdot550$	$-2\cdot295$	$+0\cdot645$	$+0\cdot475$	$+\cdot029$	$+\cdot099$	$-\cdot014$	$\cdot000$
1863	$-4\cdot381$	$-2\cdot701$	$+0\cdot885$	$+0\cdot565$	$+\cdot077$	$+\cdot049$	$-\cdot067$	$-\cdot052$
1864	$-4\cdot394$	$-3\cdot021$	$+0\cdot948$	$+0\cdot571$	$+\cdot050$	$+\cdot089$	$-\cdot065$	$-\cdot025$
1865	$-4\cdot689$	$-3\cdot171$	$+0\cdot971$	$+0\cdot672$	$+\cdot025$	$+\cdot146$	$+\cdot029$	$-\cdot048$
1866	$-4\cdot086$	$-2\cdot512$	$+0\cdot783$	$+0\cdot516$	$+\cdot077$	$+\cdot166$	$+\cdot006$	$-\cdot014$
1867	$-3\cdot986$	$-2\cdot479$	$+0\cdot759$	$+0\cdot492$	$+\cdot089$	$+\cdot119$	$-\cdot023$	$+\cdot002$
1868	$-4\cdot900$	$-3\cdot106$	$+0\cdot918$	$+0\cdot488$	$+\cdot100$	$+\cdot184$	$-\cdot027$	$-\cdot029$
Mean	$-4\cdot278$	$-2\cdot718$	$+0\cdot879$	$+0\cdot562$	$+\cdot062$	$+\cdot122$	$-\cdot043$	$-\cdot016$

DIFFERENCES of the COEFFICIENTS from the MEAN, in thousandths of a Degree.

Year	A_0	$-A_1$	$-B_1$	A_2	B_2	A_3	B_3	$-A_4$	$-B_4$
1849........	$+\ 630$	-173	-264	$-\ 57$	$-\ 64$	$+72$	$+16$	$+25$	$-\ 3$
1850........	$-\ 170$	$+\ 38$	$+\ 35$	$-\ 62$	$-\ 38$	$+39$	$+14$	$+17$	$+22$
1851........	$-\ 310$	$+\ \ 9$	$+\ 10$	$+\ 28$	$-\ 50$	$+\ 2$	$-\ 5$	$+10$	-11
1852........	$+\ 940$	$+\ 89$	$-\ \ 1$	$+\ 57$	$+\ 81$	-16	-16	$+17$	$+33$
1853........	-2150	-456	-493	-126	$-\ 68$	$+45$	$+\ 3$	$+24$	-56
1854........	$-\ 460$	$+415$	$+476$	$+170$	$+108$	-28	$+21$	$+35$	$+42$
1855........	-2520	$+\ \ 3$	$-\ 20$	$+\ 39$	$+\ 35$	-40	$+73$	-14	$+10$
1856........	$-\ 550$	-254	$-\ 85$	$+\ 84$	$-\ 61$	$+\ 3$	-26	$+22$	$-\ 2$
1857........	$+1550$	$+346$	$+164$	$+125$	$+\ 27$	$+11$	$+45$	$+35$	-19
1858........	$-\ 190$	$+358$	$+299$	$+114$	$+\ \ 9$	-56	-20	$+\ 2$	-14
1859........	$+1180$	$+374$	$+\ 75$	$+\ 12$	$+118$	-11	-20	-13	$+\ 8$
1860........	-2120	-558	-394	$-\ 85$	$+\ \ 5$	-17	-43	-16	-19
1861........	$+\ 340$	$-\ 22$	$-\ 47$	$-\ 59$	$+\ 59$	$-\ 8$	-65	$-\ 5$	-29
1862........	$+\ 250$	-728	-423	-234	$-\ 87$	-33	-23	-29	-16
1863........	$+\ 970$	$+103$	$-\ 17$	$+\ \ 6$	$+\ \ 3$	$+15$	-73	$+24$	$+36$
1864........	$-\ 730$	$+116$	$+308$	$+\ 69$	$+\ \ 9$	-12	-33	$+22$	$+\ 9$
1865........	$+1180$	$+411$	$+453$	$+\ 92$	$+110$	-37	$+24$	-72	$+32$
1866........	$+\ 710$	-192	-206	$-\ 96$	$-\ 46$	$+15$	$+44$	-49	$-\ 2$
1867........	$-\ 890$	-292	-239	-120	$-\ 70$	$+27$	$-\ 3$	-20	-18
1868........	$+2330$	$+622$	$+388$	$+\ 39$	$-\ 74$	$+38$	$+62$	-16	$+13$
Mean	1085	278	220	139	56	26	31	23	20
„ $\div\sqrt{20}$	242	62	49	31	12	6	7	5	4

The calculations were pretty carefully checked. In the steps of the calculations I never went beyond the third decimal of a degree, and therefore this figure is liable to an error amounting to a few units. With this understanding, the mean of the reductions agrees with the reduction of the mean previously given [p. 207], which was calculated quite independently, so that the two check each other.

The last line is given as allowing an estimate to be formed of the mean errors of the coefficients derived from different groups of 20 independent years, compared with the coefficients which would be obtained from an infinite number of years.

The general conclusions obtained in my former memorandum in an indirect way are borne out by the direct comparison between different years which is here made. As before observed, the minute fourth order is swallowed up, even in the mean of a 20 years' series, by the uncertainty in the amount of fluctuation, not to speak of the much larger uncertainty in the mean temperature for a day. Nevertheless, it appears from the figures that though this order is so minute, and though its effect on the mean temperature at any particular hour is trifling compared with the average variation from year to year of the mean for that same hour, it still emerges roughly with tolerable certainty from the mean for a good number of years.

If we were in possession of a theory by which mean temperatures could be calculated with the same precision as the planetary motions, there would then be some interest in working out from observation this fourth order, though so minute. But as there appears to be no prospect of such a theory, and as it seems hardly conceivable that so minute a feature of the diurnal fluctuation could assist us in the endeavour to connect observed meteorological effects with their causes, it does not seem advisable to incur the additional labour and expense that would be involved in the endeavour to elicit from observation such very small fluctuations*.

[* Sir George Airy's cordial concurrence with the practical conclusions here derived, from the theory of meteorological statistics sketched in these memoranda, is expressed, *supra*, p. 186.]

[From the *Report of the Meteorological Council*, 1880–81, pp. 25—27.]

ON THE WORKING OF THE HARMONIC ANALYSER.

By PROF. STOKES.

The harmonic analyser at the Office is now in regular use. It has been applied in the first instance to the analysis of air temperatures. The numbers on the seven cylinders give respectively for each day the mean temperature, and the coefficients of the first three pairs of terms in the variable part of the diurnal fluctuation. It has not, however, been considered necessary to read all the cylinders for each day. The mean cylinder and the pair which give the coefficients of the two terms in the component of the 24-hour period are read for each day, and the others only at the end of each period of five days and at the end of each civil month. Whether the coefficients of the two principal terms in the diurnal fluctuation will be published for each day, or only as hitherto the mean temperature, is not yet decided. In any case they would be published for suitable intervals.

The numbers on the cylinders are carried on by the machine, so that no numerical additions are required in getting the mean coefficient for any interval that may be chosen. Suppose the year divided into equal or nearly equal parts, say into months. From the mean coefficient, for each month, of any one of the terms in the diurnal fluctuation, it will be easy to deduce, either by direct numerical calculation or by the use of the machine, the harmonic expansion of the annual fluctuation of the coefficient, and by treating each coefficient in this way we may express the average temperature as a function of the time of day and time of year, by a series involving only a moderate number of coefficients derived from observation. A comparison of the more important coefficients as obtained for the same year for two different places, or for the same place for two different years, would afford a general view of the leading differences of climate of two places, or of the leading differences of weather in different years.

We are hardly yet in a condition to pronounce definitively upon the degree of accuracy to which the machine practically works. No investigation of the diurnal fluctuation of temperature has hitherto been made at the Office, with which to compare the work of the machine. The mean temperature alone admits, with the data at present in our hands, of thorough comparison, and with regard to it we are able to affirm with confidence that the work of the machine is perfectly satisfactory. In fact, the mean temperature for the year came out the same by the machine as by numerical calculation, from ordinates measured for each hour on the photograms, to the hundredth part of a degree, or thereabouts. It must be noticed, however, that there are no cross-heads with slots in which pins work involved in the government of the motion of the mean cylinder as there are for the others, so that we must not jump to the conclusion that the working of the latter is equally satisfactory.

With the view to obtain some data for a fair comparison, a book belonging to the Office was employed, in which are entered the hourly temperatures for Valencia for the year 1871, the place and year with which the work of the analyser was begun. From the entries, the mean monthly temperatures were calculated numerically for the even hours; and then these, as well as the annual means obtained from them, were expressed by the usual arithmetical process, in a harmonic series with a year for period. On comparing the coefficients thus obtained with those got by the machine, discrepancies were found amounting perhaps to $0°\cdot2$ or $0°\cdot3$ in the larger terms, but generally smaller. In the small terms, the discrepancies were usually considerably smaller. This gave an idea of the greatest error to be feared, but left the question undecided which was the more near to the truth. Independently of arithmetical mistakes, which can be guarded against by sufficient careful checking, the first process is open to the imperfection of disregarding all changes that take place between one even hour and the next. The principle of the Harmonic Analyser is perfect in this respect, but the employment of the machine is subject to the finiteness of accuracy which belongs to every mechanical process.

To give some idea of the degree to which the coefficients were liable to be vitiated in the first process by the disregard of all changes that occurred between one even hour and the next, the means for one month were taken for the odd hours only, and then expressed arithmetically in a harmonic series. The coefficients thus got from the even hours alone and the odd hours alone certainly agreed with one another a good deal better than either agreed with the coefficients got by the machine, the discrepancies between the first two sets reaching only to about $0°\cdot05$.

This seems to make it pretty clear that the chief source of error is in the use of the machine, though it is true that the basis of the induction on which this conclusion rests is rather limited. That errors of some such magnitude should occur is not to be wondered at when it is considered that the tenth of a degree, which is a convenient measure of liability to error, is represented on the photograms by only the $\frac{1}{150}$th of an inch. That the errors should be comparable with so minute a quantity, shows that the machine must have been worked with great care.

On examining the errors (assumed to belong to the coefficients got by the machine) it was noticed that in general if one coefficient of a pair belonging to the same order were too large, the other would be too small, and it was found that the square root of the sum of the squares agreed with the same quantity as obtained by measurement and calculation a good deal better than did the coefficients individually. The error in fact fell rather upon the epoch than upon the coefficient when the two terms were combined into one. The sign of the error of the epoch indicated a lagging. It is probable that the results got by the machine might be improved by introducing a correction to the epoch which may be expected to be in the proportion of the numbers 1, 2, 3 for the pairs of terms of those orders respectively.

But even if we take the results as they are, seeking for no further correction, the error is so small as not to be likely to be of any practical importance, especially as it does not affect the mean and, as to other terms, tends to be alike for different places or different years that may be compared. If the accuracy be sufficient for all practical purposes, it would surely be unreasonable to refuse to employ the machine merely because results slightly more accurate could be obtained by a vastly greater expenditure of time and labour, an expenditure so great that hitherto it has not been attempted by the Office.

[From the *Report of the Meteorological Council*, 1879–80, pp. 28—32.]

ON THE EFFECT OF SLUGGISHNESS ON THE READINGS OF MARINE BAROMETERS ON SHORE.

Mr Buchan in his report called the attention of the Council to the discrepancy in the relative readings of the portable barometer with a highly contracted tube, belonging to the Office, which he carried about with him on his tour of inspections, and of a barometer with uncontracted tube, belonging to the Scottish Meteorological Society. If there were merely a difference of index error, the difference between the readings of the two ought of course to be constant. It was found, however, that the difference varied according as the barometer was rising or falling. The character of the change was such as to indicate that the barometer with contracted tube was retarded in following the actual changes of the true barometric height. The variation was sufficient to affect slightly the second place of decimals of the height expressed in inches.

It is obvious that an effect of the kind must be produced by the resistance which the capillary part of the tube presents to the flow of the mercury through it. If this be the sole cause why the barometer with contracted tube is behindhand in its indications of barometric changes, the circumstances of the retardation may be determined by mathematical calculation.

It is known that in the case of the slow motion of liquids through narrow tubes the resistance, depending on the viscosity of the liquid, varies as the first power of the velocity. This leads to a differential equation of the simplest kind connecting the variation of the height of the mercury with the difference between the actual and true heights at the time; namely

$$\frac{dx}{dt} + q\,(x - h) = 0, \tag{1}$$

where x is the height of the mercury in the barometer with contracted tube, t the time, q a constant, and h the true height, which will be a function of t. This equation as it stands, or rather put into the form

$$h = x + \frac{1}{q}\frac{dx}{dt}, \tag{2}$$

when once the constant q is known, gives the height in terms of the observed height and its rate of variation. For the apparent height in terms of the true height we have from (1)

$$x = qe^{-qt} \int e^{qt} h dt = q \int_{-\infty}^{t} e^{-q(t-t')} h' dt', \qquad (3)$$

where h' is the same function of t' as h is of t. In this expression, as we shall see when we come to determine the value of q, the part of the integral relating to times more than 20 minutes or so earlier than the time t is practically insensible, and the inferior limit may be taken accordingly.

When the barometer is steadily rising or falling, so that the rate of change is tolerably constant for a good many minutes, h is a slowly varying function of the time, and (3) may be converted into a rapidly converging series by integration by parts. It will be sufficient to take the first two terms, giving

$$x = h - \frac{1}{q} \frac{dh}{dt}, \qquad (4)$$

which determines the error of lagging for a known rate of barometric change, just as (2) gives the correction for lagging, supposing the rate of change of the apparent height observed.

It seemed doubtful à priori whether the resistance could still be taken to vary as the velocity when we come to the, comparatively speaking, enormously rapid changes of apparent height which take place in the process of testing for sluggishness. In this the time (which I will call the testing time, and denote by T) is observed which the mercury takes to fall from 1·5 to 0·5 inch above its true height. To examine this question I made an experiment at the Office on Barometer No. 261, with Mr Scott's assistance. The mercury having been raised, one of us set the vernier to heights successively decreasing by 0·1 inch, and called out when the mercury reached the mark; the other observed the time, and wrote down the time and the height. Two experiments in which the mercury descended from 1·1 to 0·1 and from 1·2 to 0·1 above the true height, were made, and the mean of the times, which agreed well, was taken.

Supposing the law true, we have, putting h constant in (1),

$$x = h + Ce^{-qt}, \qquad (5)$$

and the logarithm of the excess of the apparent over the true height is of the form $a - bt$. On plotting the mean result of the experiments by taking the time for abscissa and the logarithm of the excess for ordinate, a curve was obtained which certainly was very nearly a straight line, from which it did not certainly differ by quantities exceeding the limits of errors of observation.

It follows that our equations may be applied even to the more rapid changes which occur in testing, and the value of q may be deduced from that of T. It will be observed that q^{-1} expresses a time; this may conveniently be called the "lagging time," and be denoted by L. It is [the time] that the mercury takes to fall from a height moderately exceeding the true height to another in which the excess is reduced to its e^{-1}th part. It follows, from the definition of T, that

$$3 = e^{qT}, \text{ whence } L = 0\cdot910 \ T.$$

The permitted limits of T being 3 m. to 6 m., those of L will be 2 m. 44 s. to 5 m. 28 s.: average, say 4 m.

The last terms in (2) and (4) are simply the lagging time into the rate of change, or what comes to the same thing if the rate be taken as uniform, the change during the lagging time. It follows that a reading taken at 8 h. a.m. gives the true height at a time earlier by the lagging time, say at 7 h. 56 m. a.m. Unless therefore the observers at the secondary stations can be trusted for great punctuality, the influence of lagging may be neglected; and if they can they have only to date the observations earlier than the times by the lagging times of their respective barometers.

Further, in comparing two barometers the more prompt should be read first, and after an interval equal to the difference of the lagging times the more sluggish, and the two readings treated as simultaneous.

We learn also that in the most sluggish barometer allowed, for which $T = 6$ m., when the barometer is set up after carriage, 39 m. from the time the mercury stood at 1·5 inch above the true height will suffice for the error of lagging, so far as depends on previous carriage, to be reduced to less than 0·001 inch. For $T = 3$ m. the time would of course be halved.

The simple expression (4) was obtained on the supposition that the rate of barometric change was sensibly uniform for a good many minutes, and it would not therefore apply to the small rapid heavings which are sometimes observed when a storm is approaching. It is obvious that the sluggishness tends to efface rapid fluctuations.

To take a simple example, suppose that x is subject to a fluctuation expressed by $c \sin nt$, so that

$$h = H + c \sin nt,$$

where H is either constant or a slowly varying function of t. From the linearity of equation (1) we may express separately the parts of x depending on H and on c, and the expression for the former has already been given. For the part of x depending on the fluctuations in h we have

$$\frac{cq}{q^2 + n^2} (q \sin nt - n \cos nt), \text{ or } \frac{cq}{\sqrt{(q^2 + n^2)}} \sin (nt - \lambda),$$

where

$$\tan n\lambda = \frac{n}{q} = \frac{2\pi L}{P},$$

P being the period of the fluctuation. We see that the phase of the inequality in the apparent height x lags behind that of the barometric inequality by the time λ, and the coefficient is reduced in the ratio of $\sqrt{(q^2 + n^2)}$ to q, or $\sqrt{(P^2 + 4\pi^2 L^2)}$ to P.

Supposing $L = 4$ m., and taking for P in succession 0 m. 30 s., 1 m., 2 m., 4 m., 8 m., 16 m., 32 m., and putting m for the multiplier of the coefficient, we have

$P = 0$ m. 30 s.	1 m.	2 m.	4 m.	8 m.	16 m.	32 m.
$\lambda = 0$ m. 7 s.	0 m. 15 s.	0 m. 28 s.	0 m. 53 s.	1 m. 36 s.	2 m. 34 s.	3 m. 22 s.
$m =$ ·020	·040	·079	·157	·303	·535	·786

As P increases, λ tends to 4 m., and m to 1, as its limit.

The experiment above referred to was calculated in the first instance on the supposition that the true height of the barometer at the time was 30·09 inches, according to data supplied by the Office. Partly as this figure did not profess to be extremely accurate, and partly in order to be independent of index error, &c., and to make the experiment self-contained, I re-calculated the result, assuming the true height as an additional unknown quantity to be determined from the experiment itself. I took as known the time for height 31·2, 30·7, 30·3, and determined all the elements from these three numbers. I found :

Calculated lagging time [see p. 219] 4 m. 12·7 s.

Whence testing time [see p. 218] 4 m. 37·7 s.

Calculated final height 30·111 inches.

I subjoin a comparison between the calculated and observed times.

| Height | Time | | Mean | Calculated |
	Experiment 1	Experiment 2		
31·3	—	– 18	– 18	– 22·2
31·2	Taken as origin		—	Assumed
31·1	27	27	27	24·3
31·0	51	53·5	52·2	51·3
30·9	73	80	76·5	81·4
30·8	119	114·5	116·7	115·7
30·7	156	154·5	155·2	Assumed
30·6	204	201	202·5	202·3
30·5	264	266·5	265·2	260·1
30·4	330	335	332·5	335·2
30·3	440	445	442·5	Assumed
30·2	619	639	629	630·2

The agreement with the observed numbers is very good. It shows that the calculation may safely be trusted, at least for heights ranging from about 0·1 to 1 inch from the true height. The calculated final height is 0·021 above that given us, *about* 30·09, but temperature and index error may account for at least part of it. The irregularity of capillarity, if it existed, would tend to make the height too great, as the mercury was descending. Perhaps a part of the excess 0·021 was due to this cause.

The effect of sluggishness properly so called is, however, mixed up with another which, if sensible, is not so easily allowed for, namely, that of an irregularity in the capillary depression depending on variability in the angle of contact of the mercury and the glass. This effect would certainly be sensible if the tube or mercury were at all dirty, and may perhaps be not insensible even when they are in the best condition. In a barometer with un-contracted tube it is obviated by tapping, which sets the mercury free to take its normal height. But in a marine barometer the effect of tapping is merely to render the angle of contact normal for the moment, altering thereby slightly the curvature at the vertex, on which depends the correction for capillarity, and con-sequently the equilibrium of the column; but when the previous equilibrium is thus disturbed the mercury does not at once take its level, but only begins to move (or moves at a different rate from what it did before if it were already in motion), and the slow rise or fall again alters the angle of contact from its normal value, and so on. On shipboard the motion of the vessel supplies what is equivalent to a constant tapping, but on land when the mercury is almost at its normal height, and is slowly rising or falling, this tendency of the edge of the mercurial column to stick where it is produces something of the nature of an additional source of sluggishness, which could be obviated indeed by incessant tapping, but is not quite prevented by tapping only now and then. The Council contemplate gradually replacing the marine barometers at the secondary stations by others in which the tubes are not contracted. The sluggishness arising from viscosity as the mercury flows through the capillary part of the tube affects a barometer on shipboard equally with one on land, and *that* can easily be allowed for if the "lagging time" be known.

[From the *Minutes of the Meteorological Council*, Jan. 24, 1880.]

Professor Stokes submitted the following Memorandum relating to the use of barometers with contracted tubes :

With reference to one passage in Mr Buchan's report, I must observe that the point which he laid before the Meteorological Council respecting the sluggishness of marine barometers, in his former report, was perfectly well understood by them; and it was precisely with a view to investigate that point that I instituted the experiments and calculations detailed in my Memorandum [*supra*, p. 217]. It is perfectly true that the experiments were made with comparatively speaking very large differences of height of the falling mercury from the equilibrium height, but the object of those experiments was to lay a foundation for a mathematical calculation from which the effect of sluggishness might be deduced in cases where the differences are such as those which Mr Buchan contemplates. The rule arrived at is very simple, namely that, except in the case of rapidly recurring fluctuations, the effect of sluggishness will be allowed for by taking an observed reading of the barometer to represent the barometric height, not at the moment of observation, but at a time earlier by a certain interval T which I called the "lagging time," which is ordinarily something like three or four minutes, and is a constant belonging to each barometer which may be easily calculated from the time the mercury took to fall from 1·5 to 0·5 inch above the true height when the barometer was tested for sluggishness.

It is only when the height of the barometer is subject to very rapid *oscillations* (whether, on the whole, the barometer be rising, falling, or stationary), that the rule fails to give that portion of the whole height which is due to the small oscillation superposed on the general rise or fall.

The error arising from sluggishness in a rising or falling barometer is, however, associated with another, not depending upon sluggishness, and equally affecting a barometer which is not contracted, namely, the irregularity in the correction for capillarity due to an irregularity in the angle of contact of the mercury with the glass. I do not believe that this would be sensible with a barometer in proper condition, though it would be otherwise if the glass were at all dirty or the mercury in the tube showed the least film on its surface. This is, however, quite distinct from the error of sluggishness; and though it affects a barometer with uncontracted tube, the latter has certainly the advantage that the error may be at once removed by tapping, which is not the case with a marine barometer, unless it be tapped at frequent intervals during the space of a few minutes.

[From the *Minutes of the Meteorological Council*, Feb. 19, 1884.]

Professor Stokes submitted the following Memorandum:

RESULTS OBTAINED BY THE APPLICATION OF THE HARMONIC ANALYSER TO THE PHOTOGRAPHIC RECORDS OF THE BAROMETER.

In relation to this subject two distinct questions present themselves:—First, how far may the instrument be trusted to give the coefficients in the harmonic expression for the height, regarded as a periodic function of the time, superseding thereby the laborious process of measuring a great number of ordinates and calculating the coefficients numerically? Secondly, how far do the coefficients obtained in either of these ways represent a true periodic inequality, as distinguished from the mere un-eliminated residues of non-periodic fluctuations?

(1) The instrument has hitherto been applied to the diurnal variation in the different months. Its accuracy for eliciting this

feature, so delicate a one in our latitudes, has been tested in the following manner.

The height of the barometer for each month of the year 1876, and for each of the seven self-recording observatories, has been expressed as a function of the time of day in a harmonic series to three orders, by numerical calculation in the usual way, from the data furnished by the hourly means calculated by Mr Eaton. The coefficients got in this manner show a very close agreement with those obtained by means of the machine. The coefficients of the variable terms, which have been calculated to four places of decimals, agree so well that the average difference is only about the thousandth of an inch. The comparison does not furnish the means of forming a judgment which of the two sets of numbers is the more accurate. The processes are, however, the same as those which had been tested with the thermometer, in which case the calculated coefficients had been deduced from the odd and even hours separately; and the results rendered it probable that the coefficients obtained by calculation were rather the more accurate. But practically the agreement is so close that either set of numbers may be accepted.

The mean height as got by the two methods shows an equally close agreement in five out of the seven observatories, while in the remaining two the discrepancies are as great as 0·020 for Valencia and 0·034 for Glasgow. This, doubtless, depends on the scale-value, which in the case of Glasgow is known to have been affected two or three times in the course of the year 1876 by alteration of focus. As the top of the column is a good way from the fiducial line, this practically amounts to an index error, which of course does not affect the variable terms.

In the numbers relating to the barometer no correction has been made for the discontinuity introduced by breaking off at the two ends of the period analysed, that is, at the beginning and end of a month, when the values of the height and of its differential coefficients of the various orders are not the same. This, however, does not sensibly affect the comparison of the coefficients obtained in the two ways, though it might sensibly affect the coefficients if taken as representing those in the true diurnal inequality.

(2) Taking the uncorrected coefficients for the several months, we observe a considerable amount of regularity in those of the

second and third orders, those of the second being tolerably constant, while those of the third show a well-marked seasonal inequality, as they vanish about the equinoxes and have opposite signs in summer and winter. The coefficients of the first order appear to be a good deal less regular. This is probably due to two causes : first, the correction for discontinuity decreases with the increase of the order; secondly, the large non-diurnal fluctuations of the barometer present rises and falls of somewhat the same frequency as those in a regular periodic inequality with a period of some few days, at any rate more than one day; and the uneliminated residues of these large non-periodic fluctuations, which affect the coefficients of the true diurnal inequality, would accordingly have greater influence the lower be the order of the term. The latter cause of irregularity is probably the more serious of the two.

It is satisfactory to find that a well-marked seasonal constancy or inequality in the diurnal fluctuation for the different months exhibits itself in the reduction of observations extending only over one year.

[From the *Minutes of the Meteorological Council*, Nov. 26, 1884.]

ON RAIN-BAND SPECTROSCOPY.

With reference to the resolution passed at the last meeting (Minutes, p. 47), Professor Stokes reported that a renewed examination of the sky with reference to the rain-band led him now to think it possible that something might be made of it, with suitable directions as to the part of the sky to aim at; but it is obviously desirable that the instrument should, if possible, be made quantitative, and as one simple method had occurred to him which might conceivably effect this object, he was unwilling to order an instrument until he should be able to try this method.

[From the *Report of the Meteorological Council*, 1885-6, pp. 22—3.]

MEMORANDUM ON CLOUD PHOTOGRAPHY.

By PROF. STOKES. *Presented June* 2, 1885.

The Council have now obtained a series of cloud photographs taken with cameras at two stations at a distance of 800 yards.

Each observation involves four photographs, one pair taken simultaneously at the two stations respectively, and another pair taken about a minute after the former. The object was to determine the height, direction of motion, and velocity of various clouds.

In the observation the cameras are set to correspond, so that the lines of collimation in taking the four photographs are in the same direction, save as to certain small corrections of the nature of index errors. A selection is made of cloud-points which can be identified, and these are marked on the four photographs. The coordinates of the marked points, referred to vertical and horizontal lines shown on the photograph, can be measured, when it becomes a question of calculation only to determine the position of a cloud-point, both for the first and second moments of observation, and consequently to determine not only its height, but the direction and velocity of its motion. At either moment the photographs determine the two lines of sight for the cloud-point, as seen from the station respectively. If the identification be correct, these lines ought to intersect, which would give a relation between the four observed quantities, namely, the altitudes and azimuths of the cloud-point as seen from the two stations; accordingly three of these observed quantities suffice for the determination, and then the fourth affords a verification.

The requisite formulæ for the reduction are readily obtained from spherical trigonometry. The calculation is perfectly straightforward, and there would be no difficulty about it if we had only a small number of cloud-points to reduce.

It is very desirable, however, to have numerous observations, and to determine, at least at first, a considerable number of cloud-points for each. The calculation, however, as above indicated, though straightforward, is by no means short; and the labour and consequent expense of making very numerous reductions in this manner would be almost prohibitive. Accordingly an endeavour has been made to simplify the process, and a method has been devised which promises to give the results with very little trouble, and at the same time very accurately.

It would be premature to describe this method at length until it shall have been actually put in practice. Suffice it to say that paper positives are taken from the glass negatives, and the cloud-points are marked on them. The four positives are laid in succession on a piece of paper on which a pair of cross-lines had previously been drawn, a correction being made for index errors by slightly shifting the positive from the position in which the cross-lines on it would be in a line with the cross-lines on the paper below. The places are then copied by pricking through. The lower paper now contains the record of the whole observation, cleared of index errors.

It is easy then to pass from the projection of the cloud-points on a plane parallel to the plates, to a projection on a horizontal plane, in the following manner. A wooden frame is prepared containing (1) a lens or combination of short focus with its axis horizontal, (2) a vertical frame, or board with a window, at a horizontal distance from the focus of the lens equal to the focal length of the objective of the camera, (3) a drawing board moveable round a horizontal axis intersecting at right angles the axis of the instrument, and capable of being clamped at any desired angle to the vertical frame.

The drawing board is set at an angle to the vertical equal to the zenith distance of the line of collimation, in the set of observations to be reduced, and the pricked paper is fastened by drawing pins to the vertical frame, the four arms of the cross having been previously marked by pricking through. A piece of paper having been pinned on to the drawing board, the sun's light is reflected into the lens, and the operator marks the centres of the spots of light corresponding to the various holes. This can be done with the utmost precision, if the pin or needle used for pricking has the proper diameter to give, as a result of diffraction, a very fine black centre in the middle of the illuminated patch.

When the four marked points corresponding to the four observations on any one cloud-point are joined, a quadrilateral is formed, two opposite sides of which, if there be no error of observation, will be parallel to each other and to the line joining the stations. If, moreover, the cloud be not moving up or down, but only in a horizontal plane, the sides above mentioned will be equal, and the quadrilateral will be a parallelogram. Two opposite sides represent the drift on the same scale on which the other two sides represent the base line of 800 yards, so that the magnitude and direction of motion are in a matter represented to the eye. Moreover, the height is given by the simple expression $P \cos z_0 \div p$; where P is the product of the base and the horizontal distance from the focus of the lens to the drawing board, which accordingly is always the same, z_0 is the zenith distance of the line of collimation, and is therefore the same for all the cloud-points belonging to the same set, and p is the base side of the parallelogram belonging to any particular cloud-point which is parallel to the base. The calculation is accordingly extremely easy ; and, moreover, an inspection of the different figures shows at once the relative heights of the different cloud-points, since the heights are inversely as the parallel sides of the corresponding parallelograms.

The photographs hold out the prospect of giving very good results. It is to be remembered that there is no object in extreme accuracy ; what we want is rather numerous observations, so as to get a general insight into the movements of the air at different heights, and under different circumstances.

[From the *Minutes of the Meteorological Council*, July 29, 1885.]

With reference to the rain-band observations, Professor Stokes reported that since the date of the resolution (Minutes, 1884, p. 79), he had been in the constant habit of observing the spectrum of the sky with a view to the rain-band, using a small direct-vision spectroscope which he generally carried about with him. As regards the appearance of a darkening on the red side of the line D, so much depends on the intensity of the light to be observed, and on the width of the slit employed, and so unsatisfactory is a comparison of what the observer actually sees with his memory of what he had seen on other occasions, that he doubted whether anything could be made of it, unless perhaps by an amateur who was free to devote much of his attention to it. Under these circumstances he was unwilling to incur expense in ordering apparatus unless again requested by the Council to do so.

Professor Stokes also explained, with reference to the resolution (Minutes, 1882, p. 31), the plan he proposed for determining the constant of the bridled anemometer, that is, the ratio of the moment of the pressure on the cups to the moment of the pressure on a pressure plate acting at a known leverage.—He was authorised by the Council to expend the sum not exceeding £15 on apparatus for the determination of the constant.

[From the *Minutes of the Meteorological Council*, Nov. 30, 1887.]

RESIGNATION OF MEMBERSHIP OF METEOROLOGICAL COUNCIL.

With reference to the resignation of Professor Stokes (Minutes, p. 80), the following resolution was adopted, and the Secretary was instructed to communicate it to Professor Stokes (P.C. 2439):

Resolved—"That the Meteorological Council, having been informed of the withdrawal of Professor Stokes from their body, desire to record their sense of the high value of his services while a member of the Council, and their regret that circumstances should have deprived them of a colleague so eminently qualified to aid them with his advice and give weight to the results of their determinations."

The Secretary was further instructed to inform Professor Stokes that the Council would be most happy to present to him in future any of their publications which he might wish to possess.

LENSFIELD COTTAGE, CAMBRIDGE,
January 10, 1888.

DEAR MR SCOTT,

I feel that my apologies are due to the Meteorological Council for the kind, but I fear only too flattering, resolution which they passed on my retirement from their body, for indeed I have felt often how little I was doing for the Office or for Meteorology: for the last, indeed, I can hardly claim anything.

The Council will, I hope, be less disposed to think that I have been unmindful of their kindness when I tell them that shortly before Christmas I went suddenly to Ireland in consequence of the serious illness of my sister.

Will you kindly convey my thanks to the Council? Yours, &c.

(Signed)   G. G. STOKES.

[From the *Report of the Meteorological Council*, 1889–90, pp. 36—48 *.]

NOTE ON EXPERIMENTS ON PRESSURE OF WIND MADE BY
MR W. H. DINES.

Mr Dines first took up this subject at the instance of the Wind
Force Committee of the Royal Meteorological Society, and his experiments were directed to a re-determination of the factor of the
Robinson cup-anemometer, the exact value of which was still
involved in some doubt.

The experiments were made with a whirling machine of 29 ft.
radius, driven by steam, erected in a level and somewhat sheltered
field at Hersham, Walton-on-Thames. The results of the experiments were published in the Quarterly Journal of the Society,
Vol. XIV. p. 253, and Vol. XVI. p. 26. Mr Dines found the factor
to be about 2·2, and thus practically confirmed the conclusion of
Sir G. G. Stokes, who deduced (*Proc. Roy. Soc.* 1881, p. 170) the
value 2·3, in place of 3 as found by Dr Robinson.

As Mr Dines had shown great skill in devising and carrying
out these experiments, and had expressed his willingness to continue his researches with his whirling machine, the Council
considered it desirable that the opportunity thus presented should
be utilised for further experiments in certain branches of aëro-
dynamics.

A grant was accordingly made, and Mr Dines proceeded to
investigate the connexion between the velocity of the wind and
the pressure on obstacles of various kinds. A full description of
the experiments and of the results was submitted to the Council
in May, 1889, and has since been published in the *Quarterly
Journal of the Royal Meteorological Society*, Vol. XV. p. 1; it will
therefore here suffice to give a summary.

The natural wind is not available for experiments on wind
pressure, because it is usually gusty and variable in direction, and
because it is not easy to measure its velocity with accuracy. It
thus becomes practically necessary to have recourse to a whirling
machine. But if such a machine be made on so small a scale that
it can be sheltered from the natural wind, then, unless the pressure

[* The following reports, published after Prof. Stokes had resigned his position
on the Council, are included here because they refer to work in which he was
closely concerned.]

plate be small, the circularity of the motion introduces a large error, and besides the possibility of the persistence of eddies in the air may be a cause of uncertainty in the results. A whirling machine should be therefore on a large scale and must accordingly be erected in the open air. Under these conditions, the natural wind, which always exists to a greater or less extent, is superposed on the artificial wind, and the total wind, to which the pressure is due, oscillates about a mean value in the course of each revolution of the whirler.

These various difficulties have, however, been obviated by Mr Dines by means of an ingenious device. It has long been known that, to a close degree of approximation, wind pressure varies as the square of the velocity of the wind. Now if a weight be swung round in a circle the restraint required to keep it in its course varies as the square of its velocity. It therefore occurred to him to balance the pressure on the experimental plate, when carried round by the arm of the whirler, by the centrifugal force, or tendency to fly outwards, of a weight swung round by the same arm. This plan was found to answer admirably, for when the right amount of centrifugal force for balancing the pressure of the artificial wind was found for any one speed of the whirler, the balance was maintained at all other speeds. It thus became unnecessary to make any accurate measurements of the rate of revolution of the whirling machine.

Mr Dines also devised an ingenious arrangement by which the amount of centrifugal force opposing the wind pressure varied automatically, until a true balance was attained. Also since the automatic adjustment did not take place instantaneously, but only by degrees, the disturbance due to natural wind was almost entirely eliminated, and thus experiments were made possible, excepting on days when there was a very high wind.

This is not the place for a full description of the apparatus by which the mechanical principle, explained above, was utilised, but a few words of general explanation are advisable.

The pressure plate was rigidly attached to a short arm, which lay in the same plane as the plate. The other end of the short arm could turn about a pivot which was attached to the long arm of the whirler. When the plate was in position for an experiment, the short arm was in line with the long arm, and the wind pressure tended of course to make the plate turn about the pivot.

The centrifugal force was due to the weight of a horizontal metal bar, at right angles to the long arm of the whirler. This bar could slide through a slot, and when the middle of the bar was at one side of the slot the centrifugal force, due to the revolution of the machine, tended to make the slot turn. The slot was geared to the short arm of the pressure plate by mechanism, which need not be described, so that the centrifugal force reacted against wind pressure. By an automatic action the bar slid through the slot, until, when the weight of the bar was winged out to a certain distance, there was a balance between wind pressure and centrifugal force. Measurement of the distance between the centre of the bar and the slot then gave the amount of force which had been called into play. The tendency of the pressure plate to turn was arrested by two stops, so near together that the plate could only turn through about 1°; it bore on one stop when the centrifugal force was insufficient for a balance, and on the other when too great. An electrical indicator was devised for detecting on which of the two stops the plate was bearing. An experiment was made by giving a few turns to the machine, and continuing the revolution until the indicator showed that the plate was bearing on neither stop, or more commonly until it sometimes bore on one stop and sometimes on the other.

For the smaller plates the maximum velocity of which the machine was capable was about 70 miles an hour, but most of the experiments were made at speeds between 20 and 30 miles an hour.

A number of plates and cups of different sizes and shapes were used, and the values given in the following table are the mean of all the experiments made with each particular plate, the number of experiments made with each being given in the last column of the table. Each experiment consisted of about three observations taken in immediate succession. As a rule no plate was tried twice on the same day. With the larger plates a single value seldom differed from the mean by more than 5 per cent., and on the whole the different values were fairly consistent. There were, however, two cases of exception to this statement, and they were the cylindrical surface and the sharp cone when placed in front of the plate. The results of five experiments out of the series of about 150 were rejected.

The particular value of 20·86 miles per hour for the velocity given in the table was chosen for the following reason. The centre of the pivot about which the pressure plates could turn was 28 ft. 1 in. from the axis of rotation of the whirling machine, and the centre of the plate 28 ft. 7 in. Taking the numerical value of the acceleration due to gravity as 32·19, these figures give 20·86 miles per hour as the velocity of the centre of the plates at which the centrifugal force acting upon the bar is equal to its weight.

Table showing the pressure upon various plates at a velocity of 20·86 miles per hour. The values are reduced to the standard temperature and pressure. The flat plates were cut out of hard wood $\frac{3}{8}$ in. thick. Allowance has been made for the arm which carried them.

Plates	Actual Pressure, in lbs.	Pressure, in lbs. per Square Foot	No. of Experiments
A square, each side 4 ins.	·17	1·51	4
A circle, 4·51 ins. diameter, same area	·17	1·51	9
A rectangle, 16 × 1 ins.	·19	1·70	7
A circle, 6 ins. diameter	·29	1·47	7
A square, each side 8 ins.	·66	1·48	8
A circle, 9·03 ins. diameter, same area	·67	1·50	12
A rectangle, 16 × 4 ins.	·70	1·58	4
A square, each side 12 ins.	1·57	1·57	7
A circle, 13·54 ins. diameter, same area	1·55	1·55	14
A rectangle, 24 × 6 ins.	1·56	1·59	6
A square, each side 16 ins.	2·70	1·52	6
A plate, 6 ins. diameter, and 4¾ ins. thick...	·28	1·45	5
A cylinder, 6 ins. diameter, and 4¼ ins. long	·18	0·92	4
A sphere, 6 ins. diameter	·13	0·67	8
A plate, 6 ins. diameter, with a blunt cone, angle 90°, at the back	·29	1·49	4
The same, with cone in front	·19	0·98	4
A plate, 6 ins. diameter, with a sharp cone, angle 30°, at the back	·30	1·54	4
The same, with cone in front	·12	0·60	4
A 5-inch Robinson cup, mounted on 8½ ins. of ½-in. rod	·28	1·68	8
The same, with its back to the wind	·12	0·73	4
A 9-in. cup, mounted on 6½ ins. of ⅝-in. rod	·82	1·75	3
The same, with its back to the wind	·28	0·60	3
A 2½-in. cup, mounted on 9¾ ins. of ¼-in. rod	·13	2·60	3
The same, with its back to the wind	·05	1·04	3
One foot of ⅝-in. circular rod	·09	1·71	9

The results of the table may be roughly summarized as follows. The pressure upon a plane area of fairly compact form is about 1½ lb. per square foot at a velocity of 21 miles per hour, or in other words, a pressure of 1 lb. per square foot is caused by a wind of a little more than 17 miles per hour.

Experiments were also made with two kinds of plates made of perforated zinc. The pressure on the first kind, which contained about 77 holes of ·08 in. diameter per square inch, was 9 per cent. less than on a solid plate, but gave the high value of 2·43 lbs. pressure per square foot of actual surface. The values for the second kind, having 11 to 12 holes of ·22 in. diameter per square inch, were 20 per cent. less than on a solid plate, and gave about 2 lbs. per square foot of actual surface.

It was necessary to stiffen these plates by turning up the edges, and it was found that this edge when presented to the wind increased the pressure by about 6 per cent.

A cone and a projecting rim, when put at the back of the foot circular plate, caused no appreciable alteration in the pressure, but when the rim projected in front the results were, an increase of 6 per cent. with a projection of $\frac{1}{8}$ in., of 10 per cent. with $\frac{2}{8}$ in., and of 14 per cent. with $\frac{3}{8}$ in.

The effect of cutting holes in the foot square plates was also tried—eight circular holes, each of one square inch area, being made, four as near the centre as possible, and one close to each corner. No difference in the pressure could be detected, whether any or all of the holes were covered or open. The eight holes together take away more than 5 per cent. of the plate, yet a difference of 1 per cent. in the pressure, had it existed, would certainly have been apparent.

The pressure upon the same area is increased by increasing the perimeter. The pressure upon any surface is but slightly altered by a cone or rim projecting at the back, a cone seeming to cause a slight increase, but a rim having apparently no effect. This result is of importance, since a pressure plate for permanent use should be of some material which will not warp, and if thin metal be used a rim is necessary to obtain the requisite stiffness. As might be expected, a cone in front greatly reduces the pressure.

Mr Dines next determined the moment of pressure upon a 5-in. Robinson cup when placed at different angles to the wind from 0° to 180°. The cup was mounted upon an arm 1 foot in length. The moment varied as the square of the velocity, so that the same mechanical device as before was again applicable.

The following table gives the moment in feet and pounds when the velocity of the wind is 20·94 miles an hour. The angles are measured, so that when the wind blew straight into the concavity of the cup its incidence is 0°; when it blew across the mouth of the cup it is 90°; and when it blew at the back it is 180°.

0°	moment	= ·284	96°	moment	= − ·014
12°	,,	= ·292	108°	,,	= − ·050
24°	,,	= ·303	120°	,,	= − ·046
36°	,,	= ·303	132°	,,	= − ·073
48°	,,	= ·317	144°	,,	= − ·110
60°	,,	= 246	156°	,,	= − ·116
72°	,,	= ·127	168°	,,	= − ·124
84°	,,	= ·048	180°	,,	= − ·127

In Sir G. Stokes's bridled anemometer the rotation of the cups is not free as in the ordinary instrument, but is restrained by the action of a spring or weight. The amount of torsion of the spindle which bears the cups is the datum from which the velocity of the wind is deduced. One of these anemometers is under trial at Holyhead.

The instrument ought of course to give constant results from whatever quarter the wind blows, and thus the torsional force required to hold the spindle bearing the cups ought to be independent of the orientation of the wind with respect to the cups. Sir G. Stokes has used Mr Dines's results with respect to a single cup to determine the inequality, according to orientation, of the action of the wind on a bridled anemometer, consisting of any number of cups from one to six. In Mr Dines's table, given above, the semi-circumference is divided into 15 equal parts, and accordingly there are materials for computing the action of the wind on an anemometer of one or several cups in 30 different azimuths. It is obvious that if there are several cups, the same torsional couple is repeated several times over; if, for example, there are two cups, there will be 15 different azimuths, if three, 10 different ones, and so on.

The mean of Mr Dines's numbers, when repeated so as to go round the whole circumference, is 80, hence the mean couple due to a single cup may be taken as 80. Sir G. Stokes then computed the couple due to pressure on a single cup for the 30 azimuths, with such repetitions as correspond to an anemometer with 1, 2, 3, 4, 5, 6 cups, and subtracted from each the mean couple 80.

For example, when there are six cups, the excess above or defect below 80 is + 12, + 1, − 5, − 12, + 1, repeated six times in

all. The sum of these five numbers, regardless of sign, is 31, and
their fifth part is 6·2. Hence the average inequality with six cups
is represented by 6·2, the mean couple being represented by 8·0.
In this way he found that, still taking the mean couple as 80, the
average inequality is as follows:

With 1 cup,	161·5	With 4 cups,	17·2
With 2 cups,	28·2	With 5 cups,	4·8
With 3 cups,	19·3	With 6 cups,	6·2

It is obvious that the smallness of these numbers is a measure
of the goodness of the instrument as a wind measurer. In design-
ing the instrument, Sir G. Stokes had had an impression that an
odd number of cups would be better than an even number, and
this appears to be borne out by the numbers. There is naturally
a diminution in the inequality with the increase in the number of
cups, but this appears to be blended with a superiority of odd to
even. Thus 3 cups come out nearly as good as 4, and 5 rather
better than 6.

If 6 cups be taken as a superior limit to the number desirable,
5 appears to be the best number to choose.

Sir G. Stokes had been induced to undertake this investigation,
which was communicated to the Council in the form of a letter,
by some further experiments by Mr Dines, to which we now return.

Mr Dines, then, tested an anemometer of five cups, exactly
like that at Holyhead, by the same method as that by which he
had previously treated the pressure plates. He found that the
magnitude of the moment due to the wind was greatly influenced
by orientation. The values for positions at intervals of 6° are
shown in the annexed diagram, each value being the mean of
several observations. The results of the experiments were fairly
consistent with one another, the observation for each angle seldom
differing by more than three or four per cent. from the mean value.

The figures show the moment, expressed in lbs. and feet,
exerted by a wind of 21 miles per hour. The direction of the
wind relatively to the cups is shown by the angle against which
the figures are placed. In all cases the moment varies as the
square of the velocity.

72°	·108	36°	·380
66°	·140	30°	·313
60°	·160	24°	·274
54°	·203	18°	·203
48°	·262	12°	·172
42°	·314	6°	·148

It was obvious from the consideration of these numbers that the great inequality in the torsional force is not due to the causes considered by Sir G. Stokes, but arises from the interference of each cup with the one partially behind it.

In consequence of these investigations, it was therefore decided to reconstruct the anemometer, retaining five cups in accordance with Sir G. Stokes's views, but re-arranging them spirally on their spindle, so that the interference detected by Mr Dines should be obviated.

Mr Dines was then requested to undertake a new series of experiments on the bridled anemometer in its modified form. On the completion of his experiments he submitted the following report to the Meteorological Council.

Determination of the Moment of the Wind Force upon the new arrangement of Cups of the Bridled Anemometer.

The instrument consists of five cups, each 5 ins. diameter, mounted with their centres $7\frac{1}{2}$ ins. from the axis. The cups are not in one plane, but are arranged so that there is 1 in. clear vertical space between the rims of two consecutive cups, and a difference of 144° azimuth.

Since the length of the axis rendered it impossible to use the method which was adopted with the old instrument, the following plan was adopted.

A thin metal disc graduated in degrees round part of its circumference was made to fit loosely on the bottom of the shaft of the anemometer, and was provided with a screw by which it could be clamped in any position. A brass weight was attached to the disc at 3 ins. from the centre, and at the same distance on the opposite side of the centre an exactly equal piece of wood was placed to neutralize the wind pressure upon the brass. The whole was fitted in a light iron frame so that it could turn freely about its axis. The frame could be attached to the long arm of the whirler, so that the axis of the instrument was vertical, and consequently the disc horizontal.

Suppose O to be the centre of the disc, OA the direction of the line joining O with the centre of the whirling machine, OB the direction of motion, G the position of the weight, and E the zero mark of the graduations, OG being at right angles to OE. Let the angle between OE and the vertical plane through one of the cup arms be α, and the angle AOE, or BOG be β.

It is clear that if a sufficient weight be placed at G, the centrifugal force upon the weight will prevent the complete

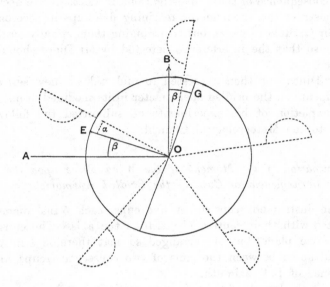

rotation of the cups, and that there will be some value of β lying between 0° and 90° which will be in a position of equilibrium.

The experiments were made thus: The disc was clamped so that α had a certain value, the whirling machine was then rotated, and soon after a steady velocity had been obtained, a brake was applied to the disc by a string running from the centre of the whirling machine, in order that no change of position should be possible until the reading had been obtained. The machine was then stopped and the value of β read off to the nearest degree. The value of $\alpha + \beta$ gave the direction of the wind relatively to the cups, the moment of the force exerted by the wind at the instant of applying the brake being proportional to $\cos \beta$.

In order that the scale might be an open one, a weight was chosen which gave the position of equilibrium with a small value of β. If W be this weight the moment due to the centrifugal force is $\dfrac{W v^2}{g\,(r + OG \sin \beta)}\, OG \cos \beta.$

As has been explained in my account of the experiments with pressure plates, the actual velocity v need not be measured, and is immaterial to the result; but in these experiments a high velocity of about 60 miles per hour has been employed to eliminate as far as possible the troublesome error due to the natural wind. Assuming that the position of equilibrium is taken up immediately, it will be seen that the natural wind may either raise or lower the moment as determined by this method. If the brake be applied when the instrument is moving with the wind the apparent moment is lowered, and conversely. On the whole, however, the tendency is to increase the moment, so that the error cannot be eliminated by taking the mean of a large number of observations.

Care has been taken that the brake should not be applied in any one part of the circle more than in another, and no experiment has been entered which was made at any time when there was the smallest perceptible wind. Still a wind of two miles per hour is barely perceptible in the open air, and such a wind might alter the moment in the ratio of $58^2 : 60^2$, that is, cause an error of from 6 to 7 per cent.

About 80 experiments have been made and worked up thus: A row of figures from 1 to 72 being prepared, the value of $\cos \beta$ was taken from a table and placed opposite to the figure indicating the value of $\alpha + \beta$ for the same experiment. In the first experiments the value of α was chosen at random, but in subsequent experiments it was chosen so as to fill in the blanks left in the list as far as possible.

When the whole set of experiments had been thus tabulated it was found that the mean value of $\cos \beta$ was ·965, the lowest value ·900, and that there was no very marked variation in the value of $\cos \beta$ in different parts of the list. The figures were pretty evenly distributed, there being no case of four consecutive numbers, without at least one value opposite to them.

The difference in weight of the brass and the facsimile piece of wood was found to be 6,259 grains, and the distance of the axis of the instrument from the axis of the whirling machine 28 feet 11 inches.

Taking these values for a first approximation we obtain $000510\,v^2$ for the value of the moment expressed in feet and lbs., v being in miles per hour.

The experiments were made at temperatures ranging from
60° to 70° F. and mostly between 60° and 65°, and at a pressure
ranging from 29·80 ins. to 30·00 ins.

One more correction is necessary; the cup which was concave
to the wind was inside, and hence the above value is rather too
low. Reference to the table of wind moments for a single cup in
various positions, given in a preceding report, will show that this
correction should be about 1·5 per cent., and hence that ·0005523 v^2
is the value of the moment due to the wind pressure upon the instrument. This result is in close agreement with that deduced
from the experiments upon a single cup, and may, I think, be
taken as very nearly correct. I think also that no greater departure than 3 per cent. from the mean value occurs in any relative
position of the wind direction and the cups.

W. H. DINES.

Upon the conclusion of these experiments the instrument at
Holyhead was dismounted, and it is being reconstructed with the
cups arranged spirally on the spindle, as explained above.

The foregoing experiments of Mr Dines have given such
valuable results that the Council have requested him to continue
his investigations, with the aid of a further grant of money. This
series of experiments was to be on the resistance of plates of
various forms at oblique incidences of the wind, and a memorandum of suggestions, as to the line which the experiments
should take, was drawn up by Professor Darwin. The account of
this work will fall into the annual report of an ensuing year, and
it must here suffice to say that the results have proved of the
highest interest, and are to be communicated to the Royal
Society*.

EXPERIMENTS WITH VIOLLE'S ACTINOMETER APPARATUS.

The *Boules conjuguées* of M. Violle are an instrument for
readily measuring the solar radiation. Like the sun thermometer,

---

* A paper containing an account of the experiments was read before the Royal
Society on June 19th, 1890. [" On wind pressure upon an inclined surface." By
W. H. Dines. Communicated by the Meteorological Council. *Roy. Soc. Proc.* vol.
xlviii. (June 12, 1890), pp. 233—258. Further experiments with the bridled
anemometer, by Mr Curtis, are described *infra*.]

as ordinarily used, it gives a relative measurement of the total radiation effect of the sun and sky, in excess of the air temperature; but according to the theory of its action as described by the inventor, with proper precautions, it may be made to furnish that of the solar intensity alone in terms of absolute units. Its principle is the same as that of the blackened and uncoated sun thermometers, used in conjunction, the theory of which has been given by Ferrel, and it is used in the same manner, viz., by exposure to the sun until its temperature has become constant.

It consists of two thin hollow copper spheres, the one coated with lamp black, the other gilt, which surround the bulbs of two thermometers precisely similar in all respects. The interior of the copper spheres and the bulbs of the enclosed thermometers are coated with lamp black. They are exposed to the sun, side by side, and when the temperatures have become constant readings of the two thermometers are taken, and at the same time the temperature of the air is observed with a sling thermometer. From these three observations the radiation coefficient is computed by a simple formula or taken from a table.

Theoretically the excess temperatures of the blackened and gilt spheres above that of the air should bear a constant proportion to each other, both varying directly with the quantity of heat received in the unit of time by radiation. The object of the tests now being applied at Kew is to ascertain whether this condition is fulfilled in practice, or whether such small variations in the coating of the spheres and the enclosed bulbs, in the circumstances of exposure, &c., as are unavoidable with different instruments, or with the same instruments at different times, seriously impair the constancy and comparability of the registers. For this purpose two pairs of spheres are now under trial, and the readings of the four thermometers will be compared with each other to see if their relative values are sensibly constant; whether, in short, the instrument is one that can be placed in the hands of meteorological observers, who have had no special training in a physical laboratory, in the expectation that the results will be more satisfactory than those yielded by the ordinary sun thermometer *.

[* There is an extensive correspondence with H. F. Blanford on these tests, and on Balfour Stewart's actinometers which were employed for some time in the Himalayas.]

ON THE WORK DONE WITH THE HARMONIC ANALYSER AT THE
METEOROLOGICAL OFFICE.

The Harmonic Analyser was designed by Sir William Thomson
in 1878. Full details of the principle of the instrument will be
found in the *Proceedings of the Royal Society*, Vols. XXIV. p. 262,
and XXVII. p. 371; while a description of the machine as constructed for this Office, and also of the method of its employment,
is given in Vol. XL. p. 382 *. Reports as to the method of employing the machine will be found in the *Reports of the Meteorological
Council* for the years 1880, 1881, 1883, and 1885.

The hope was at first entertained that the machine would
admit of the labour involved in making the hourly measurements
of the photographic traces being dispensed with, and that the harmonic constants could by means of the machine be derived directly
from the curves.

For some time after the machine had been delivered at the
Office only experimental work was carried on with it, for the
double purpose of determining the constants of the machine and
the reliability of its results, and also of familiarizing the operator
with its use.

The analysis of the thermograms for the year 1876 was first
taken in hand, because for that year the mean hourly values of
temperature had been calculated, from the published hourly values
for the seven observatories, and the data were thus supplied from
which the results obtained by the use of the machine could be
compared with results obtained by computation. As this was the
first operation in which the machine was brought into continuous
use, the results were submitted to an examination by Prof. Stokes,
who instituted "a more searching comparison between the results
obtained by the machine and those got by calculation than had
previously been possible from the available data," with the object
of ascertaining whether there was in them any indication of
systematic errors, and to what extent they might be relied upon
as correct.

[* " On the working of the Harmonic Analyser at the Meteorological Office," by
R. H. Scott and R. H. Curtis. *Roy. Soc. Proc.* vol. xl. (May 6, 1886), pp. 382–392.
It is explained that the instrument is on Lord Kelvin's model, with some details in
the mechanical arrangements that were worked out by Prof. Stokes and Mr de la Rue.
An account of a discussion of temperature curves made by the instrument is given
in the paper.]

The conclusions arrived at, together with the figures upon which they are based, were published by the Meteorological Office in 1884 as an Appendix to the *Quarterly Weather Report* for 1876, p. [39], the concluding paragraphs of which are as follows:

"Disregarding now the systematic character of some of the errors, and treating them as purely casual, we get as the average difference between the constants as got by the machine and by calculation from the 24-hourly means 0·065. It may be noticed, however, that the numbers are unusually large (and at the same time very decidedly systematic) in the case of the second cylinder of the first order, for which the average is as much as 0·125, the eighth of a degree. If b_1 be omitted, the average for the remaining cylinders of the machine is reduced to 0·047.

"We see, therefore, that with the exception perhaps of b_1, the constants got by the machine for the mean of the days constituting a month are as accurate as those got by calculation which requires considerably more time, inasmuch as the hourly lines have to be drawn on the photograms, then measured, then meaned, and the constants deduced from the means by a numerical process by no means very short."

While these preliminary trials had been going on it had not been thought expedient to discontinue making the hourly measurements of the curves and publishing the hourly values thus obtained, but hourly means were not computed. At the same time General Strachey, F.R.S., had been occupied in computing the harmonic constants of the entire series of barometric and thermometric observations made at Greenwich for the 20 years between 1849 and 1868, and as he was desirous of obtaining as complete data as possible for comparison with those he had obtained, it was considered that the satisfactory results of the treatment of the photograms of 1876 justified the extension of the use of the machine to the whole series of records of the seven observatories, which in the absence of computed hourly mean values could not otherwise be dealt with.

The analysis of the temperature records for the 12 years 1871—1882 was therefore carried out, and the results were published in 1886 as an appendix to the volume of "Hourly Readings" for the year 1883, in which the harmonic constants of the diurnal march of temperature at each of the seven observatories are given for every month and year of the period.

Subsequently the barograms for the same period were similarly dealt with, but the results have not yet been published.

With a view of obtaining as complete a knowledge as possible of the reliance that can be put on the results derived from the machine, a further set of trials has been carried out by Mr R. Curtis, which are described in the supplementary note which is annexed.

After careful consideration of the results of this further examination of the operation of the machine, the Council have been induced to modify their former opinions regarding its practical application. For since they have decided to publish the hourly numerical mean values, which are necessarily derived from the measurements of the curves, there is no room to doubt that the harmonic constants can be derived from these mean values arithmetically with less labour and with greater accuracy than by the use of the machine, while the results can be far more readily and completely checked.

The Council being of opinion that the harmonic analysis of the chief meteorological phenomena should invariably follow the computation of the requisite mean values, they have, for the reasons above stated, resolved that the analysis shall be made by arithmetical computation, while the use of the machine will be reserved for special occasions when its use may be found desirable.

[In the *Report of the Meteorological Council* 1897–8, pp. 21–30, there is a valuable "Report upon Anemometer Experiments at Holyhead" by Mr R. H. Curtis, followed by "Description of the Bridled Anemometer designed by Sir G. G. Stokes." The report institutes a comparison, on the basis of prolonged observations, between four types of anemometer, viz. pressure-tube, bridled, pressure-plate, and Robinson standard.

For the sake of completeness the greater part of this report is here reproduced*.]

Comparison of the Pressure-Tube and Robinson Cup Anemometers.

In the Report of the Office for the year 1896 an account was given of a comparison which had then recently been carried out between the records of the cup anemometer and the corresponding

[* For completion of this record reference should also be made to a series of experiments made at the National Physical Laboratory by Dr Stanton in 1904 and published in the *Proceedings of the Institution of Mechanical Engineers.*]

records of the pressure-tube anemometer on Salt Island, the result of which was to show : first, that upon the average wind velocities recorded by the cup anemometer were about 27 per cent. higher than the records of the same winds as obtained from the pressure-tube, thus pointing to the necessity for a modification of the factor employed for converting the movement of the cups into equivalent wind-velocities ; and, secondly, that in addition to this, the indications of the cup instrument were further greatly influenced by the railway sheds, grouped around the base of the lighthouse, on the top of which this particular anemometer is placed. A plan of the pier and sheds had been given in the Report for 1895, and, briefly stated, it was now shown that when the direction of the wind was such as to cause it to blow *through* the sheds, from end to end, the result was to suck down air from the anemometer, whose record was in consequence decreased and brought below that of the pressure-tube ; whilst when the wind blew *against* the closed sides of the sheds it was deflected upwards, over their roofs, causing more than the normal amount of air to pass over the cups, with the result that their record became correspondingly increased beyond that of the pressure-tube ; it was only when the direction of the wind was such that the current of air neither blew directly through, nor against, the sides of the sheds, but in such a way as to slip past them, that the records of the two anemometers agreed.

The result was of considerable practical importance, as it proved the absolute necessity for securing a perfectly free and uninfluenced exposure for anemometers, in order to obtain from them a true record of the strength of the wind, and at the same time showed that causes hitherto quite unsuspected were capable of affecting their records in a very marked degree.

The consistent character of the results obtained left very little doubt as to their accuracy, although the number of observations available for the comparison was not so large as could have been wished, owing to the pressure-tube anemometer having been at work for a few months only. Now, however, there is a fairly large mass of data available for further testing the conclusions arrived at in 1896, and therefore it has been thought desirable to repeat the comparison with observations made subsequent to the publication of the previous report. Of course the question of the proper factor for the cup anemometer is to some extent involved in the investigation, and the wind velocities by that instrument have all been dealt with upon the assumption that its true factor is not 3·0 but 2·2; but it has not been thought necessary to re-open the question further as regards this particular instrument, because if the conclusions already arrived at respecting the effect of the sheds upon it are confirmed, it follows that its exposure at Holyhead is not sufficiently good to admit of its giving a satisfactory answer upon the point*.

In the present instance one hundred observations of moderate or fresh winds were taken under each of the 16 points of the compass†, and the mean velocity as given by the two instruments was determined for each point. The results are given in the following Table I., and are also shown graphically in Fig. II. [omitted].

The results now obtained fully confirm those got by the previous comparison, the only differences worthy of note being:

(1)   That the maximum *plus* difference is shown under South, instead of South-south-east, and that it amounts to only 15 per cent. instead of 22 per cent.; and

(2)   That the *minus* difference under East is increased to 20 per cent. instead of 13 per cent.   Generally, the differences, if put in the form of a curve, are seen to be more symmetrical than before, and those for opposite points, as North and South, and East and West, which are due to a similar kind of action, now approximate more closely to each other as regards amount; these changes are doubtless due to the smoothing effect of the larger number of observations it has been possible to use on the present occasion.

The figures in this table demonstrate very clearly the necessity

---

* Various attempts have been made to obtain the factor of the Robinson cup anemometer by comparing its records with those of a pressure-tube anemometer placed by its side, and run under exactly similar conditions. Such a comparison, extending over three years, has been carried out at Rousdon Observatory, in South Devon; and another, covering a period of eight months, at the Colába Observatory, Bombay. In both these instances the cup anemometer was similar to the one in use at Holyhead. The Rousdon comparison gave as the mean factor (2·2), and the Colába comparison (2·1). The Rousdon observations covered a much larger range of wind-force than did those at Bombay, and it is probable that the relation between the speed of the cups and that of the wind is affected both by the force of the wind, and by its steadiness or otherwise; and also by the conditions attaching to each instrument *in situ* so far as they affect the question of friction and the amount of work required to be done by the cups in registration.

† So many as 100 observations could not be obtained with winds from E.S.E. and S.E.

there is for having a perfectly free exposure for anemometers, in order to get from them the true velocity of the wind. With instruments which require a mechanical connection between the recording apparatus and the parts acted upon by the wind, this is frequently a difficult condition to secure, and the question is opened up as to how far, and in what way, existing anemometrical records derived from instruments placed on buildings, as all large-sized cup-anemometers are, have been affected by the obstruction offered by those buildings to the free passage of the wind. The conditions at Holyhead are no doubt exceptional, and their effect, as shown by the diagram, is of course peculiar to that place; but it is to be feared that in other instances also the buildings, or other objects, in the close neighbourhood of the anemometers, are such as to affect their records prejudicially.

Comparison of the Pressure-Tube and the "Bridled" Anemometers.

These two instruments have now been at work side by side for nearly three years, and although during that time they have not experienced any gale of exceptional violence, yet they have afforded a good deal of material for making a comparison between their records.

A description of the bridled anemometer will be found on p. 28, and therefore it will suffice to say here that it measures the force of the wind through the resistance offered, by some weights, to the turning of a set of five cups, arranged spirally upon a vertical spindle, and exposed to the wind. The pressure-tube anemometer, on the other hand, achieves the same end by the displacement of a specially-shaped float, contained in a closed vessel of water, but communicating with the air by means of a tube, leading to a vane which is kept constantly facing the wind.

The bridled anemometer was not designed to register light winds, and therefore it is unaffected by any wind of lower velocity than 20 miles per hour, and the inertia of the instrument is not entirely overcome until the velocity has risen to fully 30 miles per hour; but with stronger winds than this its indications are found to agree very closely with those of the pressure-tube. It sometimes happens, in sudden and transient squalls, that the gust begins to abate before the float of the pressure-tube has had time to fill and rise to its proper height, in which case that instrument fails to record the maximum force which is shown by the bridled anemometer, whose cups respond at once to the wind, and indeed may occasionally, as a result of their inertia, be carried somewhat beyond the proper point. Such differences, however, seldom exceed a rate of two or three miles per hour, and only occur with very transient gusts of wind; if the gust lasts for but a few seconds, or if it is at once followed by another of equal force, as not infrequently happens, then the float has time to fill, and the two records agree.

It may therefore be said that, as regards the range of wind-force which has been covered by the observations made in this comparison, the indications of the two instruments are practically identical.

The value of the comparison lies very much in the mutual support which the records of the two instruments—each obtained in a different way, although by the application of a similar principle,—afford to each other, and in the confidence in their accuracy which their agreement supplies.

Comparison of the Pressure-Tube and the new Pressure-Plate Anemometers.

The most recently erected anemometer at Holyhead is a pressure-plate, the recording portion of which has been arranged upon a somewhat novel plan.

The plate itself is a thin circular disc of aluminium, of one square foot area, and is supported, at the height of 26 feet above the ground, upon an iron column, which stands only a few feet away from the pressure-tube and the bridled instruments. The plate turns freely in azimuth, and is kept facing the wind by a fin-shaped vane, which is placed about two feet in the rear of the plate and rises a little above its top edge; by this arrangement not only is the vane not sheltered by the plate, but any interference with the free movement of the air flowing around the edges of the plate is avoided.

The apparatus is designed to register only the *maximum* pressure exerted by the wind upon the plate since the last observation, and after every reading of the scale the index requires to be reset. Readings are therefore made only at definite times, usually once a day, except in stormy weather, when they would be made as frequently as possible; but owing to the remote position of the instrument the number of observations is necessarily restricted.

The action of the instrument will be best understood by the aid of the accompanying diagrams, Figures III. and IV. [omitted]. Figure III. shows a section of the plate (P), to the back of which is fastened a tube (T), which, by means of friction rollers, slides easily to and fro in the slightly larger tube C. To this sliding tube a chain is attached, and as the plate and tube are together driven back by the wind this chain pulls upon the spring shown in Fig. IV., p. 25, which is securely fastened inside the column at a point where it can be easily seen by the observer. The extension of the spring which is thus effected is shown by a brass bar graduated to indicate pounds, and having on each edge a ratchet and pawl, so arranged that after the spring has been extended it cannot go back again, but must remain extended until the pawl is released by the observer. By this arrangement the maximum pressure is reached gradually, pound by pound, and any error in the record due to the momentum of the plate, acting under the influence of sudden sharp gusts, is avoided.

In pressure-plates which are free to oscillate to and fro with each gust of wind, as is usually the case, the chance of error from the momentum of the plate is very great, and there is little doubt that from this cause such plates in gusty winds often travel a considerable distance in excess of the proper amount, giving rise to records of wind-pressures which are not really experienced.

Owing to a fault which became developed in the bearing of the vane, soon after its erection, the instrument had to be dismounted again, and it was not until the end of January, 1898, that it came regularly into work. The number of observations at present available is therefore not very large, and as the strongest winds of the year were experienced whilst the instrument was out of action the number does not include any very high pressures; but the results of the comparison, so far as it has gone, are very interesting.

The method adopted has been to compare with each reading of the pressure-*plate* the maximum pressure indicated by the pressure-*tube* during the interval covered by the reading. These observations were then grouped according to the pressure-plate readings, and the mean equivalent by the tube found for each pound of pressure recorded by the plate; these equivalent pressures, together with the number of observations from which they were determined, are shown in the following Table II. [omitted].

Speaking generally, it may be said that these equivalent tube values are all about 25 per cent. *above* the values actually recorded by the plate. It will be remembered that, from the construction of the instrument, increments of pressure of less than a pound are not registered; and therefore a given reading may always have been exceeded by any amount short of another pound, without such excess being registered. For example, a recorded pressure of 4 pounds must be taken as indicating that the effective pressure of the wind upon the plate not only reached that amount but may have been anything between it and 5 pounds. Bearing this in mind it will be seen from the diagram, Fig. V. [omitted] that the equivalent pressures obtained by the comparison, so far as it has gone, fall very well between the limits above stated, after adding 25 per cent. to the actual readings of the spring. The friction of the moving parts, although very slight, is probably the cause of the rather higher equivalents obtained for the lowest pressures.

The wind on striking a plane surface, such as the pressure-plate now used, does not exert all over it a uniform pressure, but a pressure which diminishes from the centre to the edge, in a ratio determined by the size and shape of the surface, and probably by the strength of the air current also *. By placing a rim round the edge of the plate, so as to stop the ready outflow of air from its surface, the pressures recorded would be considerably increased for all winds; the conditions would then approximate closely with those which exist in the pressure-tube, and it is probable that the records of the two instruments would under those conditions nearly agree.

The number of observations with strong winds is as yet far too small to warrant the drawing of any final conclusions; but so far as the results of this comparison go at present, it seems pretty clear that with pressure-plate anemometers of the ordinary construction, the momentum of the plate has been the cause of some misconception as to the frequency with which high pressures are reached; and secondly, that, even when this source of error is guarded against, it is not safe to accept the records of anemometers of that type, as being reliable records of the wind-pressures actually

---

* See *Quarterly Journal Royal Meteorological Society*, vols. viii. and ix.

experienced, until they have been corrected for the effect of the slipping of the air over their surfaces and past their edges. In the case of a circular plate of one square foot area there is reason to suppose that the loss of effective pressure upon the plate from this cause is equal to about 25 per cent. of the actual pressure the wind current is capable of exerting, but at present there are no data to show how this may be modified in the case of larger plates, or of plates of other shapes.

Description of the Bridled Anemometer designed by Sir G. G. Stokes, Bart., F.R.S.

Although the bridled anemometer has been at work ever since the year 1880 it does not appear that any detailed description of it has yet been published, the references to it which have been made from time to time in the publications of the Office having usually been very brief and general in character.

The portion of the instrument which is acted upon by the wind consists of five hemispherical copper cups, each of which is fixed by a short strong arm to a vertical spindle, around which they were originally placed equidistantly and in the same horizontal plane; after a time, however (in the year 1890), the cups were re-arranged spirally, so as to prevent any possibility of one cup sheltering another.

The special aim of Sir G. Stokes was to get a measure of the strength of the wind in gusts; and this he sought to do by attaching to the spindle a weight, which would have to be lifted as the spindle turned under the influence of the wind acting upon the cups, the weight being sufficiently heavy to prevent the cups from making a complete revolution in the strongest gust they would be likely to experience.

To achieve this the spindle is brought, through bearings fitted with friction rollers, into the room in which the recording apparatus is placed, and its lower end is fitted with a pair of snails. The load to be lifted by the cups is divided into two equal parts, and is suspended on each side of the spindle by flexible cords, which pass from the spindle over a pulley to the weight, and back over another pulley to the spindle again, where each is fastened to the root of one of the snails. As the cups turn the weight is lifted, and the cord in each case is wound upon the periphery of the snail, the extra resistance thus gained being in accordance with the well-ascertained fact that the pressure of the wind varies with the square of the velocity ; the record obtained is therefore one of velocity, and its scale is an uniform one throughout.

The record is secured in a very simple manner. Immediately below the spindle, and co-axial with it, is a cylinder, upon which a sheet of paper can be placed; and to the bottom of the spindle itself is fixed an arm, bent so as to allow it to sweep around the cylinder as the spindle turns, and having at its lower end a pen which can be kept in contact with the paper; the amplitude of every turn of the spindle is therefore shown by the length of the corresponding line the pen leaves upon the paper.

The cylinder forms the driving weight of the clock by which its descent is regulated, and the abscissæ of the trace thus become the time-scale, while the horizontal ordinates give the force of the wind. By a simple arrangement the rate of descent of the cylinder can, if desired, be doubled, and the time-scale correspondingly opened, and this enlarged scale is generally employed in strong gales ; there is also an arrangement by means of which the time-scale can be still further enlarged by substituting a hydraulic arrangement for the clock, but this is very rarely used.

PENDULUMS AND GRAVITY SURVEYS.

CORRESPONDENCE WITH SIR EDWARD SABINE.

CAMBRIDGE, *Dec.* 16*th*, 1850.

DEAR COL. SABINE,

Perhaps you may recollect speaking to me once at Prof. Miller's about pendulum experiments. In your paper " On the Reduction to a Vacuum of the Vibrations of an Invariable Pendulum," *Phil. Trans.* 1829, in speaking of the results to which you had arrived by a comparison of the vibrations in an exhausted receiver, in air at the atmospheric pressure, and in hydrogen, you add (p. 232) "Should the existence of such a distinct property of resistance, varying in the different elastic fluids, be confirmed by experiments now in progress with other gases, &c." I have not met with any further notice of these experiments, but you told me on the occasion I have mentioned that the experiments showed that the retardation could not at all be inferred from the density, in passing from one elastic fluid to another.

Now I wish to know (supposing, as I believe is the case, that the experiments have not been published) whether you would have any objection to my stating in a paper read before the Cambridge Philosophical Society that I had been informed by you that the experiments here alluded to fully established the existence of a specific action in elastic fluids quite distinct from mere variations of density.

The paper I allude to was read at the last meeting of the Camb. Phil. Soc. It contains the calculation of the resistance to a pendulum in the two cases of a sphere and of a long cylindrical rod, when the *internal friction*, as it may be called, of the fluid is taken into account. The agreement of theory with Baily's experiments is very striking. I hope in the course of a few months to be able to send you a copy.

Yours very truly,

G. G. STOKES.

CAMBRIDGE, *Jan.* 10*th*, 1851.

I am much obliged by your mention of the additional experiments on the specific action of different elastic fluids on the vibrations of a pendulum. I was not aware that they had been interrupted, and thought that you had forgotten to publish the results.

As to the existence of a specific action, I am quite prepared to expect it, and therefore believe in it even on the strength of the two hydrogen experiments alone. In fact, I find that though Baily's results are at variance with the common theory of fluid motion, in which the pressure is supposed equal in all directions in a fluid, or, which comes to the same, in which the fluid is supposed to be perfectly smooth, they agree beautifully with the formulæ to which I have been led by employing a theory in which what may be called the internal friction of the fluid is taken into account. In this theory it is supposed that a continuous sliding motion of the fluid calls into play a tangential pressure proportional to the rate of sliding. Thus, imagine a fluid to flow in horizontal layers, the velocity increasing uniformly from the ground upwards, so that the layers of fluid which are at one time arranged like ≣ after a certain time come to be arranged like ⟋; then I suppose that, if you draw any imaginary horizontal plane through the fluid, the motions of fluid above and below the plane act tangentially on one another thus ⇄. The tangential pressure for a sliding unity, referred to a unit of surface, and divided by the density, is a constant depending upon the nature of the fluid, which I propose to call the index of friction of the fluid.

I should like to be able to determine by pendulum experiments the index of friction for different gases. If your apparatus were erected at Kew, I think I might go there for the purpose some time in the summer. I should however want to make some preliminary experiments on vibrations in air, especially with the view of testing the theory by its application to the calculation of the rate of decrease of the arc of vibration. These experiments would also require a vacuum apparatus.

Believe me, Yours very truly,

G. G. STOKES.

CAMBRIDGE, *Feb. 9th*, 1863.

Thank you for the two letters, which I kept in order to enter fully into the subject, and refer to Bessel's memoir, which I have got.

The methods for eliminating the effect of the air may be divided into three, of which again the third, when considered in its details, branches off into two.

1st, There is the direct method, which you have recommended, of swinging the pendulum in vacuo. This would seem to be the most exact of all, on account of the collateral advantage that the very process which removes from the result the influence of the air on the *time* of vibration removes also its influence on the *arc* of vibration, so that the time during which the pendulum swings through an arc large enough for the observation of coincidences is greatly prolonged.

2nd, There is the method you employed of swinging the pendulum in air and in vacuo, and so determining the factor by which the correction for buoyancy must be multiplied. This method recommends itself by its extreme convenience, as the determination is made once for all at some fixed observatory. The influence, if any, of temperature on the value of the correcting factor has yet to be determined experimentally, but its effect would probably be insensible in the case of heavy pendulums, such as those used in determining the variation of gravity or its absolute value at a particular place.

3rd, There remains Bessel's method, a very elegant one in principle, to construct a pendulum which is symmetrical in figure but not in mass, and in which further the dimensions and masses of the parts are so chosen that one centre of suspension (represented by a knife-edge) shall be at the same distance from one end that the centre of oscillation (represented as approximately as may be by the other knife-edge) is from the other.

It is easy to prove that such a pendulum if rigorously convertible in air at a standard pressure and temperature would not be convertible in vacuo, and even changes in the barometer would affect its rigorous convertibility. It becomes then a question how the error of convertibility is to be dealt with.

(*a*) The pendulum might be furnished with a small weight, adjustable by a screw, by means of which it could be rendered convertible by trial. This however would destroy the rigorous invariability of the pendulum, and would involve tedious trials at each station, and ought I think decidedly to be rejected. An adjustable weight might be used in the construction of the instrument, but when once adjusted it ought not to be further touched. I should rather however that the adjustment were made by filing.

(b) If (a) be rejected it remains to have the times of vibration round the two knife-edges respectively subject to a minute discrepancy, and from them deduce the time of vibration of an invariable pendulum.

According to Bessel's notation (p. 97 of his memoir "Untersuchungen über die Länge des einfachen Sekundenpendels"), let m be the mass of the pendulum, m' that of the air displaced, μm the moment of inertia of the pendulum, K an unknown quantity, representing the square of a line, such that $m'K$ represents the moment of inertia of the fictitious mass which we must conceive added to the pendulum to allow for the inertia of the air set in motion, S the distance of the centre of gravity of the pendulum from one knife-edge, (suppose that nearest to the lighter bob,) S_1 the distance from the other knife-edge, $S', = \frac{1}{2}(S + S_1)$, the distance of the centre of gravity of the figure from either knife-edge, l, l_1 (which are very nearly equal) the lengths of the isochronous simple pendulum; then

$$l(mS - m'S') = m(\mu + S^2) + m'K,$$
$$l_1(mS_1 - m'S') = m(\mu + S_1^2) + m'K.$$

Subtracting, and observing that in the small terms multiplied by m' we may suppose $l = l_1$, we have

$$l \cdot mS - l_1 \cdot mS_1 = m(S^2 - S_1^2),$$

or
$$lS - l_1S_1 = S^2 - S_1^2 \qquad (1).$$

Let t, t_1 be the observed times of vibration, τ the time for a simple pendulum whose length $= S + S_1$ the distance between the knife-edges; then

$$\frac{l}{t^2} = \frac{l_1}{t_1^2} = \frac{S + S_1}{\tau^2}.$$

Whence substituting in (1) and dividing by $S + S_1$ we have

$$St^2 - S_1t_1^2 = (S - S_1)\tau^2.$$

Now t, t_1, τ are so very nearly equal that the squares of their differences may be neglected. We have then

$$\tau^2 = t^2 + \frac{S_1}{S - S_1}(t^2 - t_1^2) = t^2 + \frac{2S_1 t}{S - S_1}(t - t_1) \text{ nearly},$$

$$\tau = t + \frac{S_1}{S - S_1}(t - t_1) \qquad (2),$$

or
$$\tau = \frac{St - S_1 t_1}{S - S_1} \qquad (3).$$

Of course S and S_1 cannot be measured with the same accuracy as $S + S_1$, which is the distance of the knife-edges. But the quantity $t - t_1$ by which the ratio of S_1 to $S - S_1$ is multiplied in (2) is so extremely small that the error thence resulting in the value of τ is insensible.

We see from (2) that the time τ which corresponds to the distance $S + S_1$ of the knife-edges is by no means the mean $\frac{1}{2}(t + t_1)$ of the observed times. The formula (2) is equally applicable whether the difference between t and t_1 arises from an original defect of adjustment, or from a change in the state of the air, supposed however to be the same for both swings.

The formula (2) put under the form (3) shows that equal errors in t and t_1 affect τ in the proportion of S to S_1. But a given error in the observed time of a coincidence will affect the deduced time of vibration in inverse proportion to the length of the swing, which in the two cases will be as h to h_1, supposing the initial and final arcs given, and the full cause of decrement to be the resistance of the air; so that a given error in the observation of a coincidence would affect the result to the same extent whether it belonged to a swing with heavy end below or with heavy end above, notwithstanding the more rapid decrement of arc in the latter case, and consequent shortening in the time of observation.

Believe me, Yours very truly,

G. G. STOKES.

To the India Office.

The Royal Society,
February 13*th*, 1871.

Sir,

With reference to your letter dated October 3, 1870, addressed to the Secretary of the Royal Society, I have the honour to inform you, for the information of His Grace the Duke of Argyll, that the enclosure therein contained, drawn up by Colonel Walker, R.E., describing the progress of the pendulum observations now being carried on by Captain Basevi in connexion with the operations of the Great Trigonometrical Survey of India, and the measures which it is proposed to take in order to complete the same, has been communicated to several of our most eminent scientific men, with an invitation to them to make any remarks or suggestions thereon that they might think fit. It has also been read at a meeting of the Royal Society, and is now published in No. 124 of the *Proceedings*.

Great judgment appears to have been exercised in the choice of stations, which is such as fully to carry out the intentions of the Royal Society; and the observations appear to have been made with a scrupulous regard to accuracy.

In the preliminary abstract of results drawn up by Captain Basevi, a noticeable feature is one which has already been observed in comparing the results of pendulum observations made in other parts of the earth, namely, that at inland stations gravity appears to be in defect of that observed at coast stations in similar latitudes. In connexion with this law, which the present observations help to confirm, but of which the cause is still uncertain, it is very satisfactory that Colonel Walker has been enabled to include in the series an oceanic station (the island of Minicoy), where, in conformity with what the law would lead us to expect, an increase of gravity has actually been obtained.

It seems desirable that advantage should be taken of Captain Basevi's return by the Overland Route for swinging the pendulums at Aden and in Egypt, according to Colonel Walker's programme, not only because by a comparatively small expenditure of time and labour the value of gravity will be ascertained at two stations remote from any for which it has hitherto been experimentally determined, but also for the sake of comparison with the results obtained at stations of similar latitude in India.

One slight addition to Colonel Walker's programme appears to be desirable, namely, after Captain Basevi's return to swing the pendulums at Kew *as well as* at Greenwich; at Greenwich for the reasons mentioned by Colonel Walker, at Kew because it was there that the pendulums were swung immediately before departure, and a repetition of the experiment would afford a test of the invariability of the pendulums, on which the value of the whole series materially depends.

HARBOUR HOUSE, PORTPATRICK,
August 23rd, 1873.

MY DEAR SIR EDWARD [SABINE]

Dr Robinson has shown me your letter to him, enclosing copy of a letter from you to Col. Walker, and inviting remarks on the latter from him and from me.

I have talked to Dr Robinson on the subject. As I am writing in a room by myself, it will be most convenient to ask you to consider me alone as responsible for the following remarks, which I will submit to Dr Robinson for approval or otherwise, or to supplement by any additional remarks he may have to offer.

1. I think it very desirable that the elaborate series of pendulum experiments made in connexion with the great Trigonometrical Survey of India should be connected with an absolute determination.

2. I think it very desirable that the absolute value of the length of the seconds' pendulum should be ascertained from Kew, as a head station for physical experiments; such experiments as could not be carried on at the National Observatory, the duties connected with which are observational rather than experimental.

3. To check the Indian work by testing the invariability of the pendulums during use, it is essential that the pendulums should be swung at Kew after their return as they were before their departure. This is contained in the programme already agreed to.

4. Assuming No. 3, nothing is required for No. 2 beyond what is essential for No. 1.

5.   Nos. 1, 2 may be satisfied in either of two ways:—either (a) by a fresh absolute determination, or (b) by determining gravity at Kew relatively to some other station where a good absolute determination has already been made.

6.   There are two such stations, as you mention, where an absolute determination has been made by yourself, No. 2 Rutland Place, and a particular part of the Greenwich Observatory.

7.   No. 2 Rutland Place, having passed into other private hands, is presumably no longer available even for a relative determination to be made once for all.

8.   The Greenwich Observatory *is* available for such a determination.   The effect of the massive erections in connexion with the equatorial need not be feared.   I have calculated that an iron plate under the pendulum *one foot* thick and extending as far laterally as you please would affect the length of the seconds' pendulum by only about the $\frac{1}{300000}$th part of an inch, or about the $\frac{1}{8}$th part of the length of a wave of light.   And the heavy erections being in great measure of the nature of pillars, extending both above and below the pendulum, their upward and downward attractions would nearly neutralize each other.

9.   Objects 1, 2 *could* therefore be effected by a relative determination at Greenwich as well as Kew.   As gravity can be compared at neighbouring stations (where the risk of damage to an invariable pendulum in travelling is reduced to a minimum) more accurately than absolutely determined at either, the previous absolute determination at Greenwich would thus be in effect transferred to Kew as well as to the Indian stations.

10.   But as in any case the pendulums must be swung and the time-determinations made at Kew, and as all the appliances exist there for measuring the distance of the knife-edges, the only thing additional required to make the determination absolute, it would be a great pity to omit the measurements required to make the determination absolute.

11.   In favour of the comparatively small addition to the programme thus involved, it may be urged

1°.   That the work done by the Officers of the Survey would thus be rendered complete in itself, instead of having to be supplemented, so far as relates to absolute determination, by that of other observers however eminent.

2°.   That the absolute determinations made by yourself at

Greenwich and to be made by the Indian Officers at Kew will thus check each other, through the intervention of the relative determination at Greenwich as well as Kew which forms part of the programme.

12. I have taken for granted that if a fresh absolute determination be made at all it will be made at Kew. That such would be the most convenient course will not I presume be disputed.

Believe me, yours sincerely

P.S. I have read the above to Dr Robinson, who says he quite approves of it. He will however write to you himself.

Thanks for your Contributions to Terrestrial Magnetism, No. XIII. I congratulate you on having brought to a successful conclusion a work which has cost you so much labour.

CAMBRIDGE, 5th Dec., 1873.

MY DEAR SIR EDWARD [SABINE]

I am in correspondence with Capt. Heaviside about the Kater's pendulum. To two of his questions I gave unhesitating answers.

(1) There is no use in making the vibrations about the two knife-edges precisely synchronous, though they must be nearly so.

(2) Take the whole series of swings in earnest (*i.e.* excluding short provisional swings to determine the adjustments) in *one* position of the slider.

Let A, B be the knife-edges, A being that with weight below, h, h' the distances of the centre of gravity of the whole pendulum, slider included, from A, B, so that $h + h'$ $(= l$ say) is the measured distance of the knife-edges ; t, t' the times of vibration for A, B ; x the required time for strictly synchronous vibrations. Then

$$x^2 = \frac{ht^2 - h't'^2}{h - h'} \qquad (1)$$

or

$$x^2 = t^2 + \frac{h'}{h - h'} (t^2 - t'^2) \qquad (2).$$

As t, t', x are very nearly equal, this formula may practically be replaced by

$$N = n + \frac{h'}{h - h'} (n - n') \qquad (3)$$

N, n, n' being the vibrations per diem.

Now h may be determined very fairly by balancing the pendulum on its edge, or better by supporting it edgeways a shade to the heavy side of the centre of gravity, and weighing the small force at the other end required to keep it horizontal; and $h + h'$ may be deemed to be known exactly. Say we can reckon on getting the factor $\dfrac{h'}{h - h'}$ to within the $\frac{1}{200}$th part. This would require the distance of the centre of gravity to be known to $\dfrac{(h - h')^2}{h + h'} \frac{1}{200}$. In a Kater pendulum h is about 26 and h' about 13 inches, and $\frac{1}{200} \dfrac{(h - h')^2}{h + h'}$ comes out about $\frac{1}{45}$th of an inch. Suppose the error of synchronism to be 2 vibrations per diem, and the error of the factor $\dfrac{h'}{h - h'}$, consequent on the error in the determination of the centre of gravity, to be $\frac{1}{200}$th. Then the error in N consequent on the error in the determination of the centre of gravity is only $\frac{1}{200} \times 2 = 0\cdot01$ of a vibration per diem. Now $\frac{1}{45}$ inch error in place of G and 2 vibrations per diem for non-adjustment to synchronism seem pretty liberal allowances, and $0\cdot01$ vibration per diem is well below the probable errors of observation. I think therefore the correction for imperfect synchronism may be deemed perfect, so that the results thus obtained will be as good as they would have been had n, n' turned out exactly equal. The formula (3) shows that the error ΔN resulting from errors Δn, $\Delta n'$ in the vibrations about the knife-edges A, B is given by

$$\Delta N = \frac{h}{h - h'} \Delta n - \frac{h}{h - h'} \Delta n'.$$

Now in the Kater pattern h is about twice as great as h'. Therefore an error in n is twice as influential as an error in n'. Therefore the observer should not divide his time equally between the two knife-edges, but pay more attention to A. In this respect I think you were quite right, and that Baily in his criticism on your work (*Phil. Trans.* 1832, p. 471, top) was wrong. Also, it is not evident as he states at line 9 that there is an anomaly on the face of the observations. There are two points near the centre (i.e. middle point, and centre of gravity) of the pendulum between which if the centre of gravity of the slider be placed, *opposite* effects on the times t, t' are produced by a slight motion of the slider, and I am not sure but that your slider may have been between these.

I quite agree with what Baily wrote in the last sentence of p. 470; but I don't know what particular process he had in view when he wrote " proper correction, from the known principles of the pendulum."

P.S. The formula showing that n, n' are both greater or both less than the true number N belonging to synchronism,—which it is, depending on the sign of the tabulated error of synchronism. But this is of no moment.

CAMBRIDGE, 6th Dec., 1873.

If I had to construct a Kater's pendulum *de novo* I think I would dispense with the sliding weights altogether, and introduce a third knife-edge C, the pendulum being constructed so as to be approximately convertible with respect to A, B, and B, C being equidistant, or nearly so, from the centre of gravity. The distances AB, AC being measured, and the three times observed, we have the data for determining the length of the seconds' pendulum. The three swing-times being nearly equal would all be convenient for coincidences if one was, which could be easily arranged. If the pendulum were *strictly* convertible, the swing-time round C and the length AC would disappear *altogether* from the result; and as I suppose the pendulum to be very nearly convertible a rough determination of the swing-time round C and an approximate measure of AC would suffice. The pendulum is thus made *as good as mathematically convertible*, as good as a pendulum with a slider in which the adjustment of the slider was known to be mathematically perfect. There would be no need to swing the pendulum round C for longer than the interval between two consecutive coincidences.

I find the formula for the length l of a simple pendulum which swings in time x to be

$$(at'^2 t''^2 + bt''^2 t^2 + ct^2 t'^2)\, l^2 + (bc\overline{t'^2 - t''^2} + ca\overline{t''^2 - t^2} + ab\overline{t^2 - t'^2})\, lx^2$$
$$- 2abcx^4 = 0$$

where
$$\begin{array}{lll}
a = AB & t^2 = & \left.\begin{array}{l}\text{square of time} \\ \text{of vibration} \\ \text{about}\end{array}\right\} \begin{array}{l} A, \\ B, \\ C. \end{array} \\
b = BC & t'^2 = & \\
c = CA & t''^2 = &
\end{array}$$

If we put $x = 1$, l will be the length of the seconds' pendulum, and will be given by the above quadratic, which has all the coefficients known quantities, and has one of its roots irrelevant. If we prefer putting $l = c$ we shall have a quadratic in x^2, and x will be the time of strictly synchronous vibrations about A and B.

The statical method would not be less accurate, if instead of a knife-edge C for swinging we introduced near the centre of gravity a knife-edge pointing sideways for balancing the pendulum like the beam of a balance, and completed the balance of the pendulum in a horizontal position by a small riding weight. The weight and position of the rider combined with the weight of the whole pendulum gives the horizontal distance (*i.e.* distance in the direction of the length of the pendulum) of the centre of gravity from the knife-edge for balancing; and the distance of the knife-edge for balancing from either knife-edge for swinging (A or B) can be measured under the microscope, as the distance of A, B from one another.

However with care in the adjustment for synchronism, h can practically be determined nearly enough without the refinement of introducing a third knife-edge of either kind.

Nevertheless I stand up for the principle. In Kater's method of trusting to the adjustment of the slider, there are *two* unknown quantities that have to be determined by the two times of vibration, namely (α) the error of synchronism, and (β) the error in the assumed length of the seconds' pendulum, or, what is equivalent, the error in the time of vibration assumed to belong to a known length. Now by merely measuring an additional quantity that has such little influence on the result that we may deem its measured value mathematically accurate, we leave but *one* small unknown quantity β, in place of two α and β, to be determined by the observed times of vibration about A, B.

CAMBRIDGE, 9th Dec., 1873.

The only objection I have to an adjustment of the knife-edges in the actual Kater's pendulum is that it would render impossible a re-determination of his measure of the distance between the knife-edges; but I suppose it may safely be assumed that the error of his determination of the distance between the knife-edges is less than the error we should be liable to in assuming that that distance is now the same as it was when he worked with it in 1818. I don't like trusting further than we can possibly help to the effect of flexure, seeing that the pendulum is neither flexible like the fine wire of a ball pendulum nor rigid like the convertible pendulums constructed for the Russian Government, but something between. Therefore I should say, Let the knife-edges be made as truly perpendicular to the bar as may be.

CAMBRIDGE, 9th Dec., 1873.

From Capt. Heaviside's description there can be little doubt that the pendulum has received a bend since Kater worked with it. Possibly from lying for years together in a position in which it was subject to an amount of flexure quite trifling if temporary, the metal has acquired a permanent set. Since therefore, so far from our having evidence that the instrument, *quoad* the distance between the knife-edges, is now in the state in which Kater left it, the presumption is the other way, there can be no possible objection it seems to me to having the bar straightened. The formal sanction of the Council of the Royal Society cannot be obtained till Thursday week.

I should not propose to alter this pendulum by the introduction of a third knife-edge, whether for balancing or for swinging, as h (the distance of the centre of gravity from one of the knife-edges) can be obtained with abundant accuracy without altering the pendulum or incurring any expense worth mentioning. I should say

1. Trace a line (a scratch) on the side (*i.e.* face) of the bar, perpendicular to its length, and a *little* to the light-end side of the centre of gravity.

2. Rest the pendulum on an edge placed exactly under the traced line, the plane of the bar being vertical, and the bob (which tends to preponderate) resting on a support adjustable by a screw. Also put a prop under the heavy end, not quite touching the pendulum. This is merely to guard against the risk of a tumble when the rider (to be mentioned) is put on.

3. Adjust the traced line to verticality (on the edge of the bar to horizontality) by means of the screw.

4. Put on a small riding weight, and shift it till the pendulum just begins to move towards the prop.

This will give the distance of the centre of gravity to the right of the traced line.

5. Get a small metal piece with a line traced on it perpendicular to its base, to serve the purpose of a carpenter's square. Attach it temporarily to the bar so that the line traced on it cuts the line traced on the bar, and is in the middle of the pendulum.

6. Measure the distance from either knife-edge to the line on the square just as you would the distance between the two knife-edges.

From data given by Kater in the *Phil. Trans.* for 1818, I have calculated that supposing the non-adjustment to synchronism to amount to one vibration per diem, an error of *one-tenth* of an inch in the determination of h (an enormous allowance) would entail an error of only 0.000034 inch in the length of the seconds' pendulum. In other cases it would vary as the product of the non-adjustment for synchronism by the error in h; so that if there were even 10 vibrations per diem difference between the two knife-edges, and if h were determined to the $\frac{1}{34}$th of an inch instead of $\frac{1}{10}$th only, the error in the length of the seconds' pendulum consequent on the error in h would be only the $\frac{1}{10000}$th of an inch. The subsequent alteration of the tail pieces would probably render the error even less than this.

P.S. I was forgetting to mention that in speaking of the matter to Prof. Miller he much objected to the projecting knife-edges in the Kater construction, and thought there ought to be quite a short knife-edge embedded in the bar, the agate-topped support being like [bent round], which he contemplated passing in through a *lateral* hole, which would also serve as a window for measurement under the microscope of the distance between the knife-edges.

13, ASHLEY PLACE, *Dec.* 10*th*, 1873.

DEAR STOKES,

I have received yours of yesterday's date, and have sent it to Kew, where it will be carefully copied, with the aid of Captain Heaviside, and placed with your previous letters.

In your summary of the services, and consequent exposures to which Kater's apparatus has been subjected, you have omitted to notice the experiments which were made with it by myself at

Greenwich, of which an account is to be found in the *Phil. Trans.* for 1831; which is in fact, I believe, the only record of experiments to determine the Length of the Seconds' Pendulum at the Greenwich Observatory: and consequently in a locality now accessible. I refer, of course, to a determination by direct experiment, with Kater's original apparatus. I can well remember, however, that on that occasion, after the experiments were concluded, the Bar was replaced in its original case, with great care to prevent any liability to strain. It may indeed have suffered injury, by reason of its own weight, when resting, or appearing to rest on the fittings designed to receive it, either after its return to the Royal Society's house, or during its removal to Kew; but on its arrival at Kew it was removed from the wooden case, and suspended, in a glazed compartment, where it was suspended in a vertical position, its weight being borne by two loops of strong woollen material, passing round the upper knife-edge, from which it hung freely, in the direction of its length; there being no other supports. I think that your explanation of the cause of the injury is a very probable one; viz. a permanent set occasioned by its resting so many years in a case in which there may have been a liability to strain.

This forms another circumstance, however, which tends to render the construction of a new convertible Pendulum desirable; and if a new one is to be made, a reconsideration of its specialties may be quite proper, and indeed necessary. The locality for the experiments, (Kew Physical Observatory) appears to be in all respects most suitable: and the comparison will be immediate with two Russian Pendulums which are at Kew for the express purpose of comparison with the British Apparatus, and have already been subjected to the necessary experiments by Capt. Heaviside who is in all respects a most competent experimentalist. Supposing authority to be *now* given for the construction of a convertible pendulum embodying your suggestions, it would scarcely be ready before February, when the frequency of the River Fogs will have passed, and transits will become again available. But an authorisation of the expenses will be indispensable, as we have no funds available for the construction of new apparatus. Captain Heaviside will have careful and well practised assistants in the Observing Staff of the Establishment; and every aid may be assured on the part of the Committee.

The instrumental means of measuring the length of the seconds' pendulum, when the distance between the knife-edges has been determined experimentally,—viz. a Cathetometer, and a facsimile of our National Scale verified by Sheepshanks, are in readiness at Kew : where only authority to proceed is awaited for. This should proceed from the Royal Society. The chief expense is likely to be that of a new bar with its suitable knife-edges arranged under your guidance. Adie will probably be found the best person to employ for this purpose. But the primary step is the authority of the Royal Society for the proceeding.

Always sincerely yours,

EDWARD SABINE.

Dec. 19th, 1873.

I had the pleasure of seeing Capt. Heaviside a day or two ago : he encourages me to hope that the so-called permanent set in Kater's convertible pendulum may be removable by a skilled artist. I understand that it is situated in a part of the bar, near the heavy circular weight ; where there may have been a deficiency of suitable support in the box in which the pendulum was kept between 1830, when I employed it at Greenwich and the date of its removal to Kew, when there was no longer room for it, the instrument, in Burlington House. At Kew, as I have already mentioned to you, it has been suspended in a vertical position, the upper knife-edge resting on woollen loop supports, and enclosed in a chamber having a glass front.

I still think, however, that the apparatus which supplies the test of the fundamental unit of the British system of Measures (should unforeseen circumstances at any future time require a reference to such a test) ought to be in duplicate, the two being preserved in different localities. Yours and Professor Miller's suggestion of a new convertible pendulum, essentially the same in principle and use, but possibly even more simple in construction, seems to me well worthy of adoption : thus furnishing in a highly advantageous manner, the double apparatus required. We may hope that the results derived would be identical :—if not so, of course further investigation would be required.

I would suggest that no time should be lost in the construction of such a new convertible pendulum, under the authority of the Royal Society, and under the superintendence of yourself and Prof. Miller, if you should be willing to undertake such an important service to Science and to the country : and that advantage should be taken of the presence of so highly reliable an experimentalist as Capt. Heaviside, for the experiments to be made with it, as well as with Kater's original convertible pendulum. This would however require some prolongation of Capt. Heaviside's stay in England, which on a due representation from the Royal Society would, I have little doubt, be sanctioned.

CAMBRIDGE, 23 Dec., 1873.

My dear Sir Edward [Sabine]

I am two letters in your debt, but though I did not answer them I did not neglect them. I had a slight attack of bilious fever last week, not enough to keep me in bed, even for breakfast, but enough to keep me from doing more than a little work. I am thankful to say I am quite well now.

I was unable to be present at the meeting of the Council of the Royal Society on Thursday, but I sent in a written application for a grant of money to enable me to have a new pendulum constructed if I saw my way to the construction and employment of it, and the grant was made, so that as far as funds are concerned I am in a condition to proceed. I have also been looking into the literature of the subject, and have written some letters to Captain Heaviside. So that I have not been quite idle, notwithstanding my indisposition.

In reply to a proposal to Captain Heaviside to swing an additional pendulum of a different construction, he writes

"But when I have finished what I am now doing I shall have been away from India much longer than Colonel Walker intended, and I am sure he would not wish me to take up any new work of the kind."

Of course it is not for me to urge Captain Heaviside to act against his sense of duty to his superior officer. I dare say leave might be obtained from Colonel Walker for him to stay a little longer for sufficient reason. But should I be justified in asking for such leave?

If I could say, "such and such is a construction decidedly preferable to Kater's, and if carried out would lead to a more accurate, or to a more trustworthy result," then I would be prepared to urge on Colonel Walker the importance of allowing such further leave of absence to Captain Heaviside as would be requisite to complete the work in the best manner. But as the matter stands I am not prepared to take the responsibility of addressing Colonel Walker to prolong Captain Heaviside's leave of absence. As it is, he is engaged to swing the original Kater pendulum and two pendulums of the Russian pattern. These have all acquired a recognized status in science, but all beyond in the way of convertible pendulums, or indeed pendulums of any kind intended for absolute determination (with the exception of the ball pendulum of Borda and Biot, and the differential ball pendulum of Bessel) is I may say untrodden ground. I can suggest one or two changes in the Kater which *seem to me* to be improvements; but then I have had no *practical* experience in the matter, for the very light pendulums I swung at Kew were for the investigation of a special point, and don't count. And if I were actually to carry out the plans which might seem to me improvements, I know not what practical hitches I might encounter. I have been not a little deterred by a paper of Baily's in the *Philosophical Magazine* for February, 1829 (Vol. v. p. 97). Baily tried among other things a plain bar with 4 knife-edges A, B, C, D, the pendulum being rendered convertible as to A, C, and also as to B, D. He found for the length of the seconds' pendulum

by A, C 39·1386 inches.
by B, D 39·1307   „

The difference ·0079 is not far short of the $\frac{1}{100}$th of an inch, and would correspond to an error of 8·72 seconds a day in time. This cannot be accounted for by the neglect of the inertia of the air; for the error thence resulting would be almost rigorously the same for the pair A, C as for the pair B, D. The observations were all self-consistent. I can only conjecture that the discrepancy arose from the state of the knife-edges, or of one of them. Of course an observer like Baily would take good care that the knife-edges were *apparently* in good order.

In the face of discrepancies like this, which so careful an observer as Baily could not account for, were I to say to Colonel Walker "I am of opinion that such and such changes in the Kater pendulum would be an improvement: I recommend a pendulum as so constructed for adoption, and would advise you to prolong Captain Heaviside's leave of absence that he may use it," I should be disposed to apply to myself the saying "Fools rush on where angels fear to tread."

THE INDIAN PENDULUM AND GRAVITY SURVEYS.

[The early preoccupation of Sir George Stokes with the reduction of the existing classical pendulum observations, in connexion with his great memoir on the resistance offered by viscosity of the air, and the illumination which his memoirs on Clairaut's Theorem and the Figure of the Earth threw on the connexion between the form of the sea level and the distribution of gravity, naturally constituted him throughout his life the first British authority on the principles of all geodetic operations. The pendulum observations of the Great Indian Trigonometrical Survey, the results of which have taken so prominent a position in the Science of Geodesy, thus occupied a large share of his attention, both officially at the Royal Society and in the way of private discussion with the directors of the operations.

A large correspondence exists, first with General Sabine, and later with General J. T. Walker, Col. Herschel, Capt. Heaviside, and other officers of the Indian Survey,—the influence of which is to be traced in the official Memoirs of that service and in various papers in the *Phil. Trans.* and elsewhere. The selection of letters here reproduced has been made mainly with a view to the scientific value of their contents, whether as containing striking original remarks, or as giving an accurate account in ordinary language of the results that are to be found in more abstruse and technical form in Sir George Stokes' published papers.

It has been felt that any attempt at condensation in addition to omission would be inadvisable. Where similar principles are

expounded in other connexions, as for example, the comparison of
the pendulums of the British and American surveys, or the best type
of instruments for the Australian survey now in progress, the
order and treatment of the topics has been sufficiently different to
justify complete reproduction.

The Indian Pendulum Observations were instituted in 1865, and
full details of the whole of the work, up to date, were given in 1879
in a thick quarto volume, ranking as Vol. v. of the " Account of the
Operations of the Great Trigonometrical Survey of India." The
Preface and Contents, pp. i—lxii, by General J. T. Walker, R.E.,
Surveyor-General of India, narrate the inception, organisation,
and progress, of the enterprise.

At the end of his Preface, p. xlv, General Walker remarks:
" In conclusion I must acknowledge my great indebtedness to
Professor Stokes for the valuable and cordial assistance which he
has rendered, both to Captain Heaviside and to myself, on several
occasions, as is abundantly testified by the extracts from his letters
which the present volume contains."

The most recent authentic account of the operations of the
Indian Pendulum Survey, and the results brought to light by its
recent revision, will be found in the extracts from Col. Burrard's
Phil. Trans. Memoir, reprinted *infra*, p. 297.

Some preparation for an undertaking of this kind had been
made as early as 1826—30 by Col. Everest: but the Kater
pendulums which he obtained were never used, and the idea
seems to have dropped out of mind until it was revived in 1864
at the suggestion of Sir Edward Sabine, President of the Royal
Society. To confirm his own views Sir Edward obtained written
opinions from many of the most eminent British authorities on
Geodesy, which are reprinted in the Preface to the " Account of
the Operations...," pp. iv—ix. The following letter from Prof.
Stokes appears on pp. viii—ix.]

CAMBRIDGE, *June 22nd*, 1864.

DEAR GENERAL SABINE,

In reply to your letter of the 9th instant enclosing copy
of correspondence relating to proposed pendulum observations in
connexion with the Great Arc of Meridian, and stating that the
Council of the Royal Society would be glad of an expression of

my opinion as to the importance of the experiments, and as to the mode in which Colonel Walker proposes to carry them out, I would make the following remarks.

The experiments may be viewed either (1) as supplementary to the survey of the arc; or (2) as affording independent information on the Earth's figure, and on the cause and amount of local variations in the intensity of gravity.

First. It is needless to refer to the great care and thought which have been bestowed on the Grand Indian Survey, or the cost of the operation from first to last. The result is a scientific achievement worthy of the nation, and the Great Arc takes its place in the foremost rank among those on which we depend for our knowledge of the Figure of the Earth. But the results of geodetic operations of this kind are, from their very nature, beset by one source of uncertainty, that arising from local variations in the direction of the force of gravity. The numerical calculations executed by Archdeacon Pratt have shown that the amount of disturbance due to the mountains and high table-land to the north of India is much more serious than might perhaps have been anticipated; but, from a comparison with the results of the Survey, it appears that some hidden source of compensation must exist, causing the effect of the mountains to be much less than would be indicated by their mere external form. These considerations, referring especially to the Indian Arc, combined with the uncertainties as to local attraction which apply to any arc, render it highly desirable to apply an independent check to the amount and character of these disturbances, if we have the means of doing so at a cost which is trifling compared with the whole expense of the Survey. Now such a check is offered by pendulum observations, when the results obtained at various stations are combined. The pendulum no doubt indicates only the *vertical* component of the disturbing force, whereas it is the *horizontal component in the plane of the meridian* that affects the measures of arcs. At any one station, of course, a horizontal disturbance may exist without a vertical disturbance, and *vice versâ*; but in a *system* of stations disturbances of the one kind must necessarily be accompanied by disturbances of the other kind. Indeed it is theoretically possible from the vertical disturbances supposed known *actually to calculate* the horizontal disturbances, and that without assuming anything beyond the law of universal gravitation. Actually to carry this

18

out would probably require observations to be made at stations more numerous than can be thought of; but the fact of its possibility shows how severe a check pendulum observations are capable of exercising on the results of geodetic operations.

Secondly. The figure of the Earth, as you are well aware, admits of being determined by pendulum observations independently of measures of arcs. For this purpose it is not necessary that the stations should be connected in series, in which respect, as in many others, pendulum observations have a great advantage in respect of facility over measures of arcs. At the stations of an arc of meridian we have the latitudes and elevations already determined, which saves part of the labour, especially as regards the elevation in the case of an inland station. The Earth is indeed already well studded with stations at which gravity has been accurately determined, in effecting which result your own labours occupy a most prominent place. In India, however, few determinations of gravity of first-rate character have as yet been made; and besides, the stations at which gravity has hitherto been measured elsewhere are mostly situated on islands or coasts, and it would be interesting to have a good series of inland stations for comparison. Furthermore, your own observations appear to show that the observed irregular variations of gravity, which are superposed on the grand variation from the poles to the equator, are connected with the character of the formations underlying the stations; so that pendulum observations may be expected to throw light on the geology of a country.

The last point on which you requested my opinion referred to the mode in which Colonel Walker proposes to carry out these observations. As you have such great practical experience in this matter, and I have none, my opinion is of no value compared with your own. I think Colonel Walker has done wisely in leaving a good margin for stations to be chosen according as the results obtained at the principal stations may appear to make desirable.

If the nature of the country admits of it, I think it would be well to observe gravity at one station some way north of Kaliána, the northern extremity of the Arc. My reason is this. Any irregular excess or depth of matter north of that station would affect the direction of gravity, and consequently the astronomical latitude of the station, but might be situated too nearly in a horizontal direction from the station to have any sensible influence on the intensity of gravity *at the station itself.* Such excess or defect, however, if it existed, would make itself felt on the intensity of gravity at stations further north, and consequently more nearly over the region in which it occurred. The latitude and elevation of any such station would have to be ascertained; but an approximate determination, such as could be made with small portable instruments, would be quite sufficient. On account of the peculiar importance of Kaliána as being one of the extreme stations of the Arc, I think the additional trouble which might be involved in observing at a station not included in the Survey would be well bestowed.

Believe me, Yours very truly,

G. G. STOKES.

To Col. J. T. WALKER, R.E.

CAMBRIDGE, 25 *July*, 1872.

I must sincerely apologise to you for having left your letter relating to the experiments of the late Captain Basevi so long unanswered. I saw it was a letter which would require a great deal of consideration, and being pressed with business at the time I placed it with other letters of importance, and to tell you the truth the subject went out of my head till I came across your letter.

As you do not specify the contrary I presume the temperature and pressure corrections investigated by Captain Basevi were the *whole* corrections, not the residues after the hydrostatical correction for buoyancy and the temperature correction for expansion of the metal of which the pendulum was composed had been applied.

*Many years ago I investigated the problem of the resistance

---

* The remainder of the letter is published in the Survey Report, p. [75], after the following remarks: " In his letter of the 28th March [1871] Captain Basevi expressed a wish that the results of his experimental investigations of the dynamical influence of the air on the motion of his pendulums should be submitted to Professor Stokes, whose theoretical investigations on the same subject have already been alluded to in the present chapter. This was not done at the time, as there was then every prospect of Captain Basevi's returning to England shortly, when he would have an opportunity of consulting Professor Stokes in person. But after these expectations had been frustrated by his death, Colonel Walker—on whom the

of a fluid to a pendulum, taking into account the internal friction of the fluid itself. A very tough problem it was, but I succeeded in obtaining the solution in the case of a sphere, and in that of a cylindrical rod of which each element was treated as an element of an infinite cylinder vibrating with the same linear velocity and without change of direction of the axis.

I will send you by book post a copy of my paper if you have not got it, but I think you have.

In the case of the sphere the correction to the time of vibration due to the pressure of the air came out in finite terms. The time is the same as if the force got by considering the hydrostatic buoyancy acted on a mass equal to that of the sphere + k times the fluid displaced, where (eq. 148)

$$k = \frac{1}{2} + \frac{9}{2a}\sqrt{\frac{\tau}{2\pi} \cdot \frac{\mu}{\rho}},$$

where a is the radius, τ the time of vibration, ρ the density, and μ a constant, which may be called the coefficient of friction of the fluid, and which may be supposed to depend on the density and temperature, or rather pressure and temperature of the fluid.

For a cylindrical rod, k comes out a complicated transcendental function of $\frac{1}{a}\sqrt{\mu'\tau}$ or of $\frac{1}{a}\sqrt{\frac{\mu\tau}{\rho}}$. If a be not too small, or $\mu'\left(=\frac{\mu}{\rho}\right)$ too large, k may be expressed (eq. 113) by a series according to ascending powers of $\frac{1}{a}\sqrt{\frac{\mu\tau}{\rho}}$, which is at first convergent, and of which the first term is 1, so that the first term of $\mathfrak{n}\,(= 1 + k)$ is 2. Supposing therefore a and τ given, the effect of the air on the time will vary as

$$\rho\left\{2 + A\left(\frac{\mu}{\rho}\right)^{\frac{1}{2}} + C\left(\frac{\mu}{\rho}\right)^{\frac{3}{2}} + D\left(\frac{\mu}{\rho}\right)^{2} + \ldots\right\}.$$

(The terms involving odd integral powers of μ/ρ vanish.) When a is small or μ/ρ large, the expression for the transcendental function is divergent from the first, and we must use a descending series which is always convergent.

business of providing for the completion of the work had devolved, as an almost sacred duty—wrote to Professor Stokes and placed all the facts before him, and was favoured with a most valuable and interesting reply, dated 25th July 1872, from which the following extracts are taken."

Since then in the case of a cylindrical rod we meet with a very complicated transcendental function, we may infer with almost certainty that for a pendulum like an invariable pendulum, of no simple geometrical figure, the law expressing Basevi's c, as a function of p and t, is no simple one discoverable by experiment. A simple relation, approximately holding good, may however connect itself with a single term, of leading importance, in the expression for k or n. Such a term may be the second, involving $\sqrt{\mu/\rho}$ or—multiplying by the ρ outside all—involving $\sqrt{\mu\rho}$. Indeed this is the only term which would occur in the case of an infinite vibrating plane, and an element of a thin flat pendulum may roughly be treated as an element of such a plane.

We conclude then from theory,

(1), that the actual expression for c in an invariable pendulum, as a function of p and t, is complicated in the highest degree.

(2), that it is likely that a term of the simple form Constant $\times \sqrt{\mu\rho}$ may be of leading influence.

First, suppose the temperature constant. For this case Basevi found that the variable part of c seemed to vary nearly as $\sqrt{p}$, or which is the same thing, as $\sqrt{\rho}$, since t is constant. If then the leading part of c varies theoretically as $\sqrt{\mu\rho}$ and experimentally as $\sqrt{\rho}$, we must infer that μ is constant, or that the constant of friction is the same at all densities.

This is precisely Maxwell's result (*Phil. Trans.* 1866, p. 249), obtained theoretically from his theory of gases, and experimentally by his experiments on the slow oscillations of disks in their own plane.

The constant term in the expression for k multiplied by the mass of displaced fluid would vanish for an infinitely thin plane pendulum vibrating in its own plane. The pendulum employed was I suppose thin, and therefore this term would be very small.

As theory points out that the relation between c and p is a very complicated one, I doubt if there be any use in going beyond the leading term in attempting from the observations to express c in terms of p, unless it be for the purpose of obtaining an empirical formula which may be useful in reductions.

Secondly, consider the dependence of the leading term on t. As Basevi's expression for c vanishes, whatever be t, when $p = 0$,

I conclude that c is the residue after application of the correction for the expansion of the metal.

The theoretical term which has been conjectured to be the leading term varies as $\sqrt{\mu\rho}$. Now according to Maxwell (*Phil. Trans.* for 1867, p. 83), μ varies as $T + t$ when ρ is constant. This would therefore make the theoretical term vary as $\sqrt{\rho(T+t)}$. But according to Basevi the leading term varies experimentally as $\sqrt{p(T+t)}$ or as $\sqrt{\rho}(T+t)$. To make the two agree it would be necessary to suppose μ to vary as $(T+t)^2$ *.

On the other hand it is to be remembered that a function known to be excessively complicated is merely conjectured to have a leading term of the form $C\sqrt{\mu\rho}$, it being merely known that terms of this form, though they may not be the most important terms, occur for a sphere and a cylinder, and that for an infinite plane the sole term is of this form.

To Col. J. T. WALKER, R.E.

CAMBRIDGE, *9th April*, 1874.

MY DEAR SIR,

I have carefully considered your letter of the 9th of March and have made some calculations with reference to it in connexion with my paper on pendulums in the *Camb. Phil. Trans.*, and I have formed definite opinions with reference to the questions you propounded to me.

1.   †If f be the correcting factor for temperature, no doubt f may be expressed as $l + d$, where l is the part due to linear expansion of the metal, and d the part due to the thermal variation of the dynamical effect of the air. But can we throw on d the observed difference between the cold series and the 2nd

---

* See footnote to next page.

† The first part of this letter, from † to †, is published with slight omissions in the Survey Report, p. [81], after the following introduction:—"This excess, as regards the determinations anterior to Captain Basevi's swings at Dehra, has already been noticed in Section 9. The swings at Dehra, as reduced by Captain Basevi, give a mean value, for both pendulums, of $F = \cdot 4435$, which corresponds to $f = \cdot 000,010,31$, and exceeds the mean value by direct measurement by $\cdot 000,000,58$ or nearly 6 *per cent*. Professor Stokes was asked whether he thought the difference might be due, in any measure, to the dynamical effect of the air...."

hot series plotted in your figure? (I leave the less regular hot series No. 1 for the present out of consideration.) It seems to me well-nigh certain that we cannot. For

(1) The difference amounts to as much as 56 per cent. of the whole dynamical effect of the air. Now the thermal variation may be expected to be a fraction of the whole dynamical effect comparable with the fraction which the variation of the temperature t is of the whole $(T + t)$ absolute temperature. Now the variation of t is about 50° and $T + t$ is about 500, so this fraction is only $\frac{1}{10}$th.

(2) Theory shows that the dynamical effect is composed in part of a term independent of the velocity and varying as the density ρ, and therefore as $\dfrac{p}{T + t}$. In truth this term would be but small for a flat pendulum vibrating edgeways, and would vanish for a mere lamina. Still as far as it goes it implies an *increase* in the number of vibrations when the temperature rises, the variables being taken as pressure and temperature, not density and temperature.

Theory shows that in the case of an unlimited fluid the only term besides the above in the case of a vibrating sphere, and the leading term besides the above in the case of a not very thin cylindrical rod vibrating in a gas in which the viscosity is not extreme (as it appears to become for *very* low pressures), varies as $\sqrt{\mu\rho}$. Now Maxwell has shown * that μ varies as $T + t$ when ρ is constant, though ρ varies, when $T + t$ is constant. Hence $\sqrt{\mu\rho}$ varies as $\sqrt{(T + t)}\,\rho$ or as $\sqrt{p}$, which is independent of t. Omit for the present very low pressures for which the gas becomes very viscous (for if μ be constant when ρ varies, μ' or μ/ρ becomes large, and μ', not μ, determines the character of the gas, viz. whether it is to be thought of as like treacle, or like sulphuric ether, as it is called. I mean ordinary ethyllic ether). Then theory would lead us to expect that there would be little or no thermic change of dynamic effect; possibly even a minute *acceleration* for the higher temperature consequent upon our choice of pressure rather than density for the 2nd variable.

(3) Everything leads us to think that the dynamic effect of the air vanishes with the density. and consequently with the

* This conclusion, used here and elsewhere, has since been modified; for air $\mu \propto (T + t)^n$ approximately, where $n = \cdot 754$. Cf. Rayleigh, *Scientific Papers*, vol. iv. pp. 458, 482.

pressure. Hence if we take the pressure as variable and suppose two
curves drawn to represent the true dynamic effects at two constant
temperatures respectively, the abscissa being the pressure and the
ordinate the number of vibrations per diem, less a constant, *the
two curves ought to pass through the same point in the axis of
ordinates*, coming suppose something like this [omitted]. Now
assuredly your curves for 52° and 100° cannot be supposed to
meet in the axis of y. For the greater part of their course
they are almost exactly parallel.

(4) To show this more clearly I replotted your curves
measuring above and below the mean curve represented by a
straight line. In this way the parallelism is still more striking,
and there can I think be no reasonable doubt that the difference
between the two curves, at least as to its main part, is due to something *independent of the pressure* and *therefore* not referable to the
dynamic effect of the air. I assume that the firmness of the
supports of the pendulum was not affected by the high temperature. That being the case I see nothing left but to assume as
Basevi did that the expansion of the metal had been somewhat
under-estimated. The difference may be above the mere casual
errors of microscopic reading in the measures of expansion taken
at Dehra. But query, might not a small portion of heat from the
pendulum bath have been communicated by radiation to the piece
of matter (a stone slab, I suppose) on which the microscopes were
mounted ? This would give the expansion a little too small.

†The effect of humidity which comes first into consideration is
that it diminishes the correction for buoyancy, moist air being
lighter than dry. I don't know whether humidity was allowed
for in correcting for buoyancy, or the air was taken at a mean
state. I am almost disposed to think the latter, as nothing is
said about this correction, and Basevi did not observe the
humidity. If this correction were not made, its introduction
would go a certain way towards clearing up the remaining
anomalies, though not, I believe, far enough. The curve for the
1st series (hot) might thus be brought down towards the 2nd by
0·1 or possibly 0·2 of a vibration—right as far as it goes but going
only a little way. Similarly at Mussorie, in comparison of the hot
moist and cold dry swings, the hot swings being, relatively to the
other set, over-corrected for buoyancy, the two results would be
brought a little too near together. The diminution of viscosity

with moisture goes so far, observe, in the right direction, but it seems hardly credible that it should go far enough. Maxwell (*Phil. Trans.* 1866, p. 257), makes damp air over water at 70° F. about $\frac{1}{60}$th part less viscous than dry air at the same temperature. It is satisfactory that the mean of the coefficients found for pendulum No. 4 under somewhat ˙ opposite circumstances at Mussorie and Kaliana agrees almost exactly with that obtained by the numerous and careful measures taken at Dehra.

The difficulty however is to make moisture go anything like far enough to account for the discrepancies.

You remark on a chief source of error in pendulum observations, variations of temperature in the course of an experiment. It seems to me that the chief way in which this will operate is what I will immediately explain, and if so that it may in great measure be *diminished* by making a very easy determination.

As illustrating the tardiness with which masses of metal take up the surrounding temperature, Prof. Miller mentioned to me a curious experiment, performed I think by a German. Two tuning forks were carefully adjusted to unison, or to give only very long beats, I forget which. It was found that if one were touched with the fingers, heat enough was communicated to derange the beats, and it took 24 hours for the touched fork to regain its normal state.

Now when a pendulum is swung and the temperature of the adjustment rises or falls, the thermometers will take up the temperature more quickly than the pendulums. Theory shows that if the changes be gradual the pendulum will lag behind the thermometers, as to its temperature, *by a constant interval T of time.* Thus if T be half an hour, the temperature of the pendulum at 9 a.m. will be that shown by the thermometers at 8.30 a.m. Moreover T may be determined by time observations alone, by heating the pendulum up to say 100° or 150° (the temperature not being required to be known), swinging it, and observing the coincidences as it cools. The results, reduced in a certain way, will give T.

I have written to Capt. Heaviside advising him to make this simple observation on one of the invariable pendulums. I believe he is still at Kew, but there has hardly been time for an answer.

I send you a question I set on this subject in the last Smith's Prize Examination*.

2. †Next as to the claims of the graphical method *versus* the formula for completing the reduction to a vacuum.

Theory shows that for a sphere vibrating in an unlimited fluid the dynamic effect of the fluid on the time consists of only two terms, one varying as the density, say $A\rho$, and the other varying as $\sqrt{\mu\rho}$, say $B\sqrt{\mu\rho}$. The coefficients A and B are known from theory. If therefore d be the dynamic effect we shall have

$$d = A'\frac{p}{T+t} + B'\sqrt{\mu_0(T+t)\frac{p}{T+t}} = A'\frac{p}{T+t} + B'\sqrt{\mu_0 p},$$

where A', B' are constants and μ_0 is the value of μ at a standard temperature ; μ_0 being a constant may be included in B', and now omitting the accents we may put

$$d = A\frac{p}{T+t} + B\sqrt{p}.$$

At low pressure $B\sqrt{p}$ is the important term, so that we have nearly

$$d = B\sqrt{p},$$

and the curve drawn in your system would be a parabola touching the axis of y. If we were sure that such was the form, or the general form, of the curve, there would be a considerable (comparatively speaking) increase in the number of vibrations in passing from a pressure of say 0·6 inch to zero.

But this rapid increase of vibrations on diminishing the pressure to zero is bound up with the constantly increasing

* *Math. and Phys. Papers*, vol. v. p. 358, question 7: it is substantially given above in the text. See p. 286 *infra*.

† This section of the letter appears in the Survey Report, pp. [86]–[88], after the following remarks : "Thus a difference of nearly half a vibration is met with in the reductions to a vacuum by the formula and by the graphical method, and the question arises, which of the two methods should be adopted? For the purposes of comparison with the results of previous operations with these and all similar pendulums, in other parts of the globe, the graphical would evidently be the best, as all former swings have been reduced on the assumption that the correction for the combined statical and dynamical effect of the air is proportional to the pressure, or in other words that the pressure curve is a straight line. But on other hand the formula might give the true correction, or a more exact correction than that given by the graphical projection. Col. Walker referred this matter also to Prof. Stokes, who replied as follows :—"

volume of air over which the disturbance extends, and which to a certain degree accompanies the sphere in its movement. It is the greatness of this volume which causes the passage from a perfect vacuum to even a low pressure to be so telling*. Now when the pendulum vibrates in a confined space, the increasing volume of air which, as the pressure diminishes, tends to accompany it, at last becomes greater than the walls of the chamber or vacuum apparatus admit of. The motion of the air then becomes altered, and passes in the limit to something quite different. In the limit the motion of the fluid is the same as if it had no inertia ; the resistance to the pendulum varies as the pendulum's velocity, and its effect is thrown from off the time on to the arc. The actual resistance is very much increased by the confinement, but at the same time its phase is altered so as agree with the phase of the pendulum's velocity. It follows alike from the discussion of the formulæ (59), (60) in my paper on pendulums, which relate to the case of a sphere vibrating within a concentric spherical envelope, and from general considerations, which last are applicable to a pendulum of any form vibrating in a confined space also of any form†, that the most important part of the effect of the time will be of the form $C\rho$ or $C' \dfrac{p}{T+t}$, the C' however being quite a different constant from A.

Hence the curve instead of touching the axis of y will cut it at a finite angle. There is therefore no such very rapid augmentation in the number of vibrations in passing from a low pressure down to actual zero, as Basevi's formula would give.

For a cylindrical rod vibrating in an unlimited fluid, and for a large μ and therefore a small m and small m, the k is given by the first of my formulæ (115). In this case the leading term is not even of the form $A \sqrt{\rho}$ but of a different form, agreeing however with the form in giving $dy/dx = \infty$ when $x = 0$, y being the number of vibrations per diem and x the pressure. But as I have already remarked, the presence of an envelope will alter the form of the function for very low pressures, giving a curve which will cut the axis of y at the very low pressures. And there is some reason to think that even before the pressure is low enough to cause such

---

* See correspondence with Lord Rayleigh, p. 104 *supra*.

† Cf. the reduction of Sir W. Crookes' observations, *Math. and Phys. Papers*, vol. v. p. 100.

an increase of μ' (or μ/ρ) as to bring into play the effect of the envelope, the motion to which the theoretical calculation applies would be unstable, and the fluid would really break into eddies and the parts first set in motion be left in the rear forming a " wake " to the moving pendulum. The effect of this again would be to throw the effect of the resistance from off the time on to the arc.

It appears therefore from theoretical considerations that the curve will cut the axis of y at a finite angle, though till the pressure gets very low the approximately parabolic form will probably come near the actual result. Hence if a be the point for zero pressure, the curve near a, instead of being a tangent to the axis of y, as abc, will be of the form ade, Fig. 1 ; indeed I think an additional bend as in Fig. 2 highly probable. Now if the

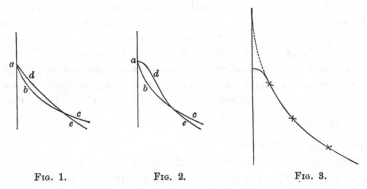

Fig. 1.   Fig. 2.   Fig. 3.

curve be of either of these forms, more especially that of Fig. 2, and if we draw an approximately parabolic curve nearly through the results of observations (represented by crosses, Fig. 3), and then on the assumption that it is approximately parabolic prolong it backwards (see dotted curve) till it meets the axis of y at a tangent, it is evident that we shall greatly exaggerate the ordinate of the real meeting point.

Seeing then that theory indicates that the curve *ought* to meet the axis of y at a finite angle, and that the curve obtained by plotting the experiments tends decidedly to do so, I have no hesitation in saying that I think the graphical result ought to be preferred to that obtained by extrapolating by means of a formula according to which $dy/dx = \infty$ when $x = 0$.

I have given up the idea of swinging the pendulum with five knife edges. First I found that Capt. Heaviside's time would be fully occupied with the work he had already in hand, and he would not feel justified in taking up this as part of the Indian work; and next I was deterred by a paper of Baily's in *Phil. Mag.* (vol. v. 1829, p. 97), who had obtained discordances between the results with one pair and another pair of knife edges in a pendulum furnished with four knife edges convertible in pairs, which he could not account for*. If this happened to so careful and experienced an experimentalist as Baily, it is clear that the subject is not yet in that state which would justify me in asking an officer of Survey (subject of course to your approval) to carry out the determination.

I am much obliged by your table of the values of l and k, which is very interesting to me, though I am not at present, at least, able to engage in the experiments.

April 10.—I had a note this morning from Capt. Heaviside answering some enquiries about dimensions of vacuum apparatus. He did not say anything about my suggestion to warm a pendulum and observe the coincidences as it cools, but he said he would write again.   I must post this before noon to catch this week's mail.

The formula you propose,

$$C = k\sqrt{p} - x_2,$$

would give $dC/dp = \infty$ for $p = 0$, and like the others would run up the vibrations greatly in passing from the lowest pressure observed to $p = 0$.

If it be wished to preserve the parabolic form of the curves and get rid of the infinite value of dy/dx for $x = 0$, it may be done by changing the formula into

$$p = \alpha C^2 + \beta C,$$

α, β being unknown constants. In comparing either this formula or Basevi's with observations there are really three (or in Basevi's formula with the correction for expansion four) constants to be determined, and not two only, for C is the *difference* between an observed result and the unknown result which would belong to $p = 0$. If C' be the difference between the correction for pressure p and any constant chosen for convenience, the formula takes this form, $p = \alpha'C'^2 + \beta'C' + \gamma'$. But I think the graphical method is to be preferred to mere empirical formulæ like these.

I am, dear Sir, Yours sincerely

* Cf. *supra*, p. 263.

CAMBRIDGE, 27*th April*, 1874.

It may be well that I should give you fuller information about the plan I proposed for ascertaining and allowing for the temperature-lagging of a pendulum*.

Let u be the temperature of a detached piece of metal, such as a pendulum, u_0 the temperature of the air and surrounding space; and suppose the difference $u - u_0$ small enough to allow us to regard the rate of cooling (or warming) of the pendulum as proportioned to the difference $u - u_0$ of the temperatures. Then

$$du = -q(u - u_0)\,dt,$$

t being the time, and q a certain constant, the reciprocal of a time; or

$$\frac{du}{dt} + qu = qu_0,$$

the integral of which equation is

$$u = qe^{-qt}\int e^{qt}\,u_0\,dt. \qquad (1)$$

Suppose the surrounding temperature u_0 to undergo periodic changes, say a diurnal fluctuation: then u_0 may be expressed by

$$u_0 = A_0 + A_1\sin(nt + \alpha) + A_2\sin(2nt + \beta),$$

where $n = 2\pi \div$ a day; or say

$$u_0 = A_0 + \Sigma A\sin(nt + \alpha),$$

n being changed to $2n$, $3n...$ in succession. This introduced into (1) gives

$$u = A_0 + \Sigma A\,\frac{q(q\sin nt - n\cos nt)}{q^2 + n^2}, \qquad (2)$$

the term Ce^{-qt} containing the arbitrary constant being omitted because it expresses what depends on the body's initial state, the effect of which I suppose to have had time to subside.

If the time the body would take, when placed in a space at a different temperature, to lose say half the difference of temperatures between itself and the surrounding space be but small compared with the period of fluctuation of u_0, or say a day,

---

* *Supra*, p. 282. Cf. also the discussion of the lagging of a marine barometer, *supra*, p. 217.

n will be very small compared with q, and neglecting n^2 in (2) we have

$$u = A + \Sigma A \left(\sin nt - \frac{n}{q} \cos nt \right) = A_0 + \Sigma A \sin n \left(t - \frac{1}{q} \right) \text{ nearly,}$$

or the temperature of the body at any time t is that of the surrounding space at a time earlier by $1/q$.

The formula (2) is true independently of the smallness of n, but as in the application of it n will be very small compared with q we may deduce the result directly from (1). Putting $u_0 = f(t)$ and integrating by parts we have

$$u = qe^{-qt} \left\{ \frac{1}{q} e^{qt} f(t) - \frac{1}{q^2} e^{qt} f'(t) + \frac{1}{q^3} e^{qt} f''(t) \ldots \right\},$$

or, as $f''(t), f'''(t) \ldots$ will be so small that we may neglect them, $u = f(t) - q^{-1} f'(t) = f(t - q^{-1})$ as before.

Suppose the body in question to be an invariable pendulum. Let it be warmed up to say 100° or 150° F. and swung in an apartment at the constant temperature u_0. If U be its initial temperature, its temperature at the time t will be

$$u_0 + (U - u_0) e^{-qt},$$

and as it cools its rate of vibration will go on increasing by a quantity proportional to e^{-qt}, so that the rate (γ) will be expressed in terms of the time by

$$\gamma = A - B e^{-qt},$$

A and B being two constants. These I treat as unknown, though A, expressing the rate for temperature u_0, will be known from previous work if the clock's rate be known. B might be expressed in terms of U, u_0, and the coefficient of expansion of the pendulum, but I suppose U unknown, and I will treat A as unknown also, which will not require a knowledge of the clock's rate.

Supposing the rate determined experimentally in terms of the time, by observing the coincidences, then taking differences of rate for equal intervals of time we have

$$\Delta \gamma = B (1 - e^{-q\Delta t}) e^{-qt},$$

$$\log \Delta \gamma = \log B + \log (1 - e^{-q\Delta t}) - Mqt,$$

M being the modulus $\cdot 43429 \ldots$, and as the interval Δt is supposed constant,

$$\log \Delta \gamma = \text{const.} - Mqt.$$

Let the results be plotted on paper, t being taken as abscissa and $\log \Delta \gamma$ as ordinate, and let a straight line be drawn by the eye through the observations, neglecting a possibly irregular portion at first, before everything has had time to settle down to its normal condition of change. The tangent of the inclination of the drawn line, divided by M, will give q the reciprocal of the lagging time τ.

Suppose now that in a regular swing (*i.e.* one taken in the course of regular work) $f(t)$ is the temperature of the air &c. around the pendulum, t being the time elapsed since the beginning of the swing; and let D be the duration of the swing. Then if $f(t)$ be reckoned as the temperature of the pendulum, the mean temperature will be reckoned as $\dfrac{1}{D}\displaystyle\int_0^D f(t)\,dt$ instead of

$$\frac{1}{D}\int_0^D f(t-\tau)\,dt = \frac{1}{D}\int_0^t \{f(t) - \tau f'(t)\ldots\}\,dt$$
$$= \frac{1}{D}\int_0^t f(t)\,dt - \frac{\tau}{D}\{f(D) - f(0)\},$$

neglecting the terms $\tfrac{1}{2}\tau^2 f''(t)\ldots$ in the expansion. Hence the correction for lagging will be

$$- \frac{\tau}{D}\{\text{temperature at end} - \text{temperature at beginning}\},$$

and depends therefore on the initial and final temperatures, which are sure to have been recorded, and not on the law of change in the interim.

I have tacitly supposed that the recorded temperature was that of the air &c. around the pendulum. But Capt. Heaviside informs me that the thermometers were placed with their bulbs in contact with a "dummy" which was presumed to have as nearly as possible the temperature of the actual pendulum. He further tells me that it has been found at the Standards Office that a thermometer not merely placed against a bar, but having its bulb protected by a brass shield, marks very accurately the temperature of the bar, and suggests the desirability of such protection in any future experiments. That such should be the result is what might be reasonably expected, as the bulb would be thus placed within an envelope which, thanks to the good conduction of the metal, would be sensibly at the temperature of the bar. A thermometer merely attached to a bar, and not

protected, would I think mark a temperature (u, say) dividing
in a constant ratio n to $1 - n$ the interval $u - u_0$ so that $\dfrac{u - u_1}{u - u_0} = n$.
The correction to the mean temperature of a swing as found above
would thus have to be multiplied by the factor n, being thus
reduced to

$$- n \frac{\tau}{D} \{ f(D) - f(0) \}.$$

May 8th.—Shortly before the despatch of last week's mail
I received from Capt. Heaviside the result of a 1st swing with
a heated pendulum. It was tantalising to have the numbers in
my hands and not to be able to tell you anything of the results,
as the discussion and calculations would take some time, and
I have now a daily lecture. I have now discussed the 1st swing.
I have not yet tried the $\Delta\gamma$ method, fearing it might be too rough,
and wishing to obtain as soon as possible a trustworthy determination of τ. So I have employed another, which if more
laborious is more certain. I first calculate the number of vibrations from the beginning, and then deduct the number which
would have been performed at the mean rate during the swing.
The difference, which of course vanishes at beginning and end
of the swing, and which I will call $m - n$, is the quantity I have
to go upon in the determination of τ. $m - n$ amounted in this
swing to a little over 0·2 vibrations at a maximum. This quantity
is got from the reduced abstractions for 22 coincidences between
the first and last (for which two it of course vanishes). My
formula should fit this when R and q are properly determined,
and I find they agree within a mean error of 0·0040 vibrations.
The lagging time τ, relatively to the apartment, comes out almost
exactly 1 hour. The pendulum was swung in the vacuum apparatus,
but at full pressure.

[In the account, Survey Report, p. 278, of Capt. Heaviside's
operations at Kew with Kater's convertible pendulum, there is
a section as follows.]

Theoretical Considerations connected with the Position of the Centre of Gravity.

Both Kater and Sabine, in their determinations of the time
of vibration of this pendulum, made the vibrations about each

axis of suspension very nearly synchronous; the former in air, the latter at low pressures. Baily, however, in the *Philosophical Transactions* for 1832, page 470, remarks with regard to Sabine's observations:—

Now, perfect synchronism I consider unattainable; or, at all events, not worth the trouble it would cost to pursue it; since the small difference which arises, in these cases, will always enable us to apply the proper correction, from the known principles of the pendulum; and which are a more sure guide on such occasions than any partial determination of the correction from actual experiment, where, in these minute enquiries, the errors of observation are sure to baffle us in our object.

Captain Heaviside wrote to Professor Stokes for advice on the subject, and received letters in reply, from which the following are extracts:—

There is no use in making the vibrations about the two knife-edges more than very roughly synchronous*. Let t, t' be the times of vibration, h, h' the distances of the centre of gravity from the knife-edges, $l (= h + h')$ the measured length, x the time of vibration for a simple pendulum of length l; then, k being the radius of gyration about a parallel axis through the centre of gravity,

$$\frac{x^2}{l} = \frac{t^2}{k^2/h + h} = \frac{t'^2}{k^2/h' + h'} = \frac{ht^2 - h't'^2}{h^2 - h'^2},$$

and multiplying by $l = h + h'$,

$$x^2 = \frac{ht^2 - h't'^2}{h - h'} = t^2 + \frac{h'}{h - h'}(t^2 - t'^2).$$

Observe it is not the mean of the times t, t' that is to be taken for x. Now, it will be easy to measure h' nearly enough for insertion in the correction $\dfrac{h'}{h - h'}(t^2 - t'^2)$: let $\delta h'$ be the error in h' and therefore $-\delta h'$ in h (for $h + h' = l$ may be deemed exact); then the proportionate error in $\dfrac{h'}{h - h'}$ will be

$$\frac{\delta h'}{h'} + \frac{2\delta h'}{h - h'} = \frac{h + h'}{h - h'}\frac{\delta h'}{h'} = \epsilon \text{ say};$$

and t, t' must be sufficiently nearly equal to render $\epsilon (t - t')$ insensible.......

I feel satisfied that it is not only far less laborious but decidedly more accurate to make the final result depend on swings taken

---

* Cf. *supra*, pp. 256, 262.

with the slider in *one* position and with measurement of h, than on swings with *two* positions of the slider with the view of deducing from *mere* observation, without having recourse to the theory of the pendulum, what would be the time of vibration had the vibrations been strictly synchronous.......

I see that in the *Philosophical Transactions* for 1832, page 471, Baily has criticised Sabine's proceeding in swinging a convertible pendulum longer on knife-edge A than on knife-edge B (A is that when the bob is below). I don't agree with Baily in this.... For the small error δx, produced by changes δt, $\delta t'$ in t, t', we have $x\delta x = \dfrac{ht\delta t - h't'\delta t'}{h - h'}$; or dividing by x, t, t', regarded as equal,

$$\delta x = \frac{h}{h-h'}\delta t - \frac{h'}{h-h'}\delta t'.$$

Now h is about twice as great as h', therefore an error in t produces twice as great an error in x as does a similar error in t'. Therefore for a given expenditure of time you get a better result by giving more observing time to A than to B, than if you divide equally between the two.......

I have consulted Professor Miller, who stands among the very first men in the country for acquaintance with the best construction of instruments of precision, and he entirely confirms me in the advice, *viz.*, (1) don't care about the vibrations being strictly synchronous : (2) let the slider once fixed remain fixed.

After these communications Captain Heaviside decided to dispense with any very close approximation to synchronism, and to swing the pendulum with the moveable weight and the slider fixed in one position, and to determine the distance of the centre of gravity of the pendulum from each knife-edge.

It was further pointed out by Professor Stokes in a letter to Captain Heaviside, dated 19th of December, 1873, that if the centre of gravity of the pendulum is not in the plane defined by the knife-edges (supposed parallel) a further correction will be required :—

Suppose the knife-edges to be parallel to each other, and let the plane of the paper passing through G, the centre of gravity, be perpendicular to the knife-edges, and cut them in A and B. Let h, h' be the distances of A and B from G; l the measured length AB; x the distance of G from AB, which we may suppose to be measured horizontally or perpendicularly to AB indifferently, the whole correction being very small: then we have, to a first approximation,

$$h + h' = l + \frac{1}{2}\left(\frac{1}{h} + \frac{1}{h'}\right) x^2 = l + \frac{l x^2}{2 h h'},$$

writing l for $h + h'$ in the small term.

Now when the heavy end (say the B end) is above, in which position the side position of the centre of gravity is most telling in making the second knife-edge lie to one side of the vertical passing through the other, let D be the distance that A lies right or left of the vertical through B; then

$$x = \frac{h'}{l} D \text{ nearly, and } h + h' = l + \frac{1}{2} \frac{h'}{h} \frac{D^2}{l},$$

and $\frac{1}{2} \frac{h'}{h} \frac{D^2}{l}$ is the small correction to be added to l, to give what would be the distance between the knife-edges, supposing it were possible to shift one knife-edge to be in a plane passing through the other and the centre of gravity, without disturbing the distribution of the mass of the pendulum, and to place it at exactly the same distance from G as it is; in which case it would give exactly the same time of vibration as it does.

To form an idea of the magnitude of the correction suppose $D = \frac{1}{10}$ of an inch; $h'/h = \frac{1}{2}$, l for round numbers $= 40$ inches. Then

$$\text{correction} = \frac{1}{2} \frac{h'}{h} \frac{D^2}{l} = \frac{1}{4} \cdot \frac{1}{100} \cdot \frac{1}{40} = 0\cdot000{,}0625 \text{ of an inch.} \text{ We}$$

see that in practical cases the correction is so extremely small that we may regard it as capable of being made with absolute accuracy.

When making an examination of Kater's pendulum before commencing the observations, Captain Heaviside found that the bar of the pendulum had become bent near the heavy end. A reference was made on the subject to Professor Stokes, who came down to Kew and saw the pendulum, and at once concluded that it should be straightened before any swings were taken with it[*]. The Council of the Royal Society having granted the requisite permission, the pendulum was re-straightened by one of Mr Adie's mechanics at the Kew Observatory.

---

[*] Cf. *supra*, p. 265.

To Col. J. T. WALKER, R.E.

CAMBRIDGE, 4th June, 1874.

I have received your letter of May 7 to which I now reply.

*As to the effect of humidity on the correction for buoyancy, the coefficient of expansion for temperature may be taken the same for dry air as for moist. It is in the density of air under standard pressure and temperature that a difference is made according as the air is dry or moist. The density of vapour of water is only 0·622 that of dry air at the same pressure and temperature; and therefore if b be the barometric height and

---

* The first part of this letter is published in the Survey Report, p. 79, and is introduced by the following statement: "Captain Basevi attributed the difference between the first and second series of hot swings with pendulum No. 4, at Dehra, to the great humidity which had prevailed during the first series, and which he thought might have caused the number of vibrations in the former to be always greater than in the latter: see the quotations from his letter on page [72]. Now the possible influence of humidity has never hitherto been taken into consideration in the reduction of the swings of a pendulum; the temperature and pressure of the air have always been observed and allowed for, but it has not been customary to recognize and allow for the influence of humidity, nor to take any steps for measuring the amount of moisture which might be present in the air during the swings. At least no such steps were ever taken throughout the operations in India, and it is believed that they have never been taken elsewhere.

"The idea appears to have originated with Captain Basevi, and it received considerable *primâ facie* support from the circumstance that, if cognizance were taken of humidity, the results of the swings which were observed at Kaliána and at Mussoorie for the determination of the temperature coefficients, see Section 4, would be brought into closer accord with each other. At Kaliána the cold swings were taken in the damp month of January, when the average humidity, as measured at the nearest Meteorological Observatory, was 0·73, complete saturation being = 1; and the hot swings were taken in the following but much drier month of May, when the corresponding humidity was 0·22; thus the difference between the results from the hot and cold swings, and consequently the values of the coefficients for temperature, would be *diminished* if the observations were reduced to a common standard of humidity.

"Conversely, at Mussoorie the hot swings were taken shortly after the termination of the rainy season, when there must have been more—it is not known how much more—moisture in the air than was the case a month later when the cold swings were taken; here the difference between the hot and the cold swings, and consequently the magnitudes of the temperature coefficients, would be *increased* by corrections for humidity. Thus the results at the two stations would certainly be brought into closer accord by correcting for the variations in the humidity; but it remained to be seen whether any possible approximation would be sufficient to account for, or to materially reduce, the existing differences."

v the vapour pressure expressed by the corresponding height of mercury, the density of the air, and therefore the correction for buoyancy will be reduced by the vapour in the proportion of b to $b - \cdot 378\, v$.

Prof. Miller, in his paper on the construction of the new standard pound, has employed the most approved modern data for the weight of moist air. He gives for the ratio of the density of air to the maximum density of water (*Phil. Trans.* for 1856, p. 785)

$$\frac{\cdot 0012930693}{1 + 0\cdot 003656\, t}\, \frac{b - \cdot 378 v}{760}\left(1 - 1\cdot 32\,\frac{z}{r}\right)(1 - \cdot 0025659\cos 2\lambda),$$

where t is the temperature (centigrade) of the air; b, v are expressed in millimetres of mercury at $0°$ C., z is the height above the level of the sea, r the earth's radius, and λ the latitude of the place.

He refers to a document signed by Biot, Regnault, and Bianchi, according to which the pressure of vapour in an apartment not artificially heated was found to be $\frac{2}{3}$ of the maximum pressure due to the temperature. This was in Paris; I should think in India, in the rainy season at least, the proportion would probably be higher.

To take a rather extreme case as to the effect of vapour. Suppose the temperature $95°$, or $35°$ C. The maximum pressure at this temperature is $41\cdot 893$ mm., and supposing the air saturated the value of $\cdot 378\, v/b$ would be $\cdot 0208$. As $\cdot 0208 \times 30 = \cdot 624$, saturated air at $95°$ and 30 in. would weigh no more than perfectly dry air at $95°$ and $29\cdot 376$ barometric height. The moisture would cause a diminution of say 2 per cent. in the correction for buoyancy. As under average conditions the vapour pressure increases with the temperature, the employment of a slightly too high coefficient of expansion for temperature would pretty well neutralize the effect of the weight of moisture in correcting for buoyancy.

You ask for the formula for the hydromechanical effect of the air. There are but two forms of pendulum for which that has hitherto been determined.

Regarding the air as a perfect fluid (*i.e.* neglecting viscosity) we must in any case conceive a mass equal to that of k times the fluid displaced, added to that of the pendulum, increasing its inertia without increasing its weight. For a fluid (like air) of small density it will suffice to multiply the correction for buoyancy by $1 + k$.

As far as the correction for a perfect fluid is concerned, moisture will affect the whole (statical and dynamical) exactly in the proportion in which it affects the statical alone.

The value of k for a sphere is $\frac{1}{2}$ and for a cylindrical rod is 1. The only form for which k has been hitherto determined is that of an ellipsoid which was given by Green in a paper published in the *Edin. Trans.* (XIII. 54) and reprinted in the 8vo vol. of his Mathematical Papers (Macmillan and Co.). The Ellipsoid includes as particular cases the sphere ($a = b = c$), the cylindrical rod ($a = b$, $c = \infty$), the bob of a pendulum regarded as an oblate spheroid ($a = b$), a thin flat bar vibrating perpendicularly to its plane ($a = 0$, $c = \infty$).

For an oblate spheroid vibrating in the equatorial plane (Green's *Works*, p. 323)...

When b is very small compared with a this becomes $k = \frac{1}{4}\pi b/a$.

But the actual result is much affected by the viscosity of air. The only forms for which the pendulum has been worked out when viscosity is taken into account are those given in my paper which you have got, the sphere and the cylindrical rod. For a sphere I find (equation 52)......

For the cylindrical rod k is a complicated transcendental function of $a^2/\mu'\tau$ which I was obliged to tabulate. When the cylinder was very narrow I found *approximately*

$$k = 1 + \frac{2}{a}\sqrt{\frac{2\mu'\tau}{\pi}}.$$

For a not too slender pendulum, for one of the invariable pendulums for instance, vibrating in a time which for this purpose we may deem constant, we should have *approximately*

$$k = F + B\sqrt{\mu'},$$

where F and B are two constants which I am not able to determine from theory. The dynamical correction would therefore be approximately of this form

$$A\rho + B\sqrt{\mu\rho}.$$

According to Maxwell μ or $\mu'\rho$ is independent of ρ and varies as $\tau + t^*$, so that the dynamical correction is *approximately* of the form

$$A' \frac{p}{\tau + t} + B' \sqrt{p}.$$

When a pendulum is thin in a direction perpendicular to the plane of vibration A' is but small, and the dynamical correction ought to be nearly independent of the temperature,—especially within error of the approximate formula (judging from analogy to the cylindrical rod) varies with the temperature in an approximate way to the first term $\frac{A'p}{\tau + t}$. This agrees very well with Basevi's results.

I have given the solution of my Smith's Prize question in a letter which crossed yours.

Captain Heaviside and I have been getting out some definite results about lagging, which I have not time to mention by this mail. Besides they are still in progress. I hope to write again when they are further advanced.

The following extracts from a memoir of Lt.-Col. Burrard, R.E., F.R.S., *Phil. Trans.* 1905, will give a view of the present state of the Indian Pendulum Survey.

Between 1865 and 1873 observations were taken at 31 stations in India by Captains Basevi and Heaviside with the Royal Society's seconds pendulums. The results were published in vol. v. of the ' Account of the Operations of the Great Trigonometrical Survey of India,' and have been subsequently discussed by many authorities†....

The physical meaning of Basevi's pendulum results was for many years the subject of controversy. The deficiency of gravity which he had found to exist in Himalayan regions was attributed by some authorities to the elevation of the level surface above the surface of the mean spheroid, and by others to the defective density of the underlying crust; by the former the surface of the geoid was held to depart largely in certain places from that of the spheroid, and by the latter the two surfaces were assumed to be almost identical. In his ' Schwerkraft im Hochgebirge,' published in 1890, Professor Helmert gave a mathematical solution of the problem, and his writings have closed the controversy....

---

* See footnote, *supra*, p. 279.

† See *Phil. Trans.* A, vol. 186, 1895; Helmert's *Die Schwerkraft im Hochgebirge*; Helmert's *Höhere Geodäsie*; Clarke's *Geodesy*; Fisher's *Physics of the Earth's Crust*.

The peninsula of India is composed of archæan and volcanic rocks; the great age of the former and the great weight of the latter would lead us to expect a high value for g; that g should be abnormally small is, from a geological point of view, surprising....

After 1874 no pendulum observations were taken in India, but the deflection of the plumb-line continued to be determined in different parts of the country. By the year 1900 the astronomical latitude of 159 stations, the astronomical azimuth at 209, and the amplitude of 55 arcs of longitude had been observed, and thus a large amount of evidence relating to the *direction* of gravity had accumulated. A discussion of the data then available showed that it would be desirable to associate determinations of the *intensity* of the force of gravity with observations of the plumb-line, and in 1902 the Indian Government sanctioned the re-opening of pendulum observations and the purchase of a new apparatus of Von Sterneck's pattern....

The new apparatus was standardised at Kew and Greenwich in the autumn of 1903, and was taken to India by Major Lenox Conyngham in November of that year. Upon its arrival he thought it advisable to commence work at some of Basevi's stations. The accuracy of Basevi's results, as given in Tables [omitted], had been questioned by Professor Helmert in his report to the International Geodetic Conference of 1901. It had been there pointed out that the observer had had no means of measuring the flexure of the pendulum stand, that during his standardisation at Kew his pendulums had not been supported on the stand subsequently used in India but between a stone pillar and a wall, and that when he visited the high Himalayan station of Moré he had substituted a light portable stand for that belonging to the Royal Society's apparatus....

Major Lenox Conyngham's first station in India was Dehra Dún; his results there were astonishing, for they showed that Basevi's value was no less than 0·103 centim. too small. Lenox Conyngham then visited Calcutta, Bombay, Madras, and Mussooree. At Calcutta observations were rendered impossible by the ceaseless vibrations of the ground, which proved sufficient to cause the pendulums, if left suspended at rest, to oscillate visibly in a few minutes; this effect on the pendulums was produced in whatever plane the latter were swung. Lenox Conyngham had therefore to abandon Calcutta without obtaining any results; that he failed where Basevi had succeeded was probably due to the half-seconds pendulums of the new apparatus being more affected by earth-vibrations than the old seconds pendulums....

In figs. 2 and 3 [omitted] the deficiency underlying Dehra Dún (38) will be reduced by almost one-half if Lenox Conyngham's value be substituted for Basevi's. Similarly the height of Mussooree (41) in fig. 3 will be almost doubled.

In the near future Basevi's other stations will possibly be visited; it seems certain that his results will everywhere be found too small, that throughout fig. 2 the curve of deficiency will have to be raised, and that in fig. 3 the line of sea-level will have to be lowered.

From Lenox Conyngham's observations at Bombay and Dehra Dún, it appears that Basevi's and Heaviside's results are not in error by any constant quantity, and that the error of each will have to be separately determined. It is not easy to account for the variation in the magnitudes of their errors; their observations were taken with a care that it is difficult for us to equal; in assuming that flexure could be prevented by the employment of a rigid stand, the old observers were following the highest authorities of their time; the only faults that have been found with their work are such as would tend to produce constant error. That their errors vary so largely can only, I think, be explained on the supposition that the flexure of the wooden stand of the Royal Society's apparatus was influenced by temperature and humidity.

The idea that gravity is exceptionally weak throughout India as compared to Europe can no longer be upheld*; the so-called "marked negative variation" of many writers has been found to rest on erroneous data.

The theory of the compensation of the Himalayas has been based to a large extent on the old pendulum results at Mussooree and Moré. The sections in figs. 2 and 3 show that a hidden deficiency of matter underlies the station of Mussooree (41) equivalent to about three-fifths of the visible excess; Lenox Conyngham's recent result reduces this hidden deficiency to one-third only of the visible excess.

Figs. 2 and 3 might lead to the belief that the Himalayas at Moré (43) are almost *entirely* compensated. The height of the visible excess is 4696 metres, the depth of the ideal deficiency 4484 metres. But Lenox Conyngham has not visited Moré, and, as Basevi employed there a special and lighter stand, it is impossible to gauge the error introduced into his result by its flexure; we have lately gained some idea of the effects of the flexure of the Royal Society's heavy stand, and we can only suppose that the light Moré stand was less rigid. That the Himalayas at Moré are compensated to a considerable extent is certain; that the error due to flexure could have affected Basevi's result to the extent of

---

* No standard value of g has as yet been adopted by the International Geodetic Association. When the absolute values of gravity at European standard stations have been finally determined, it may be found that the values at Kew and Greenwich, which we are now accepting as our standards, are not themselves normal. Both Basevi's old and Lenox Conyngham's new values will then have to be corrected by a constant quantity.

22 seconds of time is out of the question. On the other hand, it is more than probable that the compensation, that does exist, lacks that completeness, which has hitherto been considered among its most remarkable features *.

To Colonel J. HERSCHEL, R.E., F.R.S.†

Lensfield Cottage, Cambridge.
21 Oct., 1885.

My dear Sir,

I do not know whether you have heard that the India Office sent your report on pendulums to the Royal Society, to consult the Society on account of the bulk of the MS. and the consequent expense of printing it. This was some time since; but as the vacation intervened, the Royal Society Council's Report has not yet been sent in to the India Office. However I have had the opportunity of looking into the MS., and that being the case I thought that I might as well mention three points which I noticed. Two of them are small matters, which I will first mention to clear them off.

I well recollect Sir Edward Sabine's mentioning to me in conversation that some clock, and I feel pretty sure it was the one he used in his pendulum observations, might be trusted to have the same rate for three or four hours that it had on the mean of the 24. He spoke of it as a rather exceptional merit in this clock. This shows that he was fully alive to that possible source of error, though I do not know whether he has expressly mentioned it in any of his published papers.

The second point is this. You write as if the effect of the air on the time of vibration depended on nothing else than the density of the air. This is not true, though the difference in practical cases would not be much. It would be true if air were a perfect fluid, but it is not. In the case of a sphere suspended by a wire which we may deem infinitely fine, and swinging as a pendulum once in a second nearly, if the correction for buoyancy be called

* Clarke's *Geodesy*, p. 350.
† The following letters were kindly communicated by Col. Herschel.

1, the additional correction for inertia, if air were a perfect fluid, would be 0·5, and the total correction 1·5 would vary simply as the density of the air.   But in consequence of the viscosity of air the total correction exceeds 1·5, by a quantity which depends on the radius of the sphere and on the time of vibration.   If the diameter be an inch and a half, and the time of vibration about one second, the total correction is about 1·85.   Now the viscosity of air increases with the temperature, according to a law which is not I think yet satisfactorily made out, so that for air of a given density the correction would be a little higher for a higher than for a lower temperature.

These two small matters being disposed of, I come to the principal point, which is your method of observing coincidences*.

At first sight I was disposed to regard either the common method or yours as practically perfect so far as accuracy is concerned, and your method seemed to offer great, or at least considerable, facilities in observation.   For absolute determinations it would require a small correction, which however could easily be made.   But further consideration led me to doubt its trust-worthiness, even supposing this correction made.

Suppose the clock pendulum and its mark or marks projected on the plane of motion of the free pendulum, by lines drawn through the centre of the object glass of the observing telescope if no lens be used, or let its image be formed by a Carlini lens, as may be preferred.   Let a be the semi-amplitude of horizontal excursion of the projection or image of the clock pendulum at the height of the mark, b that of the fiducial edge of the free pendulum. We have only to consider the relative positions of the mark and fiducial edge; and the difference of abscissae may be represented by

$$a \sin nt - b \sin (nt + \omega t + \alpha) + c,$$

where ω is a small quantity depending on the difference of rates of the two pendulums, α a constant, c a constant depending on the distance of the mark from the fiducial edge when both pendulums are at rest.   In actual observation, the quantity ω may be 300 times or so as small as n.   We may deem ω constant; for the small variation, depending on the slow variation of the minute correction

---

* This is printed for its interest in connexion with the theory of observations; but it turned out (*infra*) that Col. Herschel's work was not sensibly vitiated by any defect of this kind.

for the finite arc of vibration, is quite too small to need taking into account for our present purpose.

The above expression may conveniently be put under the form

$$\rho \sin (nt - \phi) + c$$

where
$$\rho = \sqrt{a^2 + b^2 - 2ab \cos \overline{\omega t + \alpha}}$$

$$\phi = \text{arc tan} \, \frac{b \sin \overline{\omega t + \alpha}}{a - b \cos \overline{\omega t + \alpha}} - \alpha.$$

In this form the relative abscissa is represented (making abstraction for the moment of the constant part) as a quantity varying according to the same law as a simple pendulum, and having the period of the clock pendulum, but subject to a secular variation of amplitude and phase. The march of the variable part is rudely represented in the figure below, where the abscissa is proportional to the time.

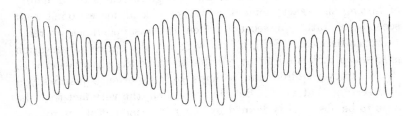

In the actual experiment, there would be probably about 300 of the subordinate periods, answering to two seconds, in one period of the joint fluctuation, answering to the interval between coincidences.

We are only concerned with the values of the relative abscissæ at the turning points; and if in the above figure we imagine them dotted, and the rest of the figure removed, we shall get two series of dots lying in two wavy curves which we may practically treat as continuous. The narrowest parts of the enclosed strip will answer to coincidences, the broadest to oppositions. As we are only concerned with the relative position, we may regard either pendulum, say the clock pendulum, as at rest, by superposing on both a motion equal and opposite to that of the clock pendulum. In the figure below, the curves represent the continuous curves in which the dots above mentioned lie, and the ordinate of the

horizontal line above the axis represents the place of the fiducial
edge of the mark observed.

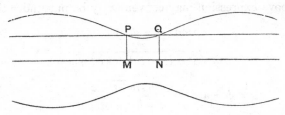

The intersections P, Q represent the critical events of disappearance and reappearance, and PM, QN being the ordinates,
the mean of the abscissæ of M and N is taken as the moment
of coincidence.

The method therefore labours under the very serious disadvantage of determining a time by observing the moment when
a certain quantity is a minimum, or in the case of an opposition
a maximum. The slowness of the change increases the difficulty
of hitting the exact moment, or rather second, for we don't split
seconds, or rather bi-seconds, when the disappearance or reappearance happens. Still, if this were all, the optical observation
is so sharp that it might be possible to make the observation with
the accuracy required, and perhaps your experience justifies you
in asserting that it is. But that is not all, the very fact that the
time to be observed is defined by being the moment of a minimum
introduces the risk of very serious disturbance from small disturbing
causes.

If we may assume a and b constant, we have for the minimum
of ρ the condition

$$\omega t + a = 0 \text{ or multiple of } 2\pi,$$

and the observation of the coincidences gives the true measure
of ω. But supposing a to remain constant, the time of minimum
will be affected by the variation of b; taking account of this
variation we have for the condition of maximum or minimum of ρ

$$(b - a \cos \Omega)\, db + \omega ab \sin \Omega . dt = 0,$$

where Ω is written for $\omega t + a$. Now the arc of the free pendulum
decreases sensibly in geometric progression, whence

$$b = \text{const. } e^{-qt}$$

and therefore $$db/b = -q\, dt;$$

and on substituting in the above equation we have for the condition of maximum or minimum

$$(b - a \cos \Omega) q - \omega a \sin \Omega = 0$$

or

$$a (q \cos \Omega + \omega \sin \Omega) = bq$$

or

$$\sin \left(\Omega + \tan^{-1} \frac{q}{\omega} \right) = \frac{q}{\sqrt{\omega^2 + q^2}} \frac{b}{a}.$$

Hence, unlike the former case the value of Ω depends on the ratio of b to a; and as that ratio is widely different near the beginning and near the end of an observation, the correction to Ω will not be the same at the end as it was at the beginning, and the total interval of time between the first and the last coincidence will not therefore be given correctly by taking the difference between the times of the observed coincidences.

As q is practically very much smaller than ω, we may without sensible error substitute for the above

$$\Omega = 2N\pi + \frac{q}{\omega} \left(\frac{b}{a} - 1 \right),$$

where N is the number of intervals of coincidence which have elapsed since the first coincidence was observed.

As the last term is constant, and therefore disappears in taking the difference, we may take $\frac{q}{\omega} \frac{b}{a}$, or in clock seconds $\frac{q}{\omega} \frac{b}{a} \frac{I}{2\pi}$ where I is the interval between coincidences, to be the correction to be applied additatively to the observed time of any coincidence. As the correction goes on decreasing with b, the total interval from first to last coincidence will be deemed too small.

On taking the numbers for q and ω from an observation of Sabine's with an invariable pendulum, given in the *Philosophical Transactions* for 1828, the error in the total interval which would have been produced by observing the coincidences in your manner came out about 15 seconds, and the corresponding error in the estimated number of vibrations per diem about 0·38. This no doubt is small, but still too large to neglect. In purely differential measurements taken by the same method, and with arc of as nearly as possible the same magnitude, the correction might no doubt pretty well be disregarded; but supposing you swung a pendulum at say Kew, and another observer, trusting to the safe keeping of the pendulum in the interim, swung the same at say New York, observing the coincidences by the ordinary method, an error would be introduced if it were assumed that the two results were comparable.

If this were all, the correction could easily be applied; and till the other day I thought that a might be assumed to be practically constant. But about four days ago a cause of possible error occurred to me that seems to me very serious.

The arc of the clock pendulum is determined on the one hand by the resistance and viscosity of the air, on the other by the periodic impulse given by the driving weight acting through the escapement. The weight has to overcome the friction &c. in the clock train. Now from a slight roughness, or slight dirtiness, or slight eccentricity, of an axle relatively to the wheel to which it belongs, it is quite possible that the amount of friction &c. may be different in different parts of the wheel as we go round. If this were the case in the early wheels, where the leverage of the friction would be comparatively speaking considerable, there might result a periodic variation of the portion of the driving force available for keeping up the arc of the pendulum, the period not being long. But as regards the possible vitiation of the place of minimum— I mean place with reference to the figure, in reality time—we are only concerned with the *rate of variation* of a. But if the period be not long the total variation of a may be too small to be observed directly against a scale, and yet its rate of variation may be sufficient to vitiate to a not insensible degree the rate of variation which comes in, in the observation of the coincidences by your method. Hence, even if the method might practically have been trusted had the clock pendulum been free, it is open to very grave doubts considering that the pendulum is connected with a train of wheels.

The common method is free from the objections which I have endeavoured to point out as belonging to yours, and is therefore, I think, more trustworthy. As regards the differential work which you have done, I do not suppose the objections I have mentioned operated to any serious degree, or the thing would have been found out in the course of the observations. Still I do not think that your method is one to be recommended for future use.

I am, dear Sir, yours very truly

CAMBRIDGE, 27 *Oct.*, 1885.

I got your letter this afternoon. I have read it carefully twice over. I quite understand your method, and had understood it before. I can picture it to my mind's eye, having been engaged once, for a fortnight, at Kew taking coincidences. The thing to be watched is much the same in your method and in the common. I am perfectly well aware that you do not observe a minimum as such, but observe a disappearance and a reappearance, as indeed I mentioned in my letter. I can quite understand that to a person using your method it might appear elegant and charming, I should be disposed to say seductive. Till you come to consider what it really means, it appears superior to the common method. It requires a correction, which however is very easily made, and supposing this made I am not prepared to say that it may not be a good method, though I think it hazardous, and to use it with assurance would require a good deal more experience.

I had noticed that the discs which gave decidedly larger intervals between disappearance and reappearance gave a mean slightly different from those which had a small interval. This I see you refer in your MS. to the decrease of the arc; and it would produce such an effect. This however I disregarded, thinking that where the interval between disappearance and reappearance was long enough to make the discrepancy worth noticing, the observation with that disc would be rejected. I assumed in fact the observation to be perfect in this respect; and it is such apparently perfect observations, where different discs would tell the same tale, that still require a correction, and *that* a considerably larger correction, in consequence of the decrement of arc operating in a different way.

P.S. I am rather disposed to think that your method would have been very good if you had chosen for observation discs which would have given an interval of 2 or 3 minutes between disappearance and reappearance, instead of only a few seconds. A correction, if required, might still have been made for the decrement of the arc, but in that case it would have been very small. I should say that being revolutionary you have not been revolutionary enough.

CAMBRIDGE, 29 Oct., 1885.

I worked out last night the correction to the observed time of coincidence arising from the diminution of the arc, supposing that the disappearance and reappearance were observed anywhere, and not merely very near a coincidence. In the particular case in which it was very near, the correction of course reduced itself to the former one. On putting it roughly into numbers, I got in a practical case a correction hardly amounting to 3 seconds, instead of 15 or so. I could not see any error, and on looking again at my former work, where the expression was reduced to number, I see that I omitted a division of 2π. This reconciles the two, and makes the correction much more minute than I supposed. It should still be included in an absolute determination; but even here the error arising from omitting it would be only about 0·06 vibration per diem.

CAMBRIDGE, 31 Oct., 1885.

In the common method of taking coincidences, the diaphragm used just takes in the tail piece of the free pendulum when that is at rest. I write of course with reference to the projections on a common plane; usually the free pendulum, the clock pendulum, and the diaphragm are so near to each other compared with the distance from whence they are viewed, that they are nearly enough in focus together; but a lens may be used if the observer pleases, or two if he is very particular, one to form an image of the clock pendulum to coincide with the free pendulum, and another to form an image of both coinciding with the diaphragm. Usually however the diaphragm is placed very near the free pendulum, and the latter lens at any rate is not required; and even the first mentioned lens is commonly dispensed with. These variations do not in any way affect the principle of the method, and for simplicity I will suppose the three are projected in viewing on to a common plane. The diaphragm as I have said just admits a view of the tail piece, and the tail piece just a little more than covers the mark. In the common method it is an essential condition that the amplitude of swing of the clock pendulum should exceed that of the free pendulum. This is not so in your method. In the common method the diaphragm is absolutely essential; in yours no diaphragm is wanted at all.

Let the figure [omitted] represent the left cheek of the screen and the left edge of the tail piece as it is advancing to the left, and on the point of squeezing (visually) to nothing the aperture between. Let the moment when the aperture is just squeezed to nothing be called the critical moment. Suppose a coincidence happened exactly at the critical moment. At that moment the mark would be invisible and a fortiori invisible as the two pendulums had been moving towards the cheek, since the clock pendulum swings through a larger arc, and therefore had been towards the tail piece like an express train towards an ordinary train moving in the same direction. But if the clock rate of vibration be a trifle greater than the free pendulum rate, after 2, 4, 6... seconds the mark will have advanced a little to the left at the critical moment, and will therefore at one of these even swings begin to appear.

Similarly as regards the journeys from left to right. In the first position of the mark it will be invisible at the critical moment, and a fortiori after it as the aperture is opened. But 2 seconds before coincidence the clock had been a trifle behind relatively to the free pendulum, and more after 4 &c., so that by going back some even number of seconds relatively to that first mentioned in speaking of this case, and consequently some odd number of seconds from the critical moment mentioned with reference to the first case, the mark would just be seen on first opening the aperture, before it had disappeared behind the tail piece.

Hence the order of things from some time before to some time after the coincidence would be:—Visibility in moving from left to right and invisibility in moving from right to left; cessation of visibility in moving from left to right, and continued invisibility in moving from right to left. Disappearance; invisibility both ways; invisibility in moving from left to right, and commencing visibility in moving from right to left. Reappearance; the two events necessarily take place at seconds of opposite parity.

Had the clock been losing on the free pendulum, the only difference is that the parity of the seconds at which disappearance and reappearance take place would have been reversed.

20—2

The critical condition which separates the *yes* from the *no* in this method is:—does the edge of the mark lie right or left of the common edge of cheek and tail piece at the critical moment? As the pendulums are moving most rapidly at the moment of observation, the difference of phase of the free and clock pendulums which accrues in 2 seconds tells as much as possible in deciding whether the answer shall be *yes* or *no*.

Your method is altogether different. In the course of 2 seconds, or one period, making abstraction for the moment of the slight change of relative phase in 2 seconds, the edge of the mark swings to and fro relatively to the edge of the tail piece, the amplitude of excursion slowly changes with the change of relative phase of the two pendulums. According to the position of the mark it may be visible all through, or invisible all through the period of 2 seconds, or may be partly visible and partly invisible. When it changes from partly visible to wholly invisible, you have a disappearance, when from wholly invisible to partly visible you have a reappearance. According to the position of the mark, the interval from disappearance to reappearance may be anything from nothing to the interval between coincidences. The mean of the two is taken as the moment of coincidence or opposition as the case may be.

This method is disadvantageous when the interval between disappearance and reappearance is very small, as the width of the portion exposed then changes very slowly from bi-second to bi-second. There is no need however to choose it small; and from the numbers you sent me in your last it appears that in some of your experiments it was not small. A small correction is required to the time of coincidence in consequence of the dying away of the arc of the free pendulum. I gave you this for the case in which the interval between disappearance and reappearance is small. I find the correction for the general case is

$$\frac{b - a\cos\left(\frac{I}{K}180°\right)}{a\sin\left(\frac{I}{K}180°\right)} \, qI \cdot \frac{K}{4\pi},$$

where b is the semi-amplitude of the free and a of the clock pendulum, I is the interval from disappearance to reappearance, K the interval between coincidences, and q such that $1 : e^{-qt}$

is the ratio in which the arc decreases in the time t. When I vanishes, the above expression takes the form 0/0 and its limiting value is the correction I gave before. The correction is so small, amounting to 3 seconds or so, that it need only be made very roughly. The trouble of making it need therefore hardly be considered. Provided a too small interval be avoided, and the small correction required on this method is introduced, I see nothing to object to in it. As so used, I believe that your method and the common method may both be regarded as perfect, the chief source of error being in other directions.

UNITED STATES GRAVITY SURVEY.

FROM PROF. C. S. PEIRCE.

ITHACA, N.Y.
Jan. 14, 1886.

MY DEAR SIR,

I desire to thank you very warmly for your letter of Dec. 28, which I received only a day or two ago.

We are ready to return the Kater pendulums; indeed, we should have done so a year ago, had I not hoped to experiment with the clock that came with them at the top and bottom of the Washington monument [550 feet high], but I have had no leisure to do so. I will see that the pendulums are sent back as soon as I return to Washington, in a few days.

They have been oscillated at various stations in the Pacific Ocean by Mr Edwin Smith, of the Coast Survey. You have, I suppose, received his report. The work was not under my direct control, but I exercised advisory supervision over it. Both the points you mention were attended to. At my instance, Mr Smith always *levelled up his knife-edge plane with a spirit-level*, and was not guided by touch.

The correction of the time of coincidence for decrement of arc was not, I believe, applied, as we wished to keep the work as homogeneous with Herschel's as we could. The swingings were all made at a constant pressure, and all began and ended at the same amplitudes, and the same spots on the disc were used at all corresponding coincidences, so that no sensible errors in the ratios of gravity could result from this cause. After seeing Colonel Herschel's plan of observing coincidences, I was led to paste a scale of millimetres on the clock-pendulum, and observe co-incidences of a needle-point on the gravity-pendulum with the divisions of this scale. I always corrected the time of coincidence for decrement of arc by means of a diagram which solved the trigonometrical problem; besides, I chose the divisions to be observed upon so that when the amplitude of the gravity-pendulum was greater than that of the clock-pendulum the correction was reduced to zero, and was a minimum in the other case. But I have now given up coincidences, and gone back to registering transits on a chronograph; for that is generally less trouble, and it has been fully shown 1st, that the probable error of a single transit, including that of reading the sheet, is only 0·014 second at the beginning of a swinging, and about half as large again at the end; 2nd, that the absolute personal equation is nil at the beginning, and at the end of 30,000 oscillations [*even if I do not change the power of the eye-piece, as I always do*] is only 0·022 second; 3rd, that the probable accidental variation of the personal equation is equal at the beginning and end, and amounts only to 0·011 second approximately.

You are very indulgent to my unfortunate paper on the effect of the flexure of a pendulum, which was printed from rough notes, without my seeing the proof-sheets. In the reversible pendulums now in use, the staff is considerably cut away to allow the passage of the tongue on which the knife-edge rests. My note was based on the idea that the flexure mainly occurred at the cut at the heavy end. I suppose the sagging to be measured when the heavy end is held firmly horizontal, simply because the distance between the knife-edges being a unit of linear measure the linear sagging measured there is instantly converted into angular sagging. But the most intelligible formulae, I suppose, are obtained by denoting by a single letter the harmonic mean of the angles of sagging when the upper and the lower pieces, respectively, are held firmly horizontal. I will put these formulae in a postscript, but since you have been so good as to take an interest in the matter, I will endeavour shortly to send you some actual numbers treated by a theory more conformed to the facts of the case.

My regular determinations of gravity have been executed with pendulums that are at once *Invariable and Reversible.* They are measured at the beginning and end of each campaign, and when any accident occurs, which is known by the alteration in the difference of periods in the two positions, they are sent back to be remeasured. I ordinarily swing two simultaneously on two independent supports, so as to multiply data, and get a check on the clock-rates. One pendulum measures a metre between the knife-edges, the other a yard, so that they cannot influence one another, and as the model of the two instruments is the same, I get a very good value for the ratio between the yard and the metre as an incidental result. I expect that my determinations will teach the geologists many things, and I wish, if it turns out practicable to do so, to draw lines of equal gravity residual over the map of the United States and its surroundings, —Canada, Mexico, the West Indies, Bermuda, the Eastern Pacific Islands—so as to find out something of the shape of the geoid within our borders. I have already, in addition to what I have published, a line of stations running from Montreal [Lat. $45\frac{1}{2}$] south to Key West [Lat. 24], and another from Cambridge, Mass., west to Madison, Wis., the difference of longitude being an hour and a quarter. Unfortunately the prospect is that Congress, influenced by a President not at all favorable to scientific expenditures, will refuse me another appropriation, and thus bring my work to an abrupt end. Yet Congress is generally ready to listen to the voice of Science, when it recognizes it, and any word of approval which you might choose to write to me or any other scientific man, would without doubt be of good service to my work. With great admiration and respect, yours very faithfully,

C. S. PEIRCE.

To PROF. C. S. PEIRCE.

LONDON,
19 *May*, 1886.

DEAR SIR,

A long time has elapsed since a letter passed between us. I am afraid the fault is mine; for though I had been expecting to hear from you, you had still better reason to expect a letter from me.

The question of making some additional pendulum observations has been entertained by a Committee of the Royal Society, and the committee has just received a vote of money which will suffice for a portion of the work.

Our first object is to make a fresh comparison between gravity at the American stations and gravity at Kew and Greenwich. For this purpose we should want the pendulums back again in the same condition in which they have been employed in the American work. But

1.   We do not want to interrupt work now going on, if such there be, in America which depends on the continued employment of the same pendulums.

2.   It may perhaps be advisable before returning the pendulums to swing them afresh at Washington or New York or both; I should be glad to know your opinion as to whether such a re-determination would be desirable.

3.   For fear of the knife-edges getting rusted on the voyage it might perhaps be well to run a drop of melted paraffin along them before packing. It could easily be cleaned off on arrival with clean warm wash-leather without risk of injuring the edges.

4.   Considering the way in which packages are liable to be tossed about on ship-board, it would seem hardly safe to send them as ordinary packages unless they were specially entrusted to the care of the captain, and placed in his cabin. We should be ready to make him a suitable remuneration for his trouble, and to send someone to the port of landing on being apprized beforehand of the name of the vessel and the time when she was due.

CAMBRIDGE,
17th *June*, 1886.

On receiving your telegram from New York I consulted one of the leading members of the Pendulum Committee of the Royal Society (though indeed I felt sure that we should all be extremely glad if your proposal could be carried out) and then I telegraphed in reply and also wrote. I had no address but "New York" for your present quarters, so I addressed the telegram and letter simply to you to New York. Shortly after that I got a message from the post office that the telegram was not delivered because the name was not known. I cannot at present tell whether my letter reached you. Your last letter was dated from Ithaca, but the telegram was dated from New York.

I said I was sure the Committee would be extremely glad if you could come over to swing the pendulums at Kew and perhaps also at Greenwich. It would seem indeed to be an almost necessary supplement to the American work. Obviously the value of this would be greatly enhanced if a thoroughly trustworthy connexion were established between American and European stations. A connexion has indeed been made by Col. Herschel but he employed a new method of adjusting the agate planes, which *à priori* would seem to be rather precarious, and at any rate has not yet received that degree of examination which would seem requisite to inspire full confidence.

It would be most desirable as you have been swinging the pendulums at American stations that you should yourself complete the connexion by swinging them in this country, as that would ensure the strict comparability of the observations on the two sides of the Atlantic Ocean.

It has been suggested to me that if I put "U.S. Coast and Geodetic Survey" on the letter it would be pretty sure to find you.

I am, dear Sir, yours very faithfully,

G. G. STOKES.

DISTURBANCE OF GRAVITY BY CONTINENTS.

To Prof. E. HULL.

26 *Feb.*, 1887.

I am afraid you will think I have forgotten the question you asked me in your letter. It is not so, but I thought that before answering it I would look into Professor Suess's book if I could find it....But if I had found it I do not know that I should have been much the wiser, for as you say he gives no numerical data as to the dimensions of the supposed continent, nor does he specify what the continent actually is if he is dealing with a real, not an ideal, continent.

In a paper " On the Variation of Gravity at the Surface of the
Earth*," which I wrote long ago, and which is published in the
Transactions of the Cambridge Philosophical Society, and in my
Collected Papers, of which as yet only two volumes have appeared,
I showed that the effect of a continent would be to make a slight
apparent diminution of gravity in continental stations as compared
with detached oceanic islands.   It operates in this way:—that
the attraction of the land causes the surface of the sea level,
the level surface, that is, which would be determined by a system
of geodetic levelling carried from the coast inwards, to stand
higher from the centre of the earth than it would have done had
the place of the continent been occupied by ocean.   The raising
of the sea level is greatest inland, but it is quite sensible, and
even important, at the coast itself of the continent.

How much the raising amounts to, depends of course on
the dimensions you attribute to the continent and the height
you give it above the undisturbed level of the sea.   To take a
numerical example, I suppose the case of a circular island, or
continent, whichever you please to call it, 1000 miles in diameter,
and elevated a quarter of a mile above the sea level.   I suppose
the depth of the ocean in which this island is supposed to be
placed to be 2 miles.   I make the usual suppositions as to the
average density of the rocks &c., in the neighbourhood of the
earth's surface, and as to the mean density of the earth, which is
fairly well ascertained.   I find the elevation of the sea level in the
interior of the island or continent (Australia), a good way from
the coast, to be about 400 feet.   Of course in a great continent it
might be considerably greater.   This would cause an apparent
diminution of gravity in continental stations, I mean of course in
gravity as reduced by the usual methods to the level of the sea.
In the first place in reducing to the level of the sea we leave out
of consideration the attraction of the stratum of earth between
the actual sea level and what the sea level would have been if
the continent had been away.   As far as this goes, corrected
gravity ought to appear too great.   But in the second place, in
reducing to the level of the sea we reduce to a point further
from the centre of the earth than we should have done if the
sea level had been unchanged, and therefore in correcting we

* *Cambridge Phil. Trans.* vol. viii. (1849), pp. 672—695.   [*Math. and Phys.
Papers*, vol. ii. (p. 155).]

don't add enough to bring it up to what it would have been if ocean had been beneath us instead of land. On this account therefore gravity should appear too small, I mean reduced gravity. The two effects on apparent gravity are antagonistic, but the second is the stronger, so that on the whole gravity ought *caeteris paribus* to appear a little less on continents than on detached islands. Sabine and Airy have pointed out that such appears to be the result of observation, but so far as I know I was the first to point out that such a result ought to follow from the attraction of a continent, by disturbing the sea level.

Far inland, the thing could only be tested in those cases where the sea level has been accurately determined by geodetic operations. We could not accordingly throw much light on the question by means of pendulum observations in Thibet.

I am, dear Sir, yours very truly

P.S. You will find Airy's discussion in his article " Figure of the Earth " in the *Encyclopaedia Metropolitana*.

An elevation of 400 feet, even if there had been no intervening attraction to reduce the resulting diminution of gravity, would only alter the number of vibrations per diem of a seconds' pendulum by about one and a half. Of course apart from disturbance the difference in the number per diem at two stations on the same parallel of latitude would be nil, and therefore the small difference of 1·5 would be infinity times that. I do not know what the term of comparison used by Fischer may be.

To E. F. J. LOVE.

7 QUEEN'S PARADE, BATH,
6 *August*, 1891.

DEAR MR LOVE,

I have come here for a week or two, and have brought your letter of June 19 with me to answer it from this house, which is that of one of my sisters.

You do not say expressly, but I take for granted that in the contemplated gravity survey [of Australia] you mean to use invariable pendulums, not Kater's pendulum or some other form available for absolute determinations. It is generally, I think, allowed that for determining the *variation* of gravity from place to place the results obtained with invariable pendulums are the more accurate. The series of determinations would be rendered absolute by transporting the pendulums to some station where gravity has been well determined absolutely and swinging them there. It will suffice if the station last mentioned be one for which gravity is accurately known absolutely by comparison, by means of invariable pendulums used by previous observers, with some other station where gravity had been determined absolutely.

At least two pendulums just like each other should be used, in order that any accidental derangement of a pendulum may be detected. Sabine said to me in conversation that there ought to be three, as that would enable you, in the event of any derangement taking place in course of transit or handling, to tell which pendulum it was that had got altered. If you had only two, and one got slightly deranged, you could only tell which it was by going back to one of the stations where they had been previously used and swinging them afresh. However I think two only have as a rule been all that have been used in gravity surveys, and I believe that with care in packing, transporting, and handling, such derangements are not likely to occur.

Before fixing on the form we must answer the question, Is the correction for the resistance going to be determined by calculation or by experiment ?

(*a*) If by calculation, we are restricted to forms for which it is possible to effect the calculation. The pendulum might be a plain cylindrical rod, or such a rod with a sphere at the end. In an invariable pendulum, soundness of casting would not be of any very great moment, the observations being strictly differential. If a rod be used, I should prefer the ends being made hemispherical, or thereabouts. The exact form is of no particular consequence, for, for a small portion of the rod near the end the calculation cannot be effected, whether the rod be left plain, or formed into a hemisphere. The calculation for a sphere would not apply to a hemisphere joined on to a cylinder. But the part of the resistance which depends on what is near the end of the rod forms only a small fraction of the whole, and if we are obliged to have recourse to estimation for that small portion, the uncertainty thence arising can only be very small, since the rod is supposed to be but narrow for its length. The alternative is to adopt the form mentioned by General Walker, a cylindrical rod with a sphere at the end. I do not think there is much to choose between these two forms. I think the latter would keep up its oscillations somewhat longer, and the former would have to be about 5 feet long (for a seconds' pendulum) which might perhaps be a little inconveniently long. I do not know however that this would be any serious inconvenience.

As to the calculation, it is to be remarked that the numerical value of the index of friction given in my paper is much too low. This arises in great measure from my having corrected for the residual air in Baily's swings at reduced pressure (about one inch of mercury) on the supposition (which seemed to be conformable to the single experiment that Sabine had made on the subject) that the coefficient of viscosity, the μ of my paper, varies as the density. We now know that Maxwell's law, according to which it is independent of the density, is very accurately true in experiment. The true coefficient is now well known for air. I have not got here books of reference, but towards the end of a paper of Tomlinson's in the *Phil. Trans.*, in which he treats of the viscosity of air, you will find collected the numerical results of various observers, himself included. The effect of reducing, in my paper, by a law as to the relation between viscosity and density, now known not to be the law of nature, was to exaggerate the effect of reduction of pressure, in other words, to underestimate the effect of the residual air, and therefore, in equating the theoretical expression for the difference between 30 inches pressure and 1 inch in the observed result, to bring out a coefficient which was decidedly too small.

The adoption however of the true law, though it raises considerably the coefficient of viscosity as got from Baily's experiments, leaves it still too small. I do not see how to account for this except on the supposition that the motion of the pendulums was not small enough to allow of a strict application of the formulas of my paper. I have remarked in my paper (at least with reference to a suspending wire, and the same would of course be true generally) that the effect of the formation of eddies would be to tend to throw the effect of the resistance from off the time on to the arc. Whether any sensible part of the resistance is due to the formation of eddies, may be tested by seeing whether the arc of vibration decreases strictly in geometric progression as the time increases in arithmetic. I examined in this way some of Sabine's experiments in the *Phil. Trans.*, and some of Bessel's experiments with the long and short pendulums. Plotting a curve with the time and the log. arc for coordinates, it came a straight line for the long pendulum, but the curve, though very nearly a straight line for the shorter pendulums, had a sensible though slight curvature. I forget whether this is mentioned in my paper. I rather think not. Yet on second thoughts it seems to me as if I recollected making a slight allusion to it. It appears therefore that with the amplitude of vibration usual in pendulum experiments, at least in the early portion of the swing, the effect of eddies is not wholly insensible, and therefore it may well be that the formula in which the motion is assumed to be small enough to be regular may not be quite applicable to the actual experiments. However beyond the discrepancy between the calculated and observed decrement of the arc of vibration, which I have mentioned in my paper, and also the decrement being not quite strictly in geometric progression, there was nothing to indicate that the formulas were in any way in fault, so very good seemed the agreement between theory and observation, until it was shown that the correction of the assumed law as to the relation between the viscosity and the density still left the numerical value of the index of friction as determined by the pendulum experiments slightly too small.

But in merely differential observations, such as those carried on with invariable pendulums, I think any uncertainty of this kind would be quite insensible, provided that care were taken that the observations should be strictly differential or very nearly so. Hence if you wish to connect a group of Australian stations with Indian stations, it is a perfectly open question whether you shall choose a pressure of say 28 inches for the Australian set or a pressure of say 4 inches (or whatever the usual Indian pressure for India was). That is on the supposition, which I gather from your letter is intended, that you mean to construct new pendulums. The pendulums being different, the two series cannot be connected till the new pendulums are swung at one of the old stations, unless you are ready to trust to a reference of each series to

an absolute determination belonging to it.   But in either case
if the higher pressure were thought the more convenient for
the Australian stations, and it were not wished to trust to a
correction for so great a difference of pressure as 28 and 4 inches,
it would merely be requisite to swing the invariable pendulums
twice in succession at the reference station, once at the higher
pressure, to connect with the Australian series, and once at a low
pressure to connect with the Indian series or with the absolute
determination as the case may be.   If the vacuum apparatus be
not quite staunch, as I fear may prove to be the case, it might be
better, as a matter of convenience and indeed accuracy, and as
Major Herschel has proposed, to use the vacuum apparatus only
for securing a constant pressure of say 27 or 28 inches, except of
course for the one set of swings at low pressure taken at the
station of reference.   However, much would depend on the
condition of the vacuum apparatus.

I will mention here, lest I should forget it, that it is well to
allow an observation (whether by a single swing, as may be done
in vacuo, or by a succession of swings does not much matter) to
extend over 24 hours, or if that be inconvenient at least from
dark to dark, through day or night as may be chosen, so as to
rate the clock by transits for the interval of time over which the
observations extend.   For you cannot trust a clock, even though
the rate from day to day be very uniform, to be quite exempt
from a diurnal inequality of rate.

(b)   Suppose now that we prefer to depend on experiment for
the correction for the air.   Then we may choose our form of
pendulum as we please.   That usually employed has the bar
somewhat thin, in a fore and aft direction, so as to be slightly
flexible.   Without this there is I believe some difficulty in
ensuring that the weight shall bear well on *both* agate planes,
so as not to run the risk of turning slightly about a vertical
axis to and fro as it swings.   I recollect some one (Sabine I
think) telling me that some one, I forget who, did not like the
flexibility, and proposed to make the pendulum stiff, and Kater
(I think it was) said, 'He'll find it will not do.'

The form having been chosen, we have to find the correction
for the air experimentally.   This demands the use of a vacuum
apparatus.   I think the most convenient plan would be to get a
fac-simile of the pendulum made of wood.   The resistance of the

air depends only on the form and time of vibration of the pendulum, I mean supposing the state of the air given; and these would be the same for the actual pendulum and for the wooden model. By avoiding specially dense wood, we might easily get the model 10 or 12 times as light as the actual pendulum, and the effect of the air on arc and time would be magnified 10 or 12 times. The whole time of the swing would be reduced in the same proportion; but this would not signify as regards having a shorter interval by which to divide any error of observation in the initial or final coincidence, for the method of coincidences is so exact that it may be deemed perfect; that is to say any error from this would be swallowed up by much larger errors from other sources; and that being the case there is a great saving of time in using the model, besides which we are less exposed to errors from variations in the clock's rate, changes of temperature, &c. However, the actual pendulum might of course be used, and probably in any case an observation or two would be taken with this for control. And besides the saving of time in taking the observations, resulting from using a wooden model, the possibility of taking swings in close proximity, merely allowing an interval sufficient to allow the disturbance of temperature consequent on the exhaustion or admission of air to subside, would I think be conducive to accuracy as securing a more near identity in the rate of the clock on the occasion of the two swings that are to be compared.

I do not see that Reynolds's paper, important though it be in its proper place, has much to say to our problem. I need not say that there is nothing new in the formation of eddies; it is a matter of common observation; and in an earlier paper I pointed out that the kind of resistance of a fluid to a moving solid that we usually consider is due to eddies. When I wrote my paper on pendulums I was keenly alive to the possibility of the formation of eddies; and indeed if I rightly recollect I was rather surprised that in the pendulum motion eddies seemed to have so little effect as manifested by the close accordance between theory and observation, though the theoretical results were calculated on the supposition that the motion was regular, *i.e.* non-eddying. The novel results obtained by Reynolds are in my mind the two following:

(1) The conditions of dynamical similarity which enable us to

pass from one system to another dynamically as well as geometrically similar, provided the motion be of the regular kind, may also be applied to the discrimination between regular and eddying motion; so that if in one case we are on the border between the two, we will also be on the border in all other cases which are dynamically as well as geometrically similar.

(2) The said conditions may further be applied to the mean effects in two cases which are dynamically as well as geometrically similar, and which lie on the irregular side of the limit.

As an example of what we could infer from Reynolds's work: suppose we had ascertained that a sphere of 1 inch diameter vibrating in a time from rest to rest of 1 second, the amplitude of vibration being 1·5 inch, was just in a transitional state from regular to eddying motion, then we could infer from Reynolds's principle that if we had a sphere vibrating in a time of 0·64 sec. instead of 1 sec., then if the diameter were 0·8 inch and the amplitude 1·2 inch, this also would be in the transitional state.

You mention the corrections for pressure and temperature. The latter depends partly on the expansion of the metal, partly on the effect of temperature in altering the state of the air, and therewith the correction on account of the air. I am not sure whether or not you meant to include the effect of the expansion of the metal.

If it is intended to keep the two parts separate, I suppose it is meant to calculate the part due to the expansion of the metal from the linear expansion either ascertained or assumed as known for the kind of metal employed. As to the air, the correction for buoyancy, and that portion of the correction for inertia which would form the whole if there were no viscosity, both one and the other vary as the density, and therefore in a known manner as regards the temperature. The rest of the correction for inertia depends in a more complicated manner on the temperature *. The whole of this residue for a sphere, and the first term and most important part of it for a not too narrow cylindrical rod, varies as $\sqrt{\mu\rho}$; ρ of course varies inversely as $1 + \alpha\theta$, θ the temperature, α the coefficient of expansion, but μ increases as the temperature rises, according to what law does not appear to be known for certain. I think experiments on transpiration gave it about as $(1 + \alpha\theta)^{0.7}$, but I am away from books of reference, and I do not remember exactly.

* See *supra*, p. 279.

If the temperature correction should be determined directly as a whole, *i.e.* effect on metal and on air together, by swinging the pendulum at two pretty widely separated temperatures, it is to be remembered that as it is made up of two different parts (effect on metal and effect on air) following different laws, so it will not be available unless some one element (say the pressure) be kept constant. The experiment would involve the use of an apartment artificially heated in an equable manner*, unless we were content to wait all the time from one season to another, say summer to winter. The temperature correction so determined for a pressure of say 28 inches would not apply (on account of that part of it which depends on the air) to a pressure of say 3 inches. It would seem to be best to correct as best may be for that part which is due to the air, so as to get the part which is due to the expansion of the metal. I think the effect of the air can be got well by using a wooden model, and altering the observed effect in the ratio of Mh to $M'h'$; and M/M' can be got by weighing, and h/h' by balancing separately the model and actual pendulum on their edges.

I shall be happy to reply to further enquiries.

Yours very truly.

* P.S. With a wooden model the effect of the air is so much larger, the time of swinging so much shorter, and the expansion of the material by heat so much smaller, that there would be little difficulty in rigging up an apartment which would serve quite well enough for that.

[In 1891 the memoir on the Effect of Viscosity on Pendulums was translated by C. Wolf for inclusion in the two volumes of *Mémoires sur le Pendule* which belong to the *Collection de Mémoires sur le Physique* published by the French Physical Society; and the translator expresses his obligation to Prof. Stokes for assistance in his task.]

J'avais eu l'intention de reproduire ici la partie du Mémoire de M. Stokes: *Sur les théories du frottement intérieur des fluides en mouvement* (*Transactions of the Cambridge Philosophical Society*, Vol. III. Part II. p. 287; 1847), dans laquelle il a établi les équations (1). Mais M. Stokes m'a fait remarquer que ces équations peuvent et doivent être considérées comme connues: 'Navier a donné bien avant moi les équations du mouvement d'un liquide où l'on tient compte de la viscosité, et Poisson a donné les équations correspondantes pour un fluide élastique ou non; ses équations contiennent deux constantes qui dépendent de la nature du fluide, et elles se réduisent à celles de Navier si le fluide est incompressible. Elles diffèrent des équations ordinaires de l'Hydrodynamique par l'introduction dans la première de termes qui peuvent se mettre sous la forme

$$\mu\left(\frac{d^2u}{dx^2} + \frac{d^2u}{dy^2} + \frac{d^2u}{dz^2}\right) + \nu\frac{d}{dx}\left(\frac{du}{dx} + \frac{dv}{dy} + \frac{dw}{dz}\right),$$

avec des termes semblables dans la deuxième et la troisième. Mes équations seraient identiques à celles de Poisson, si je n'y avais introduit que les hypothèses qui sont le fondement même de ma méthode. Mais une hypothèse additionnelle qui, sans être fondamentale, m'a paru présenter un très haut degré de probabilité, y a réduit les deux constantes à une seule, en établissant entre elles la relation

$$\mu = 3\nu.$$

Maxwell est arrivé plus tard aux mêmes équations, avec la même relation entre les deux constantes, en partant de sa théorie dynamique des gaz (*Philosophical Transactions*, 1867, p. 81). Il me semble donc sans intérêt de traduire la première partie de mon premier Mémoire; les équations différentielles partielles qu'il faut appliquer à la solution du problème de la résistance au mouvement du pendule peuvent être regardées comme connues, puisqu'elles ont été obtenues indépendamment par plusieurs géomètres, qui s'accordent tous les uns avec les autres.'

REMARQUE SUR LE MÉMOIRE DE M. G. STOKES.

Le Mémoire de M. Stokes a fait faire un progrès immense à la théorie du pendule, en définissant d'une manière précise la cause du fait observé par Du Buat et par Bessel, et en permettant d'en calculer la valeur numérique. (Voir l'*Introduction historique* que j'ai placée en tête du Tome IV de ce Recueil, p. xxv.) Mais à l'époque où parut ce travail, n'existait pas d'expériences suffisantes pour permettre de comparer avec sûreté la théorie nouvelle et l'observation. De là une lacune au sujet de laquelle je laisse la parole à M. Stokes lui-même. Voici ce qu'il m'écrivait en répondant à la demande que je lui avais adressée, de publier la traduction de son grand Mémoire.

C. W.

'Les plus importantes séries d'expériences dont je disposais pour comparer la théorie et l'observation étaient celles de Baily, qui avait cherché à déterminer l'influence de l'air sur la durée d'oscillation d'un pendule, en faisant osciller le même pendule dans l'air et dans un milieu très raréfié. Si le vide avait été absolu, ou du moins pratiquement parfait, tout aurait marché tout droit. Mais, comme la pression de l'air raréfié atteignait ordinairement un pouce environ de mercure, l'influence de l'air restant n'était nullement insensible, et il était nécessaire d'en corriger le résultat. Le calcul de cette correction exigeait la connaissance de la loi qui lie le coefficient de viscosité de l'air μ avec la densité ρ. M'appuyant sur une seule expérience de Sabine, faite avec l'hydrogène, gaz dont il est bien difficile de garantir la pureté absolue, surtout quand il n'est pas préparé par un chimiste de profession, j'ai supposé que μ variait proportionnellement à ρ, en d'autres termes, que ce que j'appelais l'*indice de frottement* μ' était indépendant de la densité. On sait aujourd'hui que c'est le coefficient de viscosité μ, et non l'indice μ', qui est indépendant de la densité. La supposition adoptée a donc eu pour conséquence de rendre trop faible l'influence de l'air restant; et, par suite, en comparant la différence observée des durées d'oscillation dans l'air à pression entière et sous faible pression à sa valeur théorique, j'en ai conclu une valeur trop petite du coefficient de viscosité.

'J'ai eu longtemps l'intention de calculer à nouveau les expériences de Baily en prenant μ, et non plus μ', indépendant de la densité, conformément à ce que nous savons aujourd'hui; et c'est là en réalité la cause qui a retardé la publication du troisième Volume de mes Mémoires*. Un calcul rapide m'a fait voir que l'effet de cette correction serait d'accroître le coefficient de viscosité d'environ 40 pour 100 de sa valeur. Cet accroissement le rapprocherait de sa valeur vraie; je dis valeur vraie, parce que ce coefficient est aujourd'hui bien connu. Cependant il serait resté encore sensiblement trop faible. Cette faiblesse provient-elle de ce que la vitesse n'a pas été assez petite pour prévenir la formation de remous; provient-elle de quelque autre cause inconnue? C'est ce que je ne puis dire aujourd'hui.'

---

* See addition in *Math. and Phys. Papers*, vol. III. pp. 136—141.

MEDINDEE, BURTON ROAD, DIDSBURY,
May 18, 1895.

DEAR SIR GEORGE STOKES,

I would have written sooner to thank you for your note about the pendulum—but various occupations, including a Roy. Soc. Council, have interfered.

I think the argument quite satisfactory as far as you press it. I have not had time to look up the literature of pendulum observations, to ascertain whether any indication of change of coefficient or law of resistance (with amplitude) can be extracted from them*. From the point of view of your letter such a change would be more likely to shew itself with oscillations of long period than of short, since there is more time for cumulative effects?

Yours very faithfully,

HORACE LAMB.

FROM M. ALFRED CORNU.

TROUVILLE SUR MER (CALVADOS).
le 21 *Juillet* 1878.
PARIS 38 R. DES ÉCOLES.

CHER ET TRÈS HONORÉ MONSIEUR,

Je vous remercie beaucoup des démarches que vous avez faites auprès de Sir G. Airy relativement aux expériences de Baily et de l'aimable lettre que vous avez bien voulu m'envoyer à ce sujet†.

---

* "It is however possible that with amplitude of vibration so large as those actually used, amounting to about 1° to start with, there may have been a very slight production of eddies, the effect of which on the time of vibration may not have been wholly insensible." *Math. and Phys. Papers*, vol. III. p. 140, appendix to memoir on pendulums added in 1901. Cf. also Lamb, *Hydrodynamics*, 1906, § 290.

† This letter relates to the revision of Baily's reduction of his observations to determine the mean density of the Earth by the method of the Cavendish experiments, which was made by M. Cornu with a view to removing the discrepancy with his own well-known determinations. There is a series of notes in the *Comptes Rendus*, LXXVI. 1878, by Cornu and Baille, on correction for the resistance of the air to the slow vibrators, showing that a linear law of resistance suffices. The following extract (p. 701, Mar. 18) explains the subject of the letter: "En vue de diminuer la durée d'une expériénce, ces deux observateurs ont profité de l'instant

J'ai bien tardé à vous répondre, mais un malheur de famille m'a frappé cruellement et depuis encore, différentes préoccupations m'ont retenu et empêché de vous écrire.

—Vous avez parfaitement défini ce qu'il y aurait à faire pour la correction des résultats de Baily: il suffirait simplement, comme vous le proposez, de substituer la deuxième moyenne (*2nd mean*) à la troisième (*3rd mean, or Resting point*) en acceptant le temps d'oscillation *N* adopté.

Quant à la valeur du temps, *N, il ne doit pas être modifié,* car les observations qui le fournissent sont comprises dans des parties de l'oscillation complètement affranchies de la cause d'erreur qui a pu vicier la première lecture de chaque série.

Le calcul de l'erreur probable serait peu utile, on pourra donc s'en dispenser.

Le travail de révision sera donc un peu long, *mais sans aucune difficulté.*

J'ai eu l'occasion à Paris, dans ces derniers temps, d'entretenir Lord Lindsay, Président de la Société Royale Astronomique de Londres, de ce sujet: j'ai donc suivi votre obligeant conseil, et Lord Lindsay me paraît très favorable au projet dont j'ai eu l'honneur de vous faire part.

Je compte lui écrire, comme Président, une lettre avec l'exposé succinct de la question, et je pense que les calculs pourront être accomplis.

Veuillez agréer, je vous prie,
Cher et très honoré Monsieur,
l'assurance de mon respectueux
dévouement,

A. CORNU.

de l'élongation, où le levier est pour ainsi dire stationnaire pendant quelques secondes, pour effectuer la manœuvre d'inversion des masses ; la même élongation peut ainsi être utilisée à deux titres, comme la dernière d'une série et la première de la série suivante." On account of unavoidable disturbances in effecting the inversion, the first swing cannot thus be utilized; and its rejection led in fact to a satisfactory agreement.

CORRESPONDENCE WITH MR C. VERNON BOYS.

S. KENSINGTON,
3rd Nov., 1890.

DEAR SIR GEORGE,

In the summer you asked me to let you have some quartz fibre. I now leave a box containing one long one wound 18 times up and down the frame. The front and back of the box are of glass so that the fibre may be examined without risk, and black paper is placed at one end in case a dark background may be required.

The glass may be taken out after first removing a screw at the end, and then the frame can be removed.

The fibre is good, in that under the prism test it shows both perfectly uniform pieces and places where the variation in diameter gives rise to small [*i.e.* short but pronounced and jagged] bends in the dark bands of the spectrum.

To see these spectra to perfection the following is the best way to hold the box and a prism : first see the colours with the naked eye [with light from a slit, the box intervening] and then hold a low-angle prism in front of the eye, *not* in the natural position of minimum deviation, but inclined so that incident light falls at about the polarising angle. It is only thus that the dark bands appear really dark.

On then moving the box along its own length so as to make successive parts of the fibres occupy the same position, you will see some of the fibres giving a constant spectrum which is the test of uniformity, while some show the irregularities described.

Also the variation of the spectrum with the angle of incidence will be evident. The more nearly the fibre is between the eye and the sun, the fewer will be the bands.

I have the honour to be your obedient servant,

C. V. BOYS.

LENSFIELD COTTAGE, CAMBRIDGE,
7 *Nov.*, 1890.

MY DEAR SIR,

I found yesterday at the Royal Society your letter and the enclosed quartz fibre that you were so good as to leave for me. I brought it home with me, but I have not yet had sunshine wherewith to examine it. You will, I doubt not, allow me to keep it somewhat longer till I have had an opportunity of examining it, which at this time of year may not be very soon.

I am not sure whether your mention of the polarising angle was meant to do more than give an idea of the angle at which it was found best in the circumstances of your experiment to hold the face of incidence towards the incident light. The explanation, I feel persuaded, has nothing* to do with polarisation.

Yours very truly,

G. G. STOKES.

8 *Nov.*, 1890.

As this is a bright day, I went to my lecture room and had a look at the bands by sunlight. As I fully expected, and as, likely enough, you may be already aware, the question of the inclination of the prism in order to see the bands distinctly is simply one of focusing. The conditions of getting a pure spectrum, with a slit at a finite distance, are that the edge of the prism shall be parallel to the slit, and the *primary* focal line corresponding to any element of the slit shall be at the distance of distinct vision. If u be the distance of the focus of incidence from the prism, v the distance of the primary focal line after refraction through the prism, ϕ, ψ the angles of incidence and emergence, ϕ', ψ' the corresponding angles in glass, we have

$$v = \frac{\cos^2\psi \, \cos^2\phi'}{\cos^2\phi \, \cos^2\psi'} \, u = \frac{\cos^2\psi \, (\mu^2 - 1 + \cos^2\phi)}{\cos^2\phi \, (\mu^2 - 1 + \cos^2\psi)} \, u,$$

and we see at once that the ratio of v to u lies between $\dfrac{\cos^2\phi}{\cos^2\psi}$ and 1. If the eye is nearer to the fibre than the distance of dis-

* But see p. 333 *infra*.

tinct vision for the colour observed (for the eye is not achromatic) you must turn the prism from its position of minimum deviation so as to increase the angle of incidence. If you put on convex spectacles the requisite inclination (which will of course depend on the distance of the fibre and the power of the spectacles) will very likely have to be at the opposite side of the position of minimum deviation.

I have too much on hand at present to examine the precise history of the light which produces the colours; but I think it probable that those seen nearly in the direction of the incident light, that is, making no very large angle with it, are due to diffraction, the same as if the fibre had been opaque, and those seen at a moderately small angle to the direction opposite to that of the incident light are those of the primary glass-bow, if I may so call a rainbow in which water is replaced by glass.

I have seen the variations you mentioned as indicative of variations of thickness.

P.S. When I said that to get a pure spectrum the slit must be parallel to the edge of the prism, I had in view the general case. In the particular case in which the spectrum is pure when the prism is in its position of minimum deviation, the pencil from any element of the slit and for any colour diverges after refraction, not from two focal lines, but from a point, and the spectrum will still be pure though the slit be not parallel to the edge.

Science and Art Department, South Kensington,
8th Nov., 1890.

Dear Sir George,

I hope if the quartz fibres are any use you will keep them. I had the box made so that what I believed would be the most interesting feature to you, viz. the optical, could be easily examined without risk.

I used the expression "polarising angle" to express the angle at which, about, the phenomena are more marked. If the prism is held at minimum deviation, as would be natural, the dark bands are *not* dark and are not even conspicuous. If however it is held at about this angle they seem quite black.

I did not feel sure that polarisation had anything to do with it; yet I believed it had for the following reasons. First I could not understand why the tilting of the prism should make the dark bands more conspicuous unless by such tilting interfering light was got rid of; as the angle was evidently about the polarising angle, it appeared as if light polarised one way was either not broken up into bands or was differently banded, and the most perfect elimination of this was important. Acting on this I have interposed a double-image prism of very small separating power, and have found that the two lights polarised at right angles to one another do produce different spectra. These I showed at the time (nearly two years ago) to Prof. Rücker.

I do not profess to understand the complete theory of these colours, but I think it possible that a careful examination of them might at any rate confirm optical principles or possibly do more.

I should have said that if the light, the fibre, and the eye are in the same plane, banded spectra are produced, but in this case the bands are much more numerous as would be expected.

I should not presume to have sent you the fibres if you had not asked me for some in the summer.

N.B. Those on the frame are about the size that I have found most suitable for the radio-micrometer.

LENSFIELD COTTAGE, CAMBRIDGE,
11th Nov., 1890.

Thanks for your letter, and the mention you made of what you observed. I will as you allow me keep the fibres somewhat longer. At this time of year one gets on slowly with work demanding sunlight, besides which it happens that I am at present extremely busy.

I had noticed that in the bands seen by what I may call reflection the number was a good deal greater than when they were seen by what I may call transmission. As I observed them, the plane containing the incident ray and the ray entering the eye was however approximately perpendicular, not parallel, to the fibre.

I mean to try the experiment you mention with a double-image prism. Meanwhile I am disposed to think* that the difference in the two spectra (by which I understand you to mean that

* See next letter.

the places of the bands were different) was due merely to the difference of direction in which the rays entering the eye left the fibre. There *would* be a difference of direction comparable to the angular separation between the two images, multiplied by the ratio of the distance from the pupil to the double-image prism to the distance from the latter to the fibre.

12th Nov., 1890.

I had not when I wrote last looked at the spectrum of the fibres with polarised light ; and when you wrote that you had viewed the light through a prism and double-image prism, and that the spectra were different, I assumed that you meant that the bands were different* (for a difference of intensity merely, to some extent, might be anticipated)—and I pictured to myself bands lying in a *somewhat* different position, and I meant to test this, using a Nicol instead of a double-image prism, or rather, to try it both ways.

It occurred to me that the correctness (or otherwise) of my supposition that the banded spectrum was that of diffracted light might admit of being tested by polarisation, that when the deviation from the course of the direct light was tolerably large, the polarisation (if that were the origin of the light) might be strong enough to show easily. I have a way of showing† that the totality of the diffracted light equals what would be the totality of the transmitted if none were sent elsewhere by reflection, and therefore (as much more is refracted than reflected) the totality of the diffracted exceeds, but does not very greatly exceed, the totality of the transmitted. The distribution however of the two respectively over the different directions of deflection is different. Last night I calculated the proportion of the intensity for lights polarised parallel and perpendicular to the fibres, for each of the two kinds separately, for each of the angles of deviation 20, 40, 60 degrees‡. Taking in each case the portion of the light of that kind (diffracted or transmitted as the case may be), which is

* As is confirmed in the postscript.
† See letter to Lord Rayleigh, p. 119 *supra*.
‡ This calculation exists, in MS. The angles as given *infra* are different.

polarised in a plane parallel to the fibre at 100, I find for the light polarised perpendicular to the fibre:—

At deviation			20°	30°	40°
in diffracted	...	...	142·6	170·4	399·9
in transmitted	...	...	94·0	78·0	56·2

or if we prefer to express them in percentages of the total light (diffracted or transmitted) in each case:

In diffracted				
polarised perpendicular		58·8	63·0	80·0
polarised parallel	...	41·2	37·0	20·0
In transmitted				
polarised perpendicular		48·5	43·8	36·0
polarised parallel	...	51·5	56·2	64·0

Hence it may very well be that when viewed, at a considerable deviation, by light polarised perpendicularly to the fibre, the diffracted light preponderates over the transmitted, but when viewed by light polarised parallel to the fibre the transmitted light preponderates over the diffracted. The less numerous and more conspicuous bands belong to the diffracted light; those of the transmitted are more numerous.

In what I may call the reflected light, that is, the light proceeding in a direction making a moderate angle with the course of the incident light reversed, the bands seemed sensibly the same whether the light was polarised perpendicular or parallel to the fibre, but it was somewhat brighter in the former case. This agrees with the origin I attributed to it, namely, that it was the light inside the primary glass-rainbow. For this there are two refractions with an intermediate reflection; but the single reflection much predominates in its effect on the polarisation over the two refractions.

I have never worked with a silvering solution, but I don't see any reason why one of these very fine fibres should not be capable of taking a silver coating. A very thin silver coating would hardly sensibly increase the thickness of the fibre, and it would stop the transmission of light, so as to allow of the diffracted light being studied apart. It would also stop the rainbow light.

In diffracted light at deflection θ, number of band-intervals in spectrum (from theory) $= D \sin \theta \left(\dfrac{1}{\lambda \text{ blue}} - \dfrac{1}{\lambda \text{ red}} \right).$

P.S. I did not go on with the experiment with a double-image prism, because when I saw the spectra I found they were quite different, now [not?] merely showing the same bands in a slightly displaced position in the one as compared with the other.

14 *Nov.*, 1890.

I finished my last in a hurry, in order to catch the post, and I felt doubtful afterwards whether I had explained that the D in the formula meant the diameter of the fibre; but even if I did not define it you probably guessed what it meant.

I have not at present attempted, except very cursorily, to calculate the bands for the transmitted light. This is a much more complicated matter than the theory of the diffracted light. In calculating the polarisation, I took the observed light as following the course assigned to it by geometrical optics. This I believe is quite near enough for the purpose, but to give a complete theory of the bands I believe that diffraction would have to be taken account of as well as refraction*.

In calculating the polarisation of the diffracted light, I assumed the truth of a result I arrived at in a paper on the dynamical theory of diffraction, published long ago in the *Transactions of the Cambridge Philosophical Society*†, and also that, in accordance with the experiments therein described, the vibrations in polarised light are perpendicular (not parallel) to the plane of polarisation. The experiments were made by means of a glass grating. In looking at the light coming from the fibres at a considerable inclination to the course of the incident light produced, I was struck with the facility with which the polarisation could be observed in what I was led to regard as the diffracted light. It struck me that the crucial experiment mentioned in that paper could be performed by much simpler appliances by using a silvered thin fibre than by means of a grating. I should not, however, have time at present to engage in anything of the kind.

* The *positions* of the supernumerary rainbow formed in a thin jet of water, as explained by Airy, were observed over a long range by W. H. Miller. The recondite calculations required to compare theory with experiment were effected by Stokes. For the actual details of comparison cf. Larmor, *Proc. Camb. Phil. Soc.* vi. Oct. 1888, and vii. Jan. 1891.

† "On the Dynamical Theory of Diffraction," especially Part ii. Section ii. *Trans. C. P. S.* vol. ix. 1849; *Math. and Phys. Papers*, vol. ii. See *addendum* in the reprint; cf. also *supra*, p. 112.

13 *Nov.*, 1890.

DEAR SIR GEORGE,

I am much obliged to you for your letters on the optical behaviour of the threads, which explain what puzzled me when observing the march of the bands during a slow and gradual change in the angle of incidence of the light (in a plane normal to the fibre). An electric arc was placed behind the slit of a large spectrometer, and the prism was replaced by a hanging fibre. The observing telescope had its object-glass taken out and the eye-piece replaced by a direct vision prism of low dispersive power. During the variation of the angle of incidence the bands did *not* march up the spectrum uniformly, but as I put it at the Royal Institution lecture on quartz fibres, moved more like the legs of a caterpillar walking. This successive movement was no doubt due to the passage of two independent sets of bands over one another. They certainly were much more clearly defined at certain angles than at others on either side.

It was with this [idea] that I added a double-image prism of low separating power, so placed that one spectrum was immediately above the other. Then, the two spectra just overlapping, the relative displacement of the bands could be easily observed.

Last spring I silvered a fine quartz fibre which I afterwards sent to Prof. FitzGerald. I used it as a conducting torsion thread, and I remember on holding it up to the light that the colours of true diffraction were very marked.

3rd December.

Many thanks for your further explanation of the cause of the bands. I am unable to try the experiment you suggest as to the cause of those bands that are always present, as the spectrometer that I used is now employed for some experiments in the astronomical department.

The reference which I could not give yesterday I have just looked up. It is "On the Electro-magnetic theory of Light," Lord Rayleigh, *Phil. Mag.*, Aug. 1881.

CAMBRIDGE, 2 *December*, 1890.

As I travelled back to Cambridge in the dark (except as to the poor light in the carriage) I thought over the spectra which you showed me.

I feel little doubt that the transverse bands in the spectrum of a violently tapering fibre have the origin I suggested, namely, that they are due to the interference of the diffracted light with the transmitted*. In a uniform fibre the bands due to the interference of these two kinds of light would be parallel to the fibre, and naturally in an intermediate case, that is, in a more gently tapering fibre, they would have an inclined position. I now think that the fine bands I saw in the spectrum of the fibres you lent me are due to the interference of the diffracted with the transmitted light, rather than to the interference of different parts of the transmitted light with one another. And this accounts for the delicacy of the test which they afford of the uniformity of the fibre.

The other bands you showed me, which you said were seen without a slit, belong, I feel almost sure, to the source of light, and arise from vaporised matter in the arc connecting the poles. They would, I think, disappear if sunlight were substituted for the arc light with which the light from the poles is mixed. Or perhaps it might be possible to get the plate impressed by pole-light alone by arranging a pendulum with a screen containing an aperture, and making the current pass through the pendulum, by means of mercurial connections so arranged that the current is either interrupted or shunted just while the aperture is passing the slit. I do not know whether the current would strike across if merely thus very briefly interrupted. Even if it would, of course a much longer exposure would be required. But the substitution of sunlight for the electric light would show at once whether the particular bands I refer to belong to the source of light, as I feel pretty sure they do.

* But see p. 337, *infra*.

3 *Dec.*, 1890.

I have thought of a verification of the view expressed in my last as to the origin of the finer bands seen in the spectra; I mean the system which in the photographs you showed me were nearly perpendicular to the direction of the fibre. It is not a sharp verification, for it merely involves a proper choice between two alternatives; and if you were to toss up for it, it is as likely you would be right as wrong. It is this. Let the vertical link *nk* denote the direction in which you travel from the image of the thinner (*n*) to the thicker (*k*) end of the fibre, and the horizontal line *rb* the direction in which you travel from red to blue. Then the direction of the bands of the system we are considering should lie somewhere within the quadrants *nb* and *kr*, not within the quadrants *nr* and *kb*. When the fibre is uniform they would be parallel to *nk*, and when it is violently tapering they would become nearly parallel to *rb*.

Dec. 4.—Thanks for your letter received this morning. It follows from the above that a local thickening in the fibre would produce a bulge in the bands turned towards the red, not the blue, end of the spectrum; and I suppose a local thickening, as a residue of the tendency to form beads, is more likely to occur than a local thinning. Hence we might expect bulges in the bands to be usually directed towards the red end of the spectrum. Some of your photographs contain, very likely, evidence on these two points, the quadrant in which inclined bands lie (provided it was noted or remembered whether top or bottom corresponds to the image of the thinner portion of the fibre), and the direction of any bulge.

Thanks for the reference to Lord Rayleigh's paper.

5th Dec.

DEAR SIR GEORGE,

I have some difficulty in seeing how the inclination of the bands as described by you can be used as evidence as to their origin.

In a fibre which gives a banded spectrum, no matter what is

the cause of the interference, if two parts have the ratio of their diameters as $n : 1$, and if two wave lengths of light have their ratios also $n : 1$, and if one happens to be such that it is destroyed, then the other must be so also, whether the interference is due to diffraction or to light that has passed through, to polarisation, or to any combination of any of these; at least I think it must be so. Therefore the bands must be inclined in the direction you suggest on any theory.

As a matter of fact I cannot now tell from the photograph whether the bands incline the right way or not.

I have examined the fibres on the frame with a prism by the electric light to-day; and on the whole I fancy the short, sharp sudden bands are on the whole towards the red, but the bands are according to my memory much more irregular than they were at first, and this I believe to be due to very fine dust, of which there is a great deal, on the fibres.

I have not been able to-day to make a test to see if this is so for certain.

CAMBRIDGE, 6 Dec., 1890.

You are quite right. Strange to say I had not thought of it, though I am so familiar with seeing, and in teaching to my class, that in diffraction and interference phenomena the fringes or other appearances are on a larger scale for the red than for the blue.

LENSFIELD COTTAGE, CAMBRIDGE,
3 November, 1897.

DEAR MR BOYS,

I feel ashamed to approach you as a beggar; but there is one thing I should like very much to have if it would not be giving you more than trifling trouble; that is, a copy (positive) of one or two of your photographs showing the effects on the air of the passage through it of a bullet moving with a velocity exceeding the velocity of propagation of sound. What I chiefly wished for was one showing the outline, resembling one branch of a hyperbola, at which there is a rapid change of density, as evidenced by the dark mark indicating the deflection of the rays due to refraction across the place of rapidly-changing density, and one showing the reflection of the wave from the plane wall of a vessel. Perhaps the same photo would show both things, which would save you trouble.

There are mathematical difficulties connected with the theory which have as yet been only very partially mastered. Possibly an inspection of the photographs might give some hint towards the way of attacking those difficulties *.

I am lecturing at present on sound, and one of my class asked me a question about the theory of your results. This has induced me to pocket my shyness about asking you for what may involve a little trouble, though I see no reason why it should involve much.

I send you my Wilde Lecture, and two earlier papers on the same subject, marked 1, 2, 3 in order of date. I have not yet distributed them as my time was otherwise occupied.

Yours very faithfully,

G. G. STOKES.

27, The Grove, Boltons, S.W.,
24th Nov., '97.

Dear Sir George,

I regret that there should have been more delay on my return in consequence of my having to get the enclosed prints.

These four fairly well show the most conspicuous features.

1. Shows a Magazine Rifle bullet speed about 2100 ft./sec., photographed just before meeting the head wire. The little spark which jumped out to meet the wire from the bullet appears as a black line on account of its being a cylindrical diverging lens. The form of the head wave above is clearly seen, being in this photograph free from splash from the head wire.

The reflection and non-reflection of the waves at different

---

* The only account published of these experiments is the illustrated Report in Nature, vol. xlvii. pp. 415—421, 440—446, of Mr Boys' lecture at the British Association, Edinburgh Meeting, 1892. This description of his results forestalls much of Prof. Stokes' remarks, and includes many additional points of interest on other photographs; it also describes arrangements previously employed by Mach, Lord Rayleigh, Jervis-Smith, and others, for spark photography.

Negatives for the six illustrations here reproduced have been kindly supplied by Messrs Newton and Co., who own the copyright.

angles from metallic plates is also seen. Also the diffraction phenomenon on the reflected and the direct wave in the lower right-hand corner.

2. Shows the same kind of bullet piercing a sheet of glass. Also particles of card from a sheet which had been pierced, following the bullet, those which are going 1100 ft./sec. or more with waves of suitable angle, those less, without.

3. Is a Martini Henry bullet 1250 ft./sec. about between reflecting plates showing form of air wave when incidence is sufficiently grazing to give no reflection.

4. Is a badly fogged plate of an aluminium bullet about 3000 ft./sec. which also had pierced a sheet of card. Particles of card with waves of appropriate angle. Head wave distorted near head by concussion of bullet striking forward wire.

If anything further will be useful pray let me know.

Yours very sincerely,

C. V. BOYS.

LENSFIELD COTTAGE, CAMBRIDGE,
26 *Nov.*, 1897.

I am extremely obliged to you for the photographs. They are to me most interesting, as they bear on the theory of a sound-bore, a theory which hitherto has been only imperfectly developed.

I found the photographs on my return from London a little before seven yesterday evening; but I had to dress, as I had friends coming to dine with me in [college] hall, so that I had not time to study the photographs last night. To-day I had a lecture, only three men were present. I showed them your photographs and letter, and pointed out some of the inferences from the photographs. I found that one of my class had assisted you in taking the photographs.

As your photographs are numbered for me, to correspond with your letter, and you probably may not recollect to what photographs these provisional numbers refer, I will briefly describe them for the sake of identification:

No. 1. Magazine rifle bullet, vel. about 2100.

No. 2. Same kind of bullet piercing glass.

22—2

No. 3.   Martini Henry, vel. 1250, between reflecting plates.

No. 4*.   Aluminium bullet, vel. 3000.

You are doubtless familiar with the bright band (bright in the positive) on one side of the dark, and most likely have noticed† the law which determines at which side, namely, that the bright is always in the rear of the dark, "rear" referring to the direction of travelling of the wave, not the direction of motion of the bullet.   This is in accord with theory, which shows that the rapid change of density indicated by the curve is such that you pass from the higher to the lower density as you travel across the bore in the direction of propagation of the wave.   The bore forms a surface of revolution, supposing everything symmetrical about the axis, as it is at least very approximately; and a section of this by a plane perpendicular to the axis gives, at the bore, an annulus of small thickness in crossing which the rapid change of density takes place.   From the source of light, draw a tangent to a circle belonging to the annulus where the change of density is very rapid.   Imagine rays of light to be incident nearly in the direction of that tangent.   Then if about the critical circle the density be diminishing as you proceed from the centre outwards, the refraction will take place as it were through a very blunt prism slightly denser than the surrounding medium, and the refraction will be inwards.   If on the other hand the density be increasing as you go outwards, the refraction will be outwards. The rule will be the same in both cases : the refraction will be towards the side of greater density.   Hence the photographs show at once which side of the bore is that of greater and which of smaller density.

Take now No. 1, which shows bores proceeding from the front and from the rear of the bullet.   The photo shows that in both cases in crossing the bore in the direction from the axis outwards the density is decreasing.   But the density a little outside the outer and inside the inner bore must be, nearly at least, the atmospheric density.   Therefore the outer bore lies in the front part of a condensing wave, and the inner bore in the rear part of a rarefying wave.

Besides the more obvious features, No. 1 shows the character of a bore that has passed the edge of an opaque screen.   For in

---

* Not here reproduced.

† On this and other points, see report of Brit. Assoc. lecture, *loc. cit.*

the rear reflected bore if we suppose a perpendicular let fall from the end of the wall on the straight part supposed produced, as the outline has begun to be slightly curved before the foot of the perpendicular is reached, we may in imagination replace the reflected bore by a straight bore passing along a clear space (the place of the mirror) and obstructed by a screen placed in the clear space where the actual wall has ended.   We see the way in which the bore curves round the edge of the (ideal) screen, dwindling away as it gets further from the normal to the plane bore (or rather rectilinear bore, as the thing is seen in section), until it becomes insensible.

No. 1 shows at the top (this feature is of course obvious) the obliteration, or partial obliteration, of one bore by another.   It is here the main front bore that is obliterated by the bore from the struck wire.

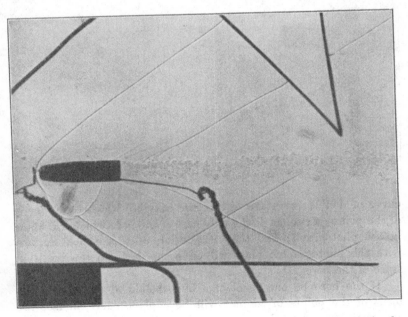

A bore is of course a surface, and if everything be perfectly regular the surface will be one of revolution about the axis of the bullet.   Regard the electric flash as a point of light.   It is only where a ray of light falls tangential to this surface or nearly so that it will be sensibly deflected from its course, so as to eave (in the positive) a dark spot in the place in the photographic plate on which it would otherwise have fallen. Hence if we make the flash, regarded as a point, the vertex of a conical surface touching the surface which represents the bore, the curve (plane curve or

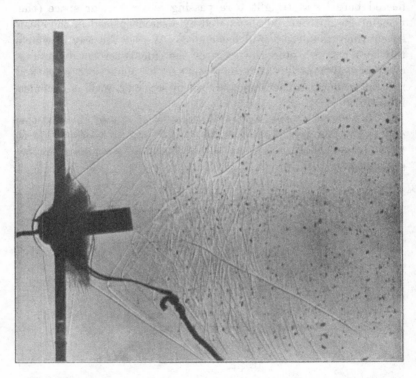

curve of double curvature as the case may be) which is the locus of the points of contact is the only part of the bore which will show in the photograph; and the intersection of this conical surface with the photographic plate will give the dark (in the positive) trace of the bore.

If the bore be one sheet of a hyperboloid of two sheets, the locus of the points of contact will be a plane curve, in fact the hyperbola (one branch of it) in which the hyperboloidal sheet is cut by the plane polar to the vertex of the cone. I should have left out the words in parentheses, since the second branch of the hyperbola would be the section of the other sheet of the hyperboloid, with which, by hypothesis, we are not concerned. Our enveloping conical surface will be an elliptic cone (*i.e.* a quadric cone) and its intersection by the photographic plate will be one branch of a hyperbola.

The intersection of the two bores (the front reflected and rear direct) near the hook in No. 1 shows some minute features which I have not discussed.

No. 2. I don't know that I have anything to add to yours.

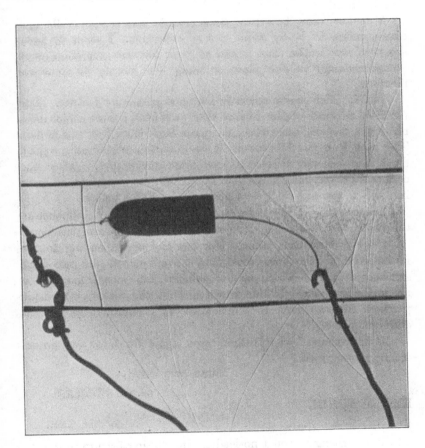

I notice that in some of the bores produced by bits of card the line of symmetry of the bore is a little inclined to the axis, or line of fire. The small inclination is to the right or left according as the bit is to the right or left of the axis, indicating that the bit was moving, not quite parallel to the axis, but in a direction deviating a little from that, to the right or left according to the position of the bit. This was to have been expected.

No. 3. The bore ahead is very strong. This may be due in part to the confinement, but I think it is due in good measure to the fact that in this case the velocity of the bullet did not greatly exceed that of sound. To judge by some of the bores due to bits of card, this would seem to be favourable to strong bores immediately ahead.

I do not know how the bores outside the copper walls got there, unless it be by some sort of reflection. I seem to have a sort of recollection that in one of your lectures you mentioned some markings on the plate as being due merely to some reflection.

No. 4. This shows specially well one particular feature. The furthest forward of the bits of card produces a bore which does not cross, but coalesces with, the main front bore from the bullet. The bore from the bit travels, in the direction of the axis, a good deal more slowly than the bore from the bullet. After the coalescence, we have a portion of straight bore somewhat more inclined to the axis than that the main bore was before.

The crossing of two bore traces without apparent influence of the bores on one another does not of itself alone prove that bores do not influence each other. For the ray corresponding to the intersection of the bore-traces may have touched the two bore surfaces at points which do not coincide, but merely happen to lie in the same straight line drawn through the vertex, so that the parts of the two bores for which a record is obtained may be in different places.

In conclusion, I wish to thank you again for these very interesting photographs.

Yours very truly,

G. G. STOKES.

Finished *Nov.* 27.

29 *Nov.*, 1897.

Since writing to you I noticed an interesting feature in No. 1, which I had either overlooked or passed by as a thing too minute to attend to. But with a slight variation in the mode of observation it becomes very distinct. It relates to the intersection of the front reflected and rear direct bores in No. 1, near the hook.

You probably may have noticed a small change of direction in the front reflected bore after its encounter with the rear direct. On viewing the photo at a very high angle of incidence, the plane of incidence containing the front reflected bore, the change of inclination of the front reflected caused by the action on it of the rear direct is rendered very evident by the foreshortening. The direction is rendered a little more nearly parallel to the axis, indicating a slightly increased velocity of propagation. This may very likely be due to its travelling in air which is rushing forwards after the bullet. If this be the cause, as seems likely, it points to the rear bore as the boundary of the portion of air which is rushing after the bullet.

1 *Dec.*, 1897.

I am a little afraid that my letters about sound-bores may be a bore to you. Still, some attention to a child of your own may not be displeasing to you.

I have reflected a good deal on what the photographs show, and have also made some measurements with a view to obtaining numerical results. Nos. 1, 2 and 4 show the front and rear bores, which tend to become straight as you go further from the bullet.

No.	Inclination		Velocity	
	front	rear	front	rear
1	34	29	1174	1019
2	$36\frac{1}{2}$	$28\frac{1}{2}$	1249	1002
4	32	$27\frac{1}{2}$	1590	1386

I laid tracing cloth on the photos, and drew on it straight lines coinciding with what appeared to be the best straight parts of the bores, continuing those straight lines onwards till they met, and measured the enclosed angles with a small protractor. The halves of these angles were taken as the inclination of the bore to the axis, from whence I calculated the velocity of propagation of the bores by multiplying the figures you gave me for the velocities of

the bullets by the sine of the inclination. Here are the results,
the inclinations being given in degrees, and the velocities in feet
per second.

Although the theory of bores has not yet been made out,
except at the first moment of their formation, yet I think we may
say from theory that at a distance from the bullet the bores
would be straight, and propagated with the velocity of sound,
and therefore the front and rear bores would be parallel to
each other. This limiting condition does not seem to have
been quite attained within the limit of the pictures. They show
the front and rear still slightly inclined to each other, and the
curvature, especially in the front bores, not yet to have quite
disappeared.

The bores for the aluminium bullet came out to have a
velocity of propagation considerably greater than that of sound.
It may be that the assumed velocity of the bullet (3000) is too
high. In any case the velocity of propagation of the front bore
would vary from that of the bullet, just at the head of the bore,
to what I think must be the velocity of sound at a distance.

In No. 3 I at first misinterpreted the strong bore just behind
the fuzzy thing, at a distance 0·63 inch in front of the flat end of
the bullet. I thought at first that it was due to the tendency
for the plate to leave a partial vacuum behind it when it was first
pushed out by the bullet striking it. But I saw afterwards that it
must be the reflection of the front bore from the glass plate;
I mean hydrodynamical not optical reflection.

Omitted to enclose, in haste to catch the post.   4/12/97.

The figure is intended to show the supposed general features
of the curve of density in a section taken across the front and

rear bores at some distance behind the bullet. The arrows are
intended to mark the direction of propagation of the bores. The
velocity of propagation of a bore probably exceeds that of sound,
and exceeds it the more the greater be the bore. The bore
probably wastes away by a continual propagation of a disturbance

backwards. In a free bore, that is, one not kept up by the
push of a bullet behind it, but abandoned to its own propagation,
the waste probably gets less and less as the bore is reduced
in magnitude, so that a bore has great longevity. The velocity
of propagation probably continually approximates to that of sound
under ordinary circumstances as the bore gets smaller and smaller.

6 Dec., 1897.

   Suppose you have an infinite mass of air at rest, the whole
having a uniform density, except a spherical portion, of radius R,
in which the density is a little greater, and now imagine the
whole left to itself. As you might expect, an annulus of
disturbance will travel outwards with the velocity of sound.
But what will be the thickness of the annulus, and what the
state of the air in it? At first sight you might suppose that
we should have an annulus of disturbance travelling outwards
with the velocity of sound, the thickness of the annulus being R,
and the air in it slightly condensed. In reality, we have an
outward-travelling annulus of which the thickness is $2R$, the
outward half, of breadth R, having condensation, and the inner

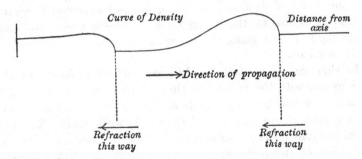

half, of breadth R, having rarefaction. This holds good for all
values of the time t for which at is greater than R, a being the
velocity of sound. What happens after starting is this:—the air
in the sphere expands, and acquires an outward velocity, except
just at the centre, where the velocity is nil throughout. The
expanding air by its inertia tends to retain its outward motion,
and to create a partial vacuum behind it; and thus it is that
we get rarefaction in the rear of the condensation.
   This is analogous to what takes place in your air which is

pushed out by the bullet. The air is condensed by the bullet, and also moved outwards; and as the air pushed outwards tends by its inertia to preserve its outward motion, it produced rare-faction in its rear; and the interior air moves outwards to fill up the partial vacuum, and thus we have a bore formed in the rarefied air, beginning with slight rarefaction, and increasing, as at a given moment we travel outwards, then increasing with great rapidity, and then rapidly changing to greater rarefaction, suddenly changing into a slowly-changing rarefaction which gradually passes into a condensation.

27, THE GROVE, BOLTONS, S.W.,
18th Dec., '97.

DEAR SIR GEORGE,

Many thanks, for the postscript especially of your last letter, which explains what I have quite failed to understand without help.

I am sending now the three more prints that I promised, the delay being due in main to the dull weather.

No. 5 corresponds to No. 2 but the bullet has now travelled about 5 inches past the plate. It has not however yet emerged from the cloud of dust and broken glass, as I expected it would. This is remarkable for the strong air wave due to the whole mass of moving lead and glass, and it shows also subsidiary waves due to odd particles.

In this the glass plate has become entirely broken. In the previous one with the bullet half through (not only just entering), which I think you have, the ripple of disturbance is still travelling *in* the glass and so it must at that time be still whole. The fact of this movement is made evident by the inclined air waves on either side of the plate with alternate light and dark exteriors, which I believe to be due to the *first* movements of the glass being alternately in opposite directions. This is consistent with the fact that an outer dark shade on one side of the plate is opposite an outer light shade on the other side of the plate.

No. 6. This shows the same event as No. 5, still later, when the bullet has travelled about 15 inches from the plate. Here the middle piece punched out by the bullet is travelling just above it and I expect the two systems of waves merge into one distorted hyperboloid with two noses and a transverse groove between.

This is the photograph which also shows the uniting of two waves so well in the upper part of the plate.

No. 7. This is the ordinary bullet going at the ordinary speed, but the atmosphere in the box was CO_2 saturated with the vapour of ether which is very heavy. The CO_2 prevented it from igniting and so risking things generally.

In connection with these solitary waves or bores did you see a letter I sent to *Nature* in the summer of this year under the heading "On the Visibility of the Shadow of a Sound Wave"? I can't now give the date but it could be found at once in the index*. If you did not see it I think the observation there described is one that would interest you.

Yours very truly,

C. V. BOYS.

* *Nature*, vol. LVI. p. 173 (June 4, 1897). The following extracts contain the main points:—

Two months ago I received the following letter from Mr E. J. Ryves.

"On Tuesday, April 6, I had occasion, while carrying out some experiments with explosives, to detonate 100 lbs. of a nitro-compound. The explosive was placed on the ground in the centre of a slight depression, and, in order to view the effect, I stationed myself at a distance of about 300 yards on the side of a neighbouring hill. The detonation was complete, and a hole was made in the ground 5 feet deep and 7 feet in diameter. A most interesting observation was made during the experiment.

"The sun was shining brightly, and at the moment of detonation the shadow of the sound wave was most distinctly seen leaving the area of disturbance. I heard the explosion as the shadow passed me, and I could follow it distinctly in its course down the valley for at least half a mile: it was so plainly visible, that I believe it would have photographed well with a suitable shutter."

....

At the time of the explosion my whole attention was concentrated upon the camera, and for the moment I had forgotten to look for the "Ryves ring," as I think it might be called; but it was so conspicuous that it forced itself upon my attention. I felt rather than heard the explosion at the moment that it passed. We stationed ourselves as near as prudence would allow at a distance of 120 yards, so that only about one-third of a second elapsed between the detonation and the passage of the shadow; but the precision of observation of coincidence when very rapid movement occurs is so great, that I am quite satisfied that the observation was correct. The actual appearance of the ring was that of a strong black circular line, opening out with terrific speed from the point of explosion as a centre. It is impossible to judge of the thickness of the black shadow; it may have been 3 feet, or it may have been more at first, and have gradually become less in thickness or, possibly, in depth of shade....

Now, in the case of a hemispherical explosive wave it is clear that the sunlight can only be tangential over a semicircle, and that the shadow of such a wave should be a semi-ellipse, the eccentricity of which would depend upon the altitude of the sun.

LENSFIELD COTTAGE, CAMBRIDGE,
22 *Dec.*, 1897.

DEAR MR BOYS,

I am much obliged to you for the additional photographs, which show further features of interest. I should have written to you before only that I wanted to see your article on the visibility of a sound shadow before replying. I felt sure the number of *Nature* was in the house, for Mrs Robinson, Lady Stokes's step-mother, who is living here, takes it in. But she is very old (92¾) and is ill, in all probability not far from her end, so

that I could not get it. I looked yesterday in the University Library, which closed yesterday for a few days, but the volume was not on the shelf; probably it was with the binder, or stored by for binding. I tried also the Library of the Philosophical Society, but the attendant was away on holiday, and I could not find it. So I got a copy of the number to-day. I have read it with great interest, but I will take the photographs first.

In No. 5 I think that the great magnitude (or rather intensity) of the bore, while due in part to the glass which accompanies the bullet, is partly also due to the reduction of the velocity of the bullet by passing through the glass. For it seemed to me on comparing other cases of bores that a velocity of the projectile

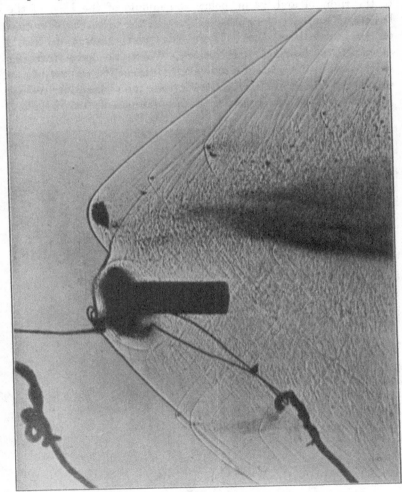

not greatly exceeding that of sound produced a bore which gave a stronger impression on the photographic plate than when the velocity of the projectile greatly exceeded that of sound.

No. 6 gives, as you mentioned, a capital example of two bores

moderately inclined to each other, and travelling in the same
direction, which coalesce on meeting, forming a single stronger
bore after meeting. The direction of the latter is intermediate
between those of the two former.

The example I mentioned before is in No. 1, about 2·05 in.
to the right of and 2·4 in. below the middle of the base of the
bullet. I suppose the photos held so that the direction of motion
of the bullet is upwards. There is a good example in No. 2,
about 2·9 to the right and 2 below. When two bores travelling
in the same direction are too much inclined to one another to
coalesce, their mutual inclination appears to be less after crossing
than before. There seems to be an example of this in No. 2,

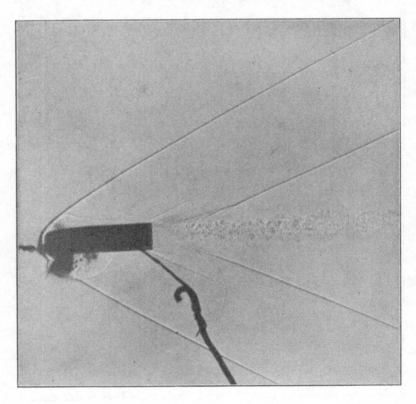

about 2 to the right and 1 below. In No. 1 on the contrary,
where bores travelling in opposite directions cross, the mutual
inclination is slightly increased after crossing.

I had heard you in a lecture explain the waves in No. 2 due to the motion of the glass when struck by the bullet but not yet broken. No. 6 shows also the mutual destruction of bores travelling in opposite directions. The hyperboloidal bores would meet at first below, and if they had not destroyed each other in good measure they ought to continue to be seen. For the rays falling tangentially on the one would pierce the other at an angle by no means small, and therefore might be expected to continue to show themselves.

I think the mutual destruction on crossing would partly depend on the magnitude of the disturbance. If, *caeteris paribus*, the disturbance was but small, we should expect them to cross. But these bores would be strong when they first met.

No. 7 is very interesting in connection with the gases used.

Now for your very interesting paper. I think you have gone in the right direction for explaining the completeness of the ring. But I should be disposed slightly to modify the ideas. I think there is highly condensed gas projected upwards with a very high velocity. It tends to expand; but having already got some way up it is able to expand in all directions. To take the extreme limit of what the condition is slightly tending towards, suppose you had a shell containing highly condensed gas, which after it had got a little way up exploded. It would produce a spherical wave which would cast a shadow, the intersection of which with the ground, supposed horizontal, would be an ellipse. If the altitude of the sun were 45 deg. the ratio of the axes of the ellipse would be about 7 to 10. I should not think that in such a rapid phenomenon you could well detect its difference from a circle.

Yours very truly,

G. G. STOKES.

25 *Dec.*, 1897.

The animated photograph you were so kind as to send me arrived by the morning's post (the only one to-day) and so came as a most acceptable Christmas gift. It is very interesting.

I did not at first see the interpretation of the rays which are seen, especially about the middle of the series, emanating from the place of the explosion. They are doubtless trails of stones or earth moving with extremely high velocity, and illuminated by

the sun. As the camera was probably in a somewhat elevated
position, and pointed downwards, the background on which the
rays are seen was probably grass, which accounts for the stones
&c. being more luminous than the background.

I am not sure whether the luminosity in the first 3 or 4
is to be attributed to the flame, or to the illumination of the
smoke by the sun. In the first 3 it *may* be due, or partly due,
to the illumination of the smoke. After the fourth it seems to
be due wholly to illumination by the sun.

The photographs are favourable to the view I threw out that
the highly condensed gas is projected upwards and gets some
distance before it has had time to expand, and that that accounts
for the dark ring being complete all the way round.

27, THE GROVE, BOLTONS, S.W.,
23rd Dec., 1897.

DEAR SIR GEORGE,

As is evident I have followed your example and
bought a type-writer.

I must thank you for your further aid in the matter of
understanding my bullet photographs. I have but little to say
now. I am so sorry that you should have had so much trouble
in raising that copy of *Nature*: if I had thought this would
have been the case I would of course have got one and sent it
myself. The conclusion that you have come to that a bullet
travelling more slowly [makes a greater bore] so long as it is
more than 1100 ft./sec. is hardly consistent with a good many
photographs that I have taken. The bore in front of a Martini
bullet, even though the bullet is larger, is much less conspicuous
than that in front of a magazine rifle bullet. The aluminium
bullet has a more conspicuous bore, bearing in mind the fogging
of the plate.

In further elucidation of the explosion experiment I enclose
a print of the animatograph film. You will see first the quick-
match burning at about a hundred feet a second in three of the
pictures, before the box containing the explosive is reached.
Then the sudden burst is by no means uniform or symmetrically
distributed. Certainly this is the case with regard to visible
matter. That curiously violent projection along the ground to
the left was very evident in its effect upon the grass. The other

local rushes of gas were also very clearly marked by torn-up grass and streaks of stones and dust. These are really rushing out more rapidly than the vertical column is ascending.

If this ocular evidence is worth anything at all it would tell against the suggestion in your letter of vertical projection and in effect radiation from a point in mid-air.

I believe the cause of the want of symmetry in such an explosion as this is to be found in the very unstable behaviour of the nitro-explosives. Thus, the rate of burning depends very largely upon the pressure, so much so that they will not keep alight in a moderate vacuum; at atmospheric pressure they burn feebly, but as the pressure rises the rate of burning rapidly increases, and so the production of gas and the rise of pressure is also increased, reacting on the rate of burning. So where the combustion starts most rapidly, there and in that direction it most rapidly becomes an explosion and a detonation.

There seems to be fair evidence of this kind of thing, which they please to call wave action, in the bores of guns.

The effect of pressure on the rate of burning of time fuses is well known and is allowed for at high altitudes. An R.A. officer told me that my photographs had been useful in explaining an anomalous rate of burning, where the exit of the time fuse gases was where he judged from the photographs the highest pressure should be, and that on putting it outside the pressure region the rate became normal.

Yours very sincerely,
C. V. BOYS.

LENSFIELD COTTAGE, CAMBRIDGE,
27 *Dec.*, 1897.

DEAR MR BOYS,
Your "enclosure" reached me on Christmas Day, but the letter itself not till Sunday. Hence when I wrote I had only seen the photograph, not the explanation. The explanation cleared up some things that had puzzled me. I had taken the light in the first three to be the commencement of the explosion, and as a dark cubical thing was evidently beyond the base of the light, I took it for something casually on the ground. I think I should have guessed that the explosion had been made by electricity. It certainly puzzled me why in No. 4 there should have been so rapid a change in the phase of the explosion.

We must distinguish between the outrush of gas resulting from the chemical changes and the outrush of elastic fluid, gas or air as the case might be. The photograph would show the former, but only that portion of the latter which was due to gas, the portion due to air remaining invisible in the photograph. The photo shows that the explosive was fired at the base of the box. Hence at the first moment of the explosion the explosive would be confined by the ground below and the explosive above, and there is a horizontal outrush of gas extending far out. The bulk of the explosive, whether in its original state or in the state of highly compressed gas formed by the chemical change that had taken place on the original, would be violently projected upwards, and would very soon consist of gas only. In expanding, this would condense and move outwards the air around it. But being no longer confined as at first, it would expand in all directions, and the gas would therefore not extend near so far from its original place as it did at the base. There would be a great outrush of condensed air outside the place of the gas, but this would not show in the photograph.

Suppose we had a sphere of gas in still air, the gas being condensed to 729 times the density it would have at atmospheric pressure. Suppose now the gas left to itself. It would expand and drive outwards the air outside it. But when the expansion was over and the gas in its natural state, it would fill a sphere only 9 times the original in radius. And if (from smoke mixed with it) the gas were visible it would not be seen to have gone near so far from where it had been as if it had been confined, say in a cannon, and so allowed to escape, with a rush, in one direction only. I think the bulk of the charge in the Maxim experiment was somewhat in the condition of the sphere of compressed gas in the above illustration.

What I took for photographic trails of rapidly moving stones or gravel in sunshine seem too persistent for such an explanation. They may have been smoke left by projected bits of the explosive which burnt as they travelled. If this were the explanation, they should have been visible to the eye.

Yours very truly,

G. G. STOKES.

27 *Dec.*, 1897.

The fact that the upright column of smoke is not wider than it is, shows, I think, that the ignition of the material or the expansion of the gas did not take place sensibly in one spot, but successively as the material was shot upwards.

The two thicker rays especially are very persistent. I doubted all along on this account whether they could be trails, but I thought they might be trails of gravel shot out in succession in certain definite directions. You having seen the thing will probably be able to say what the rays are, whether the two most conspicuous ones or the numerous very fine ones well shown about the middle of the series.

LENSFIELD COTTAGE, CAMBRIDGE,
20 *March*, 1900.

DEAR MR BOYS,

Professor Turner tells me you got no indication of radiation of heat from the dark part of the moon. This I can quite understand; for the crust being probably rock, accordingly a rather bad conductor, not a good conductor like metal, the surface would pretty rapidly cool by radiation.

A lunar eclipse would give a far better opportunity of trying, as the dark part of the moon would be caught shortly after it had been exposed to full sunshine. You may perhaps have tried it already, though if you have I do not recollect that it came before me. Lord Rosse (*Phil. Trans.* for 1873, p. 619) mentions a trial, but the circumstances were unfavourable, and besides he examined the moon as a whole, whereas you, I think, can work with quite a small portion.

There is but one lunar eclipse this year, and that only just an eclipse as regards the umbra. The middle of the eclipse is about 3.30 a.m. on June 13. At first sight the smallness of the eclipse seems disheartening. Yet in one respect it is a positive advantage. Of course in so small an eclipse we could only work with the penumbra. Now if we take two points on the moon's surface situated symmetrically with respect to the plane *SME* (the plane passing through the centres of the sun, moon and earth) for one the shadow would be advancing and for the other passing off. Hence one would have been longer under partial shelter from the sun's rays than the other, and therefore we might expect the former to emit less heat than the other, while as regards reflection the two would be alike. Had the eclipse been going to be total, the exposure at two symmetrically situated points would have been the same. Hence the eclipse of June 13 lends itself specially well for an investigation of want of symmetry in the heat coming from pairs of points symmetrically situated with reference to the plane *SME*. And the observation has the great advantage of being almost strictly differential, except as regards a possible difference of character of different patches of the moon's surface.

If you have already made observations on the eclipsed moon I should be glad to know where the results are to be found. Yours very truly,

G. G. STOKES.

23 *March*, 1900.

When I said that the lunar eclipse of June 13 was specially favourable for an endeavour to obtain evidence of the heat of the moon, I forgot one thing. In working with the penumbra we are tied down by having to compare points equidistant from the shadow, and in a grazing eclipse, like that of June 13, the most favourable choice is to take a pair of points symmetrically situated on opposite sides of the line of symmetry of the shadow. Now I forgot that in working on the umbra we are set free from the restriction of being obliged to compare two points at equal distances from the edge of the shadow, because we have no reflected heat to contend with, and we may choose the points where we please, and take them so that the line joining them shall be, at least roughly, parallel to the path of the shadow relatively to the moon. This would give us the greatest difference in the lengths of time during which the two points have respectively been in shade. So probably after all the small eclipse of June 13 would not be so good as a bigger one.

I suppose I may assume that you contemplate working with a reflecting telescope if you mean to attempt the observation on June 13. A refractor would be almost useless, as nearly the whole of the heat emanating from the moon would be stopped by the glass. Working with a reflector, Lord Rosse found that only 8 or 10 per cent. of the total heat from the moon was able to get through glass, even a fairly thin piece of plate glass; and as the total included the reflected heat, which might be expected to get through glass in much the same proportion as solar heat, the emitted heat from the moon would probably be almost wholly stopped.

66, Victoria Street, Westminster, S.W.,
26th March, 1900.

Dear Sir George,

My experiments in the heat or radiation reaching us from the moon were made with a 16-inch reflector stopped down to 6 inches. With a 12-inch stop, and using about, as far as I remember, $\frac{1}{50}$ part of the moon's visible surface, the radiation drove the instrument at once right off the scale.

I did not examine an eclipsed moon, as during the time that I had my apparatus set up there was either no eclipse or if there was one I could not arrange to observe it.

I did observe the curves of radiation, which are in my Roy. Soc. paper of about 1890, by allowing the image to pass over the small sensitive surface and taking at equal intervals of time, during the two minutes or so, the reading. These I found to be, so far as the method was capable of showing, perfectly symmetrical in the case of a full moon, but to show the maximum as we should expect at the part of the moon when the sun would be in the zenith, e.g. at the limb in the case of a half moon.

The perfect symmetry in the case of a full moon surprised me at first, because in that case one side is exposed to a rising sun after a fortnight's night while the other is exposed to a setting sun of equal altitude after nearly a fortnight's baking. Of course the very free radiation and low conductivity account for this.

The eclipsed moon is of course much better on account of the more rapid change of sun's radiation, and the observation would be worth making. I do not know whether F. W. Very at Alleghany has done anything of the kind.

I do not yet know if I can go to Algiers to observe the corona at the total eclipse. The astronomers want me to do so. Dr Common would let me have a *mirror* equatorially mounted, 20 inches diameter and only 45 inches focal length. This would give a far hotter image of anything than that which I obtained with the 16-inch mirror of Dr Huggins, which was about 60 inches focus, and so might be more suitable if the corona only radiates little heat.

My chief difficulty will be to know what part of the image is on the instrument. I have not yet gone into these details.

If the corona only radiates one 100th of the heat compared with the light of the full moon, a radio-micrometer made somewhat insensitive so as to be very quick should give with such an image good deflections, of which several should be observable in the one minute of the eclipse.

If I can get this in good order and conveniently workable, the same arrangement might be used on the moon at the succeeding eclipse.

If all this should work out so as to be possible, and the astronomer at Algiers will allow me to use the observatory for so long, it might be worth while to remain the fortnight and take such observations of the moon on intermediate days as are possible.

The greatest difficulty here is the mounting of the radio-micrometer, which must be, with an equatorial mounting, carried on a separate column at the right position and height to catch the image, and the axis of the rays must be sent into it always horizontally.

But all is very much in the clouds at present.

Yours sincerely,

C. V. BOYS.

30 *March*, 1900.

DEAR PROFESSOR BOYS,

I made some calculations relative to the radiant heat received from the moon. In the changes corresponding to phase, you found no sensible heat emission due to previous roasting, which as you notice is natural enough, as in these slow changes there would be plenty of time to cool by radiation.

Supposing for simplicity that the incident radiation is expressed by the sine or cosine of an angle proportional to the time, it turns out that the residual effect of roasting varies as the square root of the frequency. Supposing we rudely assimilate the changes in a lunar eclipse to a periodic change having its period the 1/400th of a month instead of a month, the residual effect of roasting might be something comparable with 20 times what there is from change of phase; and I should think it might be detectable in an eclipse, though if we have only the penumbra to work on we are restricted to comparing points equidistant from the shadow, in order to eliminate the portion due to reflection.

As to the ordinary conditions, *i.e.* no eclipse interfering, my views led me to think that the heat received would depend only on the lunar zenith distance of the patch observed, *i.e.* the angle between a normal to the lunar surface and a line drawn to the sun, and not on whether the patch were chosen from a dark region (one of the so-called seas) or a bright region. Shortly after I arrived at this result, I seemed to have a sort of glimmering recollection of having heard it noticed as remarkable that in a sweep over the surface of the moon it did not seem to matter whether one observed on a bright or a dark region. I thought perhaps it was in your paper giving the result of your experiments, but on reading it through I don't see it there. I think what I fancied I recollected must have been some remark made *vivâ voce* by you after the reading of your paper, which was on April 24, 1890. Whether it is a real or an invented memory, I cannot feel sure, but I seem to have a sort of recollection of your speaking it.

What a curious thing memory is! That a thing like this, which I had not attended to in the meantime, should have lain dormant for 10 years, and then turned up again. As a psychological curiosity, I should like to know (in case you recollect) whether you did say anything about it.

In an eclipsed moon I think the state of the surface *would* make some difference.

There would be a good opportunity next year in a total lunar eclipse in November. For the earlier one in May next year, the moon does not rise at Greenwich till the umbra has almost passed off. Yours very truly,

G. G. STOKES.

CORRESPONDENCE WITH SIR WILLIAM CROOKES, F.R.S.

In reply to a request, made some years ago, for information regarding his collaboration with Prof. Stokes, Sir William Crookes promptly sent a privately printed correspondence extending from Mar. 1876 to April 1879, relating mainly to the radiometer and the viscosity of ramified gases, and also various packets of letters relating to other subjects. His permission to make use of these documents has been fully taken advantage of, only such letters and passages as were of merely temporary interest having been omitted. In spite of the disclaimer in the letter next following, the reader will fully appreciate the value and the joint results of this combination of Sir W. Crookes' unrivalled experimental skill and bold intuition, with the refined theoretical insight of Sir George Stokes.

Oct. 16, 1904.

MY DEAR LARMOR,

In '76 to '79 I was corresponding with Sir G. Stokes frequently, and as I had the greatest difficulty in making out his writing I sent the letters to my printing office and got the head printer there (who could decipher almost anything) to set them in type. Then after collation and correction so as to get them correct I had a few copies printed off. *No one but myself* has ever seen them in this form, but now I send them to you—the second man of science who sees them.

These are only a selection; many of minor importance came between. But they serve to show the enormous indebtedness I, in common with most scientific men who corresponded with him, owe to Stokes. I have not read them for 20 years, and the impression they leave on my mind is that I knew very little of the subjects in question, and if what I owe to Stokes is deducted from my work there will be precious little left I can claim for my own!

I have kept religiously every letter I have received from Stokes, and they all are at your service.

Believe me, very sincerely yours,

WILLIAM CROOKES.

Nov. 1, 1889.

MY DEAR PROFESSOR STOKES,

I have only just been told of your selection to fill the presidency of the Royal Society.

Let me offer you my warmest congratulations on your thus succeeding to Newton's chair. The pleasure this gives me is, however, somewhat marred by the thought that we shall lose you as Secretary, and what a great loss this will be, will come home to every author of a paper in the *Phil. Trans.*, and to no one more than to myself.

Believe me, very sincerely yours,

WILLIAM CROOKES.

The correspondence which follows is arranged in order of time.

PEMBROKE COLLEGE, CAMBRIDGE,
March 2, 1856.

MY DEAR SIR,—It is now some months since Mr Spiller gave me your beautiful photographs of calc.-spar, nitre, and unannealed glass. As he told me he had also a spectrum for me, with a duplicate which you wished me to mark, I delayed thanking you for the first set of photographs till I should have got the spectrum. Mr Spiller was however too much occupied for a long time to mount the spectrum, so that it was only quite recently that I received it, or I should rather say them, as there are 5 spectra besides the loose one. I knew the lines at once, having repeatedly viewed them on fluorescent substances. I return you the loose spectrum with the lines marked in continuation of the first part which you have yourself done correctly. The new map referred to is merely a drawing I have made but not published* of the lines seen with a complete quartz train. I give also some measures taken with a pair of measuring compasses and a diagonal scale in which the unit is $\frac{1}{2}$ inch. I make H the origin when the contrary is not expressed, and measure positively in the direction of increasing refrangibility.

The following measurements refer to spectrum No. 2:

$-1\cdot24$ F Fraunhofer,
$-0\cdot60$ G do.
$0\cdot00$ H do.

* Published by Bunsen and Roscoe in *Phil. Trans.* 1859. See *supra*, vol. ii. p. 80.

$+ 0.08$ k, Stokes,

$+ 0.31$ to 0.35 l do.

$+ 0.46$ to 0.63 group m, Stokes,

$+ 0.55$ m, Stokes,

$+ 0.51$ to 0.63 Becquerel's group M,

$+ 0.81$ to 0.93 $\left\{ \begin{array}{l} \text{Becquerel's group } N \\ \text{first four lines of group } n\text{, Stokes} \end{array} \right\}$,

$+ 1.01$ $\left\{ \begin{array}{l} o, \text{ Becquerel} \\ n, \text{ Stokes} \end{array} \right\}$.

The following measures refer to spectrum No. 1 :

$- 0.45$ G,

$+ 0.72$ n, Stokes,

$+ 0.94$ o, Stokes; P (?) Becquerel,

$+ 1.06$ Q Stokes new map, p old map,

$+ 1.25$ R Stokes,

$+ 1.65$ S.

In Fig. III. I will measure from T instead of H. T is a narrow well-defined line with a pair of lines between it and the group S on the left and a broad line U to its right :

$- 0.24$ to $- 0.13$ group S,

$+ 0.19$ U,

$+ 0.30$ V,

$+ 0.36$ edge of photograph,

$+ 0.41$, about (from memory) W.

When the spectrum is sufficiently pure and extended R is seen to be a group of 4 nearly equidistant lines, resembling the first 4 of the group n. The two more refrangible are blacker than the two others. S is a group of 3 lines, the middle one dividing the interval between the extreme ones unequally, in the ratio perhaps of 3 to 2. The spaces between the lines of the group are somewhat dark, which I have attempted to represent by shading [*omitted*]. T is black and well-defined, not broad. U is a double line. The fluorescent light is copious about U, and V is seen without the least difficulty, but W is so excessively faint that it is only by catching the extreme end of the spectrum on the fluorescent substance, assuming all the rest of the rays to pass by, and then by looking away at a dark object (black velvet), and then looking at the fluorescent light that W is seen just on first looking. It is only during the summer months that this extreme part of the spectrum can be seen.

Should you have any difficulty about the identification I shall have great pleasure in giving you any additional information.

I received my quartz prisms in August 1852*. There were one or two very fine days towards the end of the month, and I was able to see as far as I have since seen even in the middle of summer; but I have never seen the more highly refrangible lines in winter.

I am, dear Sir, yours very truly,
G. G. STOKES.

20 MORNINGTON ROAD, N.W.
April 9, 1862.

DEAR SIR,—Some years ago you were good enough to name some of the principal lines in the ultra violet portion of the Solar Spectrum which I had photographed.

Since that time I have succeeded in obtaining the lines much sharper and also to a greater extent than I believe they have ever been got before, but now I meet with the same difficulty of recognising them owing to the altered appearance which the increased sharpness communicates to the spectrum. In this difficulty I think you will pardon my again troubling you. I enclose two copies of the ultra violet lines built up from different negatives as the foci of the lines differ in each, and shall be greatly obliged if you will write on one of them the principal lines and return it to me. The other I will beg your acceptance of. One difficulty which I have met with is that in different maps the lines are of different names, and few are sufficiently detailed to be of service when compared with my photographs.

Believe me, truly yours,
WILLIAM CROOKES.

LENSFIELD COTTAGE, CAMBRIDGE,
12 April, 1862.

DEAR SIR,—Allow me again to thank you for the very beautiful photograph of the solar spectrum as formed by a quartz prism and lens. I had little difficulty in recognizing the lines, for it was just about the degree of purity of the spectrum I commonly form

---

* Sir W. Crookes states that his own photographs were taken with this very train, which he bought from Darker two years later and still possesses. Cf. *Roy. Soc. Proc.* vol. lxxiv. (1905), p. 526.

when I wish to look at the lines by means of fluorescence, except near the very end, where from the falling off of the light I am obliged to use a somewhat wider slit, and therefore get the spectrum less pure. Most of the lines I recognize readily, in fact at a glance, for I have the spectrum pretty well by heart.

You will have seen by the table I sent this morning that you have not gone further by photography than I went by fluorescence. The reason is that we have got to the end of the solar spectrum. I stated at an evening lecture I gave before the British Association in 1852 that I conceived I had got evidence that I had now arrived at the end of the solar spectrum; and the only link in the chain of evidence which was left was shortly after supplied in the course of preparation for a lecture I gave at the Royal Institution in Feb. 1853. However as photography has an advantage over fluorescence, namely that faintness, or I should rather say, want of intensity, in the incident rays may be to a certain extent compensated by length of exposure, it was conceivable that photography might possibly push the spectrum a little further than fluorescence. That it could not go much further I have long known, as my friend Kingsley took some photographs shortly after I got my prisms and lens of quartz. The lens was meant for fluorescence, and was rather short-focused for photography, but still I was able to identify most of the lines.

I looked at the lines to-day about 1½ o'clock I think, and though the sun has still no great elevation I was able to see U readily, though it was a little faint. I saw light beyond U, but it hardly extended to V.

The line you have selected next beyond T is sharper than T, but if you view the photograph from a little distance such as a couple of feet with the eyes partially closed, you will see T much more conspicuous than the sharp line. In viewing the lines by fluorescence in this part of the spectrum, the slit cannot well be very narrow, the light beginning to be somewhat faint, and therefore for purposes of fluorescence T is a better line to select than the one next beyond.

I happen to have a revise of the plate of Bunsen and Roscoe's paper which Basire sent me as secretary*. I have got the plate in the volume of the *Phil. Trans.*, so I send you the revise with the lines you have selected marked, beginning with I.

The group S was at first represented as 3 distinct lines.

* See *supra*, p. 363.

I wished Basire to put a little shading between, and he left it out between the 2nd and 3rd lines, and put it in so strong between the first and second that it makes them appear like one broad band.

I am, dear Sir, yours very truly,

G. G. STOKES.

P.S.—As far as Q ($= p$) inclusive the map in the *Phil. Trans.* for 1852 is better than the one I send.

12 *July*, 1862.

I have had a good deal of correspondence with Mr Crookes on the subject of the photography of the solar spectrum, and have received from him some most admirable photographs of the spectrum, including that part which it requires quartz apparatus to show. Except as to some photographs taken by my friend Mr Kingsley with my apparatus, which have never been mentioned in print, Mr Crookes was I believe the first to apply photography to the investigation of the new region of the spectrum shown by a quartz prism, and to the delineation of its fixed lines. By studying photographs of the spectrum taken under a variety of circumstances, he arrived independently at conclusions respecting the influence of the Earth's atmosphere on the more refrangible rays agreeing with those which I obtained myself by working in a totally different manner.

As evidence of the ingenuity which Mr Crookes brings to bear on the most recent advances of science, I may refer to his beautiful and instructive experiment in which a candle-flame containing soda is seen surrounded by a black mantle when viewed against the flame of a Bunsen's burner also coloured by sodium.

Of his merits as a chemist I do not feel myself competent to express an opinion, as I have not paid particular attention to that science. I am well acquainted, however, with his labours on Thallium. By combining spectral observations with chemical reactions, as described in a paper published some time since, he obtained conclusive evidence of the existence of a new element, although the quantities with which he necessarily operated were very minute. He has now obtained the substance in larger quantities; and in a paper recently read before the Royal Society he has described both the element itself, which he has isolated, and several of its compounds. His name will therefore always remain in science in connexion with the discovery of this element.

31st March, 1876.

The torsion suspension you suggested, with a mounting which you have already employed, will, I find, answer perfectly to discriminate between the friction of the pivot (or the torsion which takes its place) and the viscosity of the minute quantity of the residual air or gas.

Let the fly be suspended by a fibre from a stopper with your heated indiarubber lubrication in the instrument fixed as in the figure, and the whole mounted so that you can readily turn it round its vertical axis. When all is at rest give 3 or 4 turns to the whole, and notice the *rate* at which the fly turns. When all is at rest again, give—at the same rate as before, as nearly as can be guessed—3 or 4 turns to the stopper *alone*, and notice the rate of the fly. If viscosity be the chief thing which sets the fly in motion, it ought to begin to turn much quicker when the whole vessel goes round than when the top alone goes round. The ultimate set, depending only on the torsion of the fibre, must of course be the same.

Aliter. Turn the vessel round, keeping the stopper fixed in space. If there be no viscosity the fly won't move; if there be, it will turn in the same direction as the vessel, though not so much, and, after stopping the vessel, will presently set in its old position relatively to space.

3rd April, 1876.

It is needless to say, if you carry out the experiment, the suspending fibre should be as fine as is consistent with safety, that the effect of viscosity may be more apparent. Save for the sake of using the apparatus, when made, for other experiments, it would be more convenient that the vanes should *not* be blackened on one side, lest the instrument (the fly of it) should be set rotating by the heat of the body.

13th April, 1876.

I have modified the opening sentence thus:—" During the discussion which followed the reading of Prof. R.'s and Dr S.'s papers, at the last meeting of the R.S., I mentioned an experiment *which confirmed* [BEARING ON] the observations of Dr S. I have since tried this in a *somewhat* modified form, and, as the results are very decided,...before the Society."

My reason is this:—My recollection of the results you mentioned on March 23rd is that you had obtained motions of a floating case, but they were sometimes one way and sometimes the other, and their law had not as yet been disentangled from the individual results. This was afterwards done by the very beautiful method of fixing the fly by a magnet, and *now*—*i.e.* in the modified experiment as described and shown on March 30th—all was clear and harmonious, and the results confirmed very beautifully Dr Schuster's experiment.

Consequently it can hardly be claimed for the experiments on a floating case up to March 30 that they confirmed the observations of Dr Schuster. It was the modification of the experiment (too important to be qualified by a "somewhat"), bringing it to what was shown on March 30, that made the whole thing clear; and NOW it was an obvious and very beautiful confirmation of Dr Schuster's experiment*.

17th April, 1876.

It would be desirable to register not merely the amplitude of the first swing, but the readings for the first 5 swings or so. This would afford a good value of the logarithmic decrement (the decrement per swing of the logarithm of the amplitude measured from the *new* zero, or reading at which the image sets) which is the constant most desirable to know.

17th April, 1876.

The initial arc in your experiments depends not solely upon the viscosity, but upon other circumstances as well, though the viscosity doubtless plays a leading part. That the initial swing at first *increased* as the air was exhausted I take to be due not to a greater viscosity, but to a diminished inertia. According to Maxwell the viscosity (in the sense of the resistance offered to a given motion of gliding, or, as engineers call it, of shearing) is independent of the density. Therefore the viscosity—in the sense of treacle-like character of the motion, depending therefore on the

* *Roy. Soc. Proc.* 1876.

S. B. II. 24

viscosity in the former sense directly, and the inertia of a given volume inversely—continually increases as the density diminishes. That is to say, with reference to a solid moving in the air, the opposition to motion (arising from this cause) is independent of the density, but the effect of viscosity *on the air itself* continually increases as the rarefaction progresses, the same force having to change the motion of a smaller and smaller quantity of matter. When I said the viscosity is independent of the density, there is of course a limit, and that limit is when the attenuation becomes so excessive that we are no longer at liberty to treat the number of molecules as practically infinite.

That the viscosity *appeared* at first to increase as the air was rarefied, I take to be due to this,—that when the bulb was turned the tangential force had not time to set the comparatively dense air in motion. If you turn a cup containing tea rapidly round, the liquid close to the side is moved round, but that near the centre remains nearly at rest in consequence of its inertia. If you turn the cup round through a given angle, a slower turn would be more effective in producing motion in parts about the centre than a very quick turn. So in your experiment the amplitude will depend upon the manipulation,—will be different according as you turn quickly or more slowly. And even if you turn at the same rate, in experiments with different degrees of expansion, there will be no simple relation between the arc swung through and the viscosity. What *is* simply related to the viscosity is the logarithmic decrement of the arc of oscillation. You have given me the data for finding this in the case of swings in air at full pressure. From the swings I derive the readings, and thence, by taking the difference between the reading and 255·3, the reading referred to the new zero, thence the logarithms, and the decrements of the logarithms.

Swing.	Reading.	Less 255·3.	Logs.	Diff.
+404	+404	+148·7	2·1724	
				0·3014
−223	+181	− 74·3	1·8710	
				0·2614
+115	+296	+ 40·7	1·6096	
				0·3021
− 61	+235	− 20·3	1·3075	
				0·2393
+ 32	+267	+ 11·7	1·0682	
			Mean	0·2760

The log decrement might also be deduced directly from the swings. Thus—

Swing.	Logs.	Diff.
404	2·6064	
		0·2581
223	2·3483	
		0·2876
115	2·0607	
		0·2754
61	1·7853	
		0·2802
32	1·5051	
	Mean	0·2753

The log decrement per oscillation comes out 0·276. This ought to be independent of the mode of starting the oscillations, though in some cases the first swing would hardly be nearly enough in regular series. It might be better to start an oscillation greater than you mean to observe, so that the oscillation should be fairly established when you begin to preserve the scale readings. Also for finding the log decrements it might be useful to move the stopper: then, when the plate first comes to rest, to move it back to its old position. In this way we avoid the shifting of the zero and secure a larger arc, and we may probably be able to dispense with the first arc, which is hardly in regular series,—I mean the first arc *after* the double motion of the stopper.

It is desirable, doubtless, to record the first arc, but the discussion should, I think, turn on the logarithmic decrement of the arc.

The logarithmic decrement will involve, of course, the viscosity of the glass. But glass is so nearly perfectly elastic, and the fibre so very thin, that this will be practically insensible.

The motion produced by the rotation of the bulb alone has the advantage of exhibiting palpably to the eye that there *is* a viscosity between the suspended body and the vessel; but once admit that the viscosity of the glass is practically insensible, and the experiment with the stopper moved is as good.

When the time of oscillation sensibly changes with the rarefaction, it is as well to divide the log decrement of the arc by the time of oscillation, so as to get the log decrement per second instead of per oscillation. The log decrements are mostly alternately greater and less than the mean. It is quite conceivable that this may be real, and not accidental, as the body was moving opposite ways—just as the resistance to a ship would be different according as she went prow foremost or stern foremost. Therefore it is well to take an *odd* number of arcs for final discussion, leaving an *even* number of intervals, so that the difference, if any, between fore and aft motion should be eliminated.

18th April, 1876.

It is needless to mention the satisfaction to the mind arising from plotting one's observations on paper. I did so for the numbers you sent for the swings in air, taking equal intervals for abscissae, and ordinates proportional to the logarithms of the arcs measured from the new zero. The dots lay very nearly in a straight line. There appeared to be a slight curvature to the axis of the abscissae, so slight that it hardly emerged with certainty from casual errors. On theoretical grounds there ought to be a slight deviation in this direction, which would diminish with the density. As it is hardly sensible for full pressure, it would probably be utterly insensible for anything approaching high rarefaction.

When arcs decrease in geometric progression, and the log decrement is to be found from the initial and final arcs, it is not of course good to go very far in the way of smallness of arc, for then an error in the reading of the arc tells too much. To use the initial and final arcs only comes of course to the same thing as taking the means, as I did in my last, for the sake of comparing the individual results with the mean of the whole. For to take a particular instance, say there are 5 arcs, a_1, a_2, to a_5. The mean log dec. is—

$$\tfrac{1}{4}\{(\log a_1 - \log a_2) + (\log a_2 - \log a_3) + (\log a_3 - \log a_4) \\ + (\log a_4 - \log a_5)\},$$

which is the same thing as $\tfrac{1}{4}(\log a_1 - \log a_5)$, and the same is evidently true whatever be the number of arcs.

If we enquire what is the best arc to stop with, so that a given error in the observation of the small arc shall produce a minimum error in the deduced log dec., we find it (the last arc) should be 1/eth part of the first arc used in the computation, e being the base of the Nap. logs, namely 2·71828....As the small arcs can be observed a little better than the large arcs, on account of the slowness of the motion, we may go a little lower, say to $\frac{1}{3}$ or $\frac{1}{4}$th of the first arc preserved for calculation.

Your beautiful experiment with the little pith pendulum shows very plainly the presence of residual gas. To keep the pendulum permanently up to the large swing I saw, at Burlington House, as the rate of rotation of the radiometer was slowly falling, it would require a nice adjustment of the distance of the candle, so that the rate of rotation should be almost exactly the critical one for synchronism corresponding theoretically to a rate at which one arm passes for each complete oscillation.

I confess that now the evidence to my mind is very strong, almost overwhelmingly strong, that the rotation of the radiometer is due to an action of heat between the radiometer (or more correctly the fly, the "radiometer" being the instrument as a whole) and the case, through the intervention of the residual gas. But the action is none the less a perfectly new one. No one, so far as I know, had made the slightest approach to discovering it experimentally. No one had dreamt of its occurrence as a matter of theoretical prediction. And even now its theoretical explanation is not an application of well-ascertained laws, but the following out of a certain speculation as to the ultimate constitution of matter and the nature of heat; and your discovery, from the thorough novelty of the action, cannot but exercise an important influence on the progress of our knowledge.

For getting the log dec. of the arc it is sufficient to set the mica in vibration anyhow, and the bulb and suspension might be turned together; and I should not think there would be much difficulty in arranging for that by a sufficiently flexible tube of glass.

20th April, 1876.

I notice the same slight curvature in your figure for swings in air at full pressure that I observed in my own.

The other figures relate to high exhaustions, for which, as I said, the curve for log decs. might be expected to be sensibly straight. We have not at present the data for intermediate exhaustions, and I cannot therefore judge whether the curvature for full air is real or accidental. If real it is very small and is altogether a subordinate feature of the experiment.

It will be very important, I think, to get the log decs. when the apparatus is free, for a moderate series of exhaustions. If it proves nearly constant (at least when corrected for the time) till the expansion becomes very great, and then goes on diminishing, that will, in the first place, confirm Maxwell's theory, and, in the second place, will show that you have reached such extreme exhaustions that the mean length of path of the molecules between their collisions is no longer very small compared with the dimensions of the apparatus—a condition supposed in the statement of the result that the viscosity is independent of the density*.

If the result be as I mentioned, namely, that the log dec. does not begin to fall off notably till extreme exhaustions are reached, that will, I think, afford evidence that you are really drawing towards the condition of a perfect vacuum. By plotting a curve with the number of bottles full of mercury for abscissae, and the log dec. for ordinate, some indication might be afforded of the rate at which you were approaching a perfect vacuum.

The *proportional* error of observation of a small arc is of course much greater than that of a large one, but the *absolute* error is probably, if anything, rather less. In concluding that the final arc had best be about 1/eth of the initial, I supposed the *absolute* errors the same, and the proportional error of course greater for the smaller arc.

24*th April*, 1876.

Make the fly for the radiometer with disks thus prepared [of roasted mica], and lastly blacken one face of each with a smoky flame.

It seems to me this plan would combine the very advantages we want, namely :—

1. Good absorption on one face.
2. Good reflection on the other, for heat as well as light.
3. Bad conduction.
4. Absence of organic matter.
5. Extreme lightness.

I first thought of using mica painted white on one side and smoked on the other. For the which it might be safest to use precipitated and dried silica, as we know that quartz is fairly diathermanous, and the rays have got to get through glass any-

---

* Cf. Prof. Stokes' reduction of Sir W. Crookes' observations, *Phil. Trans.* 1881, *Math. and Phys. Papers*, vol. v. pp. 100—116.

how. Probably the purest pipe-clay would be safe. If the material of a tobacco-pipe were turned into a thin lamina, probably this would be very good, and I suppose it could readily be managed by enclosing a bit of the clay in a pasty condition between two pieces of smooth paper, and pressing it then by squeezing it between flat surfaces, such as pieces of plate glass, removing one piece of glass to let it dry; then, when quite dry, heating before the blowpipe or in a furnace, which would at the same time burn off the paper. But if the roasted mica be sufficiently protected against breakage by the unroasted rim, it seems to me to combine in a pre-eminent degree the very conditions we want. I know you have tried mica, but there is a great difference I take it between clear mica and roasted mica. The effect of the heating I take to be its grand merit.

If the mica be very thin it will become pearly nearly up to the gripping-point; if thicker, it will remain clear for a little way inwards, the heat being conducted to the metal. I have tried this.

I think carbonate of lead might be used for the white paint, but not a sulphate. I am disposed to think carbonate of lead would be very good.

28th April, 1876.

I was much struck by the experiment described in your lecture, in which a metal (silver, I think) radiometer, when covered with a heated glass shade, went black side foremost. I can understand a metal radiometer going this way <i>when cooling</i>; but to explain its going this way when subjected to the radiant heat from a glass shade, I feel obliged to suppose that the lamp-black was snow-white with respect to the radiant heat of low refrangibility, while the other surface was only silver-white.

If this be so, I should expect even a pith radiometer with the opposed surfaces coloured, one by lamp-black, the other by a salt of copper (say oxalate, to take a pale one), would go opposite ways, according as it were acted on by light alone or by invisible heat alone of low refrangibility. For light the lamp-black would be black and the oxalate of copper nearly white, but for heat of low refrangibility the lamp-black would be white and the oxalate of copper black. It is conceivable that the interposition of water might even accelerate the motion produced by radiation from a luminous source.

1st May, 1876.

You asked me if I could suggest a way of measuring the friction of the point of support of a radiometer. I don't see my way to doing it directly; but if we may assume, as Reynolds does, and as we know is true for sliding contact, that the friction is independent of the velocity (I don't know whether experiments have actually been made for a body revolving round an axis passing through a point of contact), then, as I think, we have evidence that for a radiant source not too close the other resistance is as the velocity, and the force inversely as the square of the distance. Assuming this the thing may be done.

Place a standard candle at such a distance (D) that the radiometer *only just* revolves uniformly, and again a good deal nearer, but not too close, and determine the velocities (v, v') of rotation, and the second distance (D') as well as the first. Then, by our suppositions for any distance D and velocity v,

$$\frac{C}{D^2} = Av + B,$$

where C depends on the strength of the candle, which I suppose constant, B on the friction of the support, and A on the resistance of the residual air, or whatever it be that causes a resistance proportional to the velocity. The equations

$$\frac{C}{D^2} = Av + B, \qquad \frac{C}{D'^2} = Av' + B,$$

give

$$B = C\,\frac{1/vD^2 - 1/v'D'^2}{1/v - 1/v'}.$$

For example, suppose a candle at a distance 8 feet gave 3·6 turns per minute; at distance 30 inches gave 40 turns a minute; then...the friction would be balanced by the force of the candle at 327·1 inches, or 27 feet 3·1 inches. To *start* the radiometer the candle would have to be brought nearer than this, because experience shows that it requires more force to start a body resisted by friction than to keep up the motion when it is once started.

The thing might probably be got from the results which you have sent in to the R.S., drawn on paper, with $1/D^2$ for abscissae and v for ordinate. If the equation be true we ought to get a straight line till you come to higher velocities, where the square of the velocity would come in. Only the straight line would not pass through the origin, but cut the axis of x a little to the right of it.

3rd May, 1876.

In return for your kindness in presenting me with a radio-meter I have a small contribution to your subject, which I think of some interest*.   I have repeated the experiment over and over again to make sure of the result.

I take a tumbler or ale glass, with a foot, which serves for a handle.   I cover with it the spout of a kettle on the fire out of which steam is rushing, and hold it there for some time till it gets hot.   Then hold it over the radiometer, of which the bulb is buried in it, but without touching.   Almost immediately a nega-tive rotation is set up (*i.e.* black foremost), which soon becomes pretty lively.   Before very long, however, it stops, and gives place to a positive, which continues long.

If the tumbler is heated as before, but dried with a cloth before inverting it on the radiometer, there is no negative rotation, but a tardy positive one, which lasts long.

If the wet hot glass be removed while the rotation is still lively negative, the fly almost immediately stops, and then rotates positively.   The positive rotation is distinctly increased by blow-ing on the moist bulb or fanning it.

If the tumbler, whether wet or dry, be removed after the fly has been rotating positively for some time, a positive rotation con-tinues for a good while.   When the bulb is moist (at least in the earlier stages), perhaps also when it is dry, blowing on the bulb seems to favour positive rotation.

The explanation seems to me to be this:—Water radiates and absorbs a kind of heat which passes pretty freely through glass, and with respect to which pith bears to lamp-black the relation of black to white.   But with respect to the mean of what glass radiates and absorbs, lamp-black is blacker than pith.   Hence, at first with the wet tumbler, the water radiation penetrating the glass drives the pith, the glass radiation being in great measure stopped by the bulb,—some, even, by the water in liquid film and vapour.   Presently the bulb gets heated, and *then* the radiation from the bulb carries the day.

When the tumbler is removed the radiation of the bulb prevails over that of the small quantity of water in the shape of

* Cf. Prof. Stokes, "On certain Movements of Radiometers," *Roy. Soc. Proc.* 1877, *Math. and Phys. Papers*, vol. v. pp. 24—35.

dew, and on blowing the cooling of the water by evaporation favours positive rotation.

It is clearly not enough to specify the temperature of a radiating body : its nature has a most material influence on the character of the radiation.

5th May, 1876.

I have tried your experiment of breathing on the radiometer. The motions all fall in with the explanation I gave you. In my experiment and in yours the surface was bedewed. If my explanation is right the effect does not depend on the deposition of moisture, and I mean to try if I cannot produce it independently of this.

8th May, 1876.

Your figures for "repulsion by candle" are very remarkable. They denote, I presume, statical measures; for of course the first swing, when a candle is let on, is influenced by the diminished resistance, independently of any change in the force.

12th May, 1876.

Here's the formula for least squares if you care for it, but it certainly would not be worth the trouble it would cost to use it....

But though I don't recommend this, on account of the trouble, it would be easy to group a long series into intervals of, say, 4 arcs, and take the log decs. for comparison with one another. Thus, let $a_1, a_2, a_3, \dots$ be the arcs; then we may take

$$\tfrac{1}{4}(\log a_1 - \log a_5), \quad \tfrac{1}{4}(\log a_5 - \log a_9), \quad \tfrac{1}{4}(\log a_9 - \log a_{13}),$$
$$\tfrac{1}{4}(\log a_{13} - \log a_{17}),$$

and compare them with one another. They would be more regular than the log decs. for single intervals only, as the errors of observation would be divided by 4.

12th May, 1876.

I think you will be able to get some interesting results from the candle-swings without much trouble of reduction.

Register the stopping-points when the candle is let on in successive swings; and in an adjacent experiment find the log decs., without the candle. If this experiment is made after the other it will only be necessary to give time for the thermal disturbance produced by the candle to subside; you need not wait till the mica has come to rest, but give it a fresh impulse.

Let l be the log dec., N the number whose log is l. Divide

the successive arcs by $N + 1$, and apply the results to the readings
of the second ends of the arcs, to get the equilibrium readings for
the force arising from the candle supposed constant during that
swing, and equal to what it was in the middle of the swing.
These readings will increase and approach a limit. The results
may then be plotted, taking for abscissae 1, 3, 5, 7, ... and for
ordinates the equilibrium readings. You will probably get a
curve like—[omitted].

This would indicate not a direct force of the candle, but a
change of state brought about by the candle....

12th May, 1876.

By way of experiment I have tried the method I mentioned
on the numbers you gave me in your letter of April 17. For the
log dec. I take the arcs 116 and the 4th after, 27.

Log 116 = 2·0645, log 27 = 1·6721,

say log dec. = 0·098. The point is to see how near the equilibrium
reading, calculated with this log dec., giving log $(N + 1) = 0·353$,
for *each arc separately*, comes to zero.

Reading.	Arc.	Log.	Less 0·353.	Number.	Calculated Equilibrium Reading.
+65					
−51	116	2·064	1·711	51·4	+0·4
+40	91	1·959	1·606	40·4	−0·4
−31	71	1·851	1·498	31·5	+0·5
+27	58	1·763	1·410	25·7	+1·3
−20	47	1·672	1·319	20·8	+0·8
+14	34	1·531	1·178	15·1	−1·1
−13	27	1·431	1·078	12·0	−1·0
+10	23	1·362	1·009	10·2	−0·2

As to the calculation, the whole affair is the work of a few
minutes. The mean deviation from zero, in the last column, is
only 0·7 of a division. We may conclude that when the equilibrium reading (*i.e.* what would be the reading if the mica were
not swinging) is slowly changing, the method will apply without
sensible error. Whether the change of zero is *sufficiently* slow, or
whether it will be necessary to have recourse to a more refined
method, remains to be seen when one or two series are reduced. I
suspect it *will* be sufficient, so that we shall readily find how the
deflecting force grows, as well as what it attains to.

When the log dec. is small it would suffice, without even
calculating the log dec., to take the readings in threes: take the

mean of the extremes, then the mean of that and the middle, and consider this as belonging to the time of the middle rest, not to the middle of a swing. I give the result of the last example treated so. This of course does not show a progressive change in the zero of the swings, as there was no candle.

Mean.	Middle.	Mean of Last Line.
$+52\cdot5$	-51	$+1\cdot2$
$-41\cdot0$	$+40$	$-0\cdot5$
$+33\cdot5$	-31	$+1\cdot2$
$-25\cdot5$	$+27$	$-0\cdot7$
$+20\cdot5$	-20	$+0\cdot2$
$-16\cdot5$	$+14$	$-1\cdot2$
$+12\cdot0$	-13	$-0\cdot5$

The mean of the last col., without regard to sign, is $0\cdot8$ division.

13th May, 1876.

I omitted to mention how you may get the log dec. when it becomes so small that the observations as hitherto practised would be tedious.

Take 3 or 5 readings (giving 2 or 4 arcs), noting the h. m. s. of the last by your watch. Then write or do anything till the arc has fallen to $0\cdot3$ or $0\cdot4$ of what it was, and take 3 or 5 readings more, noting the moment of the first. (Or say 3 at beginning and 7 at end; the division would then be $n + 1 + 3$, or $n + 4$.) The mean of the log arcs for the first set gives the log arc for the moment of middle rest (i.e. the 2nd or 3rd reading as you took 3 or 5 in all), and the same for the last set. The difference of noted times divided by the time of one swing gives the number of swings (n say) that passed between the observations, and the difference between the mean logs, divided by $n + 2$ or $n + 4$, as you took sets of 3 or 5, gives the log dec.

Indeed this is about the best way of treating even a moderately long set of swings, only you count directly instead of using your watch as a counter. Take the mean of, say, 2 log arcs at the beginning, and of 4 or 6 when the arc has fallen to $0\cdot3$ or $0\cdot4$ of what it was, and divide the difference between the mean logs by the number of swings between the middles of the sets.

14*th May*, 1876.

I am much obliged for the communication of your very interesting results. I intend to send you the numerical results of the method I proposed to-morrow, but I write to-day to propose a couple of experiments, lest if I waited the vacuum should have got too complete. Meanwhile I will just observe that your experiments show clearly—

(1) That the force deflecting the mica (not including the torsion) is by no means constant, but mounts up during a very appreciable time.

(2) That it is well nigh at its full value even by the end of the first swing.

(1) pretty well proves by itself alone that the force is not due to radiation *directly*, but *indirectly*.

From (2) it appears that these experiments won't serve to show the law of increase of the force, as it is almost full grown by the time the first observation is taken.

And now for the experiments :—

Exp. 1.—To try if the force lasts after the candle is cut off, and whether, if it does, it falls fast or slowly. Adjust the mica to rest under the influence of the candle; now cut off the candle, and take the readings for the ends of the swings.

If the force ceases when the candle is cut off, the swings will be to and fro about zero, the point the candle started from being in series with the turning-points.

If it falls as it rose the swings will be like—[*omitted*].

If it falls more slowly they will be like—[*omitted*],

i.e. hovering about a point which gradually sinks to zero.

Exp. 2.—The growth of the force proves to be too quick to allow of making out its law by the method I mentioned. It may be ascertained indirectly in this way :—

Adjust the mica to rest, with candle cut off. Then raise the screen for, say, $\frac{1}{2}$ s., and in subsequent experiments for 1 s., $1\frac{1}{2}$ s., 2 s., $2\frac{1}{2}$ s., 3 s., $3\frac{1}{2}$ s., and note in each case how far the mica swings. I don't think anything more than the first swing will be wanted. The times of exposure might be taken 0·8, 1·6, 2·4, 3·2, 4·0, if a watch giving 5 ticks in 2 s. be used.

Exp. 931.   Series I.

Log Dec. 0·135 ; L. $\overline{n+1}$ 0·374.

Reading.	Arc.	Log Arc.	Less 0·374.	Number.	Zero for Swing.
64	64	1·806	1·432	27·0	37·0
26	38	1·580	1·206	16·1	42·1
54	28	1·447	1·073	11·8	42·2
34	20	1·301	0·927	8·5	42·5
50	16	1·204	0·830	6·8	43·2
36	14	1·146	0·772	5·9	41·9
42	6	0·778	0·404	2·5	39·5

Series VI.

Log Dec. 0·066 ; L. $\overline{n+1}$ 0·335.

Reading.	Arc.	Log Arc.	Less 0·335.	Number.	Zero for Swing.
88	88	1·944	1·609	40·6	47·4
29	59	1·771	1·436	27·3	56·3
83	54	1·732	1·397	24·8	58·2
32	51	1·708	1·373	23·6	55·6
75	43	1·633	1·298	19·9	55·1
40	35	1·544	1·209	16·2	56·2
68	28	1·447	1·112	12·9	55·1

Series XIII.

Log Dec. 0·017 ; L. $\overline{n+1}$ 0·310.

Reading.	Arc.	Log Arc.	Less 0·310.	Number.	Zero for Swing.
42	42	1·623	1·313	20·6	21·4
15	27	1·431	1·121	13·2	28·2
44	29	1·462	1·151	14·2	29·8
14	30	1·477	1·167	14·7	28·7
41	27			13·2	27·8
14	27			13·2	27·2
40	26	1·415	1·105	12·7	27·3
15	25	1·398	1·088	12·2	27·2
39	24	1·380	1·070	11·7	27·3
16	23	1·362	1·052	11·3	27·3
39	23			11·3	27·7
17	22	1·342	1·032	10·7	27·7
39	22			10·7	28·3

The numbers in Col. 6 are got by applying Col. 5 to Col. 1, by subtraction and addition alternately. The first figures are marked off with a line, because the method gives only a rude approximation for the first swing, as the whole of the force, very nearly, has grown during this swing. There seems to be some indication of a minute decrease of the force after a time. This may be real, and may depend on the conveyance of heat by conduction to the hinder face, which then acts antagonistically.

15th May, 1876.

If you attempt the experiment with the subdivided time, *i.e.* with the force let on for a short time only, the things wanted will be—

1 (and chiefly). The first arc the body swings through; and as independent matters—

2. The log decs. for the exhaustion in question.

3. The equilibrium reading for candle on, which may be got just as in weighing, as the log decs. are now so small. The swings, if inconveniently large, may be damped or destroyed by the hand screen.

4. The time of vibration, candle off or on,—say off.

2 and 3 will of course vary with the exhaustion; 4 is practically constant, but I mention it as it might be as well once for all, now that the log decs. are so small, to get a thoroughly good value by finding the time of, say, 50 vibrations.

With such small log decs. the equilibrium position, whether with candle permanently off or permanently on, may be got as in weighing, and the more elaborate formula I gave (with the $1 + N$), which would be useful with a larger log dec., is not required. The variation of the force is not traceable by the swings, as it is well-nigh accomplished within the first.

16th May, 1876.

I was working last night at the mathematics of the swings with a view to get the variation of the force with the time. I see it *would* be desirable to take 6 or 7 swings, and not the first alone. A determination, even moderately accurate, of the time of the *first* swing,—*i.e.* time from candle let on to first rest of spot of light,—would also be desirable. This would require either a second observer or a second experiment. I dare say, however, if you call "Now, Now" to your assistant, he could observe the time and you the arc, so as to do the two together. The swing-time being, say, 4 s., the *first* swing I should suppose would take 6 s. or so.

16th May, 1876.

You ask how to get the real zero, but I must first ask what you mean by the real zero. If your apparatus had no inertia, and had no resistance to overcome, it would obey the force at once, and move without swinging. The true curve would then be something like this,...the force rising at a decreasing rate till it is sensibly constant.

Actually the mica swings, and we may think of its motion as of one swinging about a point which itself changes more or less slowly with the time. The rate of change would seem to be such that the point a, or thereabouts, would correspond to the end of the first swing.

If by the real zero you mean that for the force when it has become sensibly constant, it is clear that we must reject that point where the force is distinctly varying. The simple method I gave you will apply without sensible error when the force, though not absolutely constant, varies but slowly, and may give useful information as to slow changes in the force. But when the variation is rapid, as in the first swing, a more refined method must be used.

Hence, to find the zero for the force supposed constant, the first one, or possibly two or three results, must be discarded. I marked off the first with a red line as giving but a rough approximation, and that to a changing zero. The second is so nearly in series that I cannot possibly say that it is out. Hence it will suffice, I think, to take the mean of the rest after striking out the first. In Series I. the last is out of line, for whatever reason, but whether we include it or not will make little difference in the mean. With it we get for mean 41·9, without 42·4, a difference of only half a division of the scale.

If the small falling off in the force be normal, and not accidental, the time curve would be like the following :—[omitted], and we must distinguish between the maximum force corresponding to the tangent fg and the ultimate asymptote hk. But until this feature be established as normal, if normal it be, it is not worth while to attend to it.

Instead of using the log dec. for the whole series of swings, we may get the slowly varying zero by taking the geometric means between consecutive arcs, and applying half the result to the middle readings ; and where the log dec. is small the geometric means won't sensibly differ from the arithmetic. I treat, for an example, Series VI. in this way....

The two [methods] differ by only half a division.

I don't expect the log dec. will be changed whether the candle is off or on. The log dec. depends for a given swinging body on the degree of exhaustion; the zero about which the swings take place, on the force of the candle.

Though the log dec. will vary with the suspended body, I expect the *ratio* of the log dec. to the log dec. for the same body, with the same gas, while at a still measurable pressure, depends only on the pressure. That is, if for air at 50 mm. pressure we get a log dec. 0·300 with a swinging body *A*, and only 0·030 at some high unknown exhaustion ; and if a swinging body *B* has in hydrogen at 10 mm. a log dec. 0·200, and if the hydrogen be exhausted till the log dec. sinks to 0·020, I expect the degree of exhaustion in the hydrogen and air would be the same. I am not sure of this, but anyhow the *ratio* of the log dec. at a high exhaustion to the log dec. at a standard pressure, such as 50 mm., will, I think, depend only on, and therefore serve to define, the degree of exhaustion.

19th May, 1876.

Did I understand you rightly to say that the time from letting on the candle to the first rest was less than the time subsequently from rest to rest,—*i.e.* less than *half* the time of a complete oscillation ? If so, I own I am puzzled; I don't see how to account for it. I should have expected the time to be $1\frac{1}{3}$ or $1\frac{1}{2}$ time of one vibration,—*i.e.* $\frac{2}{3}$ or $\frac{3}{4}$ of the time of a complete oscillation.

I am struck by the extreme regularity of the swings when the candle is cut off; probably minute variations of intensity of the candle prevent them from being *quite* so regular when the candle is on.

I shall feel curious to see whether the roasted mica radiometer will behave similarly to the other with a heated wet tumbler. I expect it won't.

20th May, 1876.

Many thanks for the communication of the results relative to the temporary action of a candle. My "puzzle" I think is now cleared up. When you told me at the R.S. that the first swing took less time than the regular swing-time, I thought you were speaking with reference to experiments in which the candle was let on and *left on*, whereas you were really speaking of experiments in which the candle was let on and then cut off for 1 s. or so.

The supposed diminution of time of swing in the former case I could not understand; in the latter it is intelligible enough. The large effect on the time of the second swing shows that even 3 s. after the candle is cut off the force is far from having disappeared. Hence the second swing at least, as well as the first, ought to be rejected in finding the log dec. That is to say, it could only be properly used by the aid of a very refined mathematical analysis: the simple methods I gave you, which apply when the state of things is fairly established, would hardly be accurate enough to use till after the second swing, or possibly later.

I have little time for calculations while my lectures are going on, but when they are made we shall be able to see within what limits the simple formulae may be used.

The behaviour of the mica (roasted) and lamp-black radiometer is altogether different from what I had ventured to prophesy in relation to the hot tumbler. I find my ideas as to the behaviour of mica rectified. The mica radiometer agrees with the silver and lamp-black one you mentioned in your lecture.

27th May, 1876.

I have been making some calculations as to the dynamics of one of the swings. I chose Series XIII. I assumed a law of decrease of the force calculated to represent it in its main features, and determined two disposable constants to make them fit. The constants I used were, besides the log dec. which you gave me,—(1) The amplitude of the first arc, and (2) the reading at which the image tends to set.

According to my calculations it takes about 1·487 s., or, say, $1\frac{1}{2}$ s. for the force to reach half its full value, and by the end of the first swing it would still be about $2\frac{1}{2}$ scale divisions behind its full value. I make the time of describing the first arc about 5·75 s., instead of 4·33 s., to which it tends.

The mica can hardly be assumed to come to its normal condition of swing by the end of the first swing, so that it would be well to reject the first two arcs in getting the log dec., &c.

13th July, 1876.

I see no reason why the law you mention should not go on, save for one of two things :—

(1) Glass is known to condense vapour of water on its surface in an invisible state, and platinum is known, or at least believed, to have similar power as to permanent gases, such as oxygen and hydrogen. It is *conceivable* that glass might have a similar power as to permanent gases, and that the condensed air or gas might be let out slowly into the gaseous form. Perhaps this is not very probable.

(2) You must have *some* liquid in contact with your vacuum when you use the Sprengel. This is actually, or is presumed to be, oil of vitriol, or rather chemically pure sulphuric acid. But is not this certain to contain a small excess either of water or of sulphuric anhydride, and may not this excess give rise to a vapour of its own kind of a very minute tension? Nay, even if you had chemically pure sulphuric acid, with no excess of either constituent, might not a minute portion of it in contact with a vacuum suffer dissociation into water and sulphuric anhydride, one constituent being held in solution, and the other evolved as vapour? The vapour volume of sulphuric acid seems to indicate that in the process of distillation it *is* temporarily dissociated. May not there be a minute dissociation even at ordinary temperatures?

From your figure I should say that your law probably holds up to the ninth or tenth bottle, or say a pressure of 1—1,000,000th of an atmosphere, when some action previously negligible begins to tell. If there were no such action I don't see why the evacuation should not go on indefinitely, or why it should be possible (though it might be quicker) to get a higher vacuum by chemical means than by the Sprengel.

18th November, 1876.

I can now explain to my own satisfaction the opposite behaviour of the two mica radiometers to a heated glass shade. It arises, I feel no doubt, from the greater opacity of the one you gave me long ago, arising partly, apparently, from a greater

25—2

thickness (to judge by the time it takes to get under way), and still more, perhaps, from the higher roasting, causing a more complete splitting up, and thereby multiplying the reflections. The aluminium radiometer shows little or no effect of difference of temperature of the two faces, but the motion is rapid when the bulb is hot and the fly cool, or the converse. The action depends on the more favourable presentation of the convex surface to the glass of the bulb. When you present a candle to either face, convex or concave, the metal is warmed slowly by absorption, and the face which is the more favourably placed as to the glass prevails. When you blacken the metal the black surface absorbs heat readily, and tends to become the warmer, and *so far* tends to be repelled. On the other hand, for equal temperatures of the faces the convex side, being more favourably presented to the glass, prevails and is repelled (the fly being warmer than the case), and the two distinct causes of motion conspire or oppose each other as the case may be.

The effect of shape depends not on curvature as such, but on favourable presentation....

And now for the experiment I had to suggest. I think it would be possible to make a perfectly flat radiometer, I mean with a perfectly flat fly alike on both faces, revolve *by throwing the obliquity from off the fly on to the case.* I will suppose the fly to be two-armed like your recent ones. Let vertical partitions of thin clear mica be fixed in the bulb, with their planes not passing through the axis of rotation, but inclined as shown in the figure. I think it would be well to have as many as three partitions 120° apart, and then one or other of the two opposite arms would come near a partition at every 60° of rotation, which would be favourable to the avoidance of dead points....If more than three oblique partitions could be introduced it might be an advantage, but I don't know what amount of trouble might be involved, and I think three would be sufficient for a fair trial.

As to the material of the disks (or squares, or rectangles), a metal should be employed. I should propose thin copper or silver as having good conducting power. The bright metal would absorb so little that it might be desirable it should be made to absorb a little better. I don't wish to use lamp-black, partly for fear the blacking should be overdone, and effects thus introduced depending on a difference of temperature of the two faces; partly for fear the coating of lamp-black, though thin, should be vastly inferior to the metal in conducting power, and thereby introduce a difference of temperature between the body of the metal and the outer surface of the lamp-black. I should propose to hang the fly under a bell-glass inverted over materials suitable for the slow generation of hydro-sulphuric acid, and wait till a moderate tarnish, alike on both faces of the disks, was produced, or dip it in a solution containing a very little hydro-sulphuric acid. A higher blackening would not be objectionable provided it be understood that the instrument is to be used with three (or more) candles at equal intervals all round.

I think I see my way to making a fly revolve, which consists of a horizontal circular disk of mica blackened above, and mounted so as to be movable about its axis. But it would be well to try the other experiment first.

It is obvious that the direction of rotation to be expected when the fly is heated by radiation is *against* the points.

On second thoughts I am disposed to avoid blackening altogether, and use the metallic surface. Aluminium or very thin copper might be used.

22nd November, 1876.

I find the most effective way to make the aluminium radio-meter revolve positively without first making it go negatively by heating the case, is to cover it with a tumbler (which I prefer to a shade as being of thicker glass), and hold a heater, such as a hot poker, outside, presenting it to different sides for fear of heating one side of the bulb. The office of the tumbler is to absorb the rays capable of ready absorption by glass, so as to avoid heating the bulb. If the heater is used for more than a short time, a spare tumbler should be at hand, and the first replaced by the cool one before it has had time to get sensibly heated inside. The heater and tumbler being then removed, the rotation continues for a long time, as the metal cools but slowly.

22nd November, 1876.

I may as well mention now the experiment I was thinking of for making a disk revolve about its axis.

The disk is circular and horizontal, mounted like the fly of a radiometer. For lightness sake it may be of mica, which need not be roasted, blackened above. Fixed to the bulb above the disk are three or four flat pieces of clear mica. Each extends from the side of the bulb to near the centre—there is no use going very near, as the leverage there is small—and ends below in a straight horizontal edge, leaving just space enough for the disk to revolve without risk of scraping. The edge is in a radial direction, and the plane of the plate inclined about 45° to the horizon, of course in the same direction for them all.

With a candle the rotation should be against the edge, which I will call the positive direction, with a warm glass shade negative, becoming positive when the shade is removed after it has been on some time. If the shade be hot there may be a little positive rotation just at first. A hot *wet* shade would probably produce a *little* positive rotation, quickly changing to negative; and blowing on the warm dewed bulb would produce positive rotation.

12th January, 1877.

As to par. 2 of § 191, I may observe that the curve of intensity for a diffraction spectrum may be deduced very nearly indeed from the curve for a refraction spectrum by dividing the ordinates of the latter by λ^3, λ being the wave-length.

§ 192, par. 4. I don't think the difference depends on wave-length so much, but on the greater intrinsic energy of the part of the spectrum about the red than the part about the violet. A radiometer or thermopile, except in so far as lamp-black may become deficient in blackness or pith in whiteness for the invisible rays, indicates the total energy; the eye or a collodion plate indicates nothing of the kind.

§ 206. I suspect the incoherence of the effloresced salt had more to do with it than increased absorption.

§ 207, par. 2. I suspect that you get a different result from Dewar *because* you used so thin a glass cover : that the effect you obtained was due almost wholly to the heating *of the cover* by the radiation.

13th January, 1877.

The explanation you mention of the effect of touching the bulb with the finger has not, so far as I know, been published. It occurred to me a few weeks ago. When I wrote to Prof. Righi I adopted your original explanation, namely, that with regard to the radiation from the heated part of the glass the two faces of the disks, the lamp-black and the pith, were black alike, and so the arms set at 45°. Shortly afterwards I saw my mistake, namely, that the warmed bit of the bulb did not act by radiation at all, but was itself the repelling body, of course through the intervention of the molecules rebounding from it with a greater velocity than that with which they struck it. I was thinking of writing to Prof. Righi to correct my mistake, but I have not done so yet.

But this does not get over all the difficulty of the vapour of water experiment. The radiator was a candle, some 2 or 3 inches off at least, which surely must have warmed the lamp-blacked surface more than the white, and more one would think than the wall of the bulb. Besides, if watery vapour is so adverse to the action that the blackened pith exerts no repelling action, how is it that the heated wall *does* exert a repelling action ?

Still I dare say the explanation may be in the direction you have indicated, and that you, who are familiar with all the details of the experiment, may be able to perceive the answer to the difficulties I mentioned.

You long ago mentioned in conversation, and I presume you published, your former explanation of the effect of a local heating of the bulb, and it may be as well, in an appendix or note at the end, to give what is doubtless the true explanation.

20th January, 1877.

I have for some time had in my head an experiment which I hesitated to communicate to you lest I should only be troubling you in a case in which I hardly expected a positive result. Yet inasmuch as the result *if* positive might lead to the solution of a great astronomical mystery, I think I may as well mention it.

The experiment is this. Let two or three (better three) thermometers be taken, as nearly alike as may be. The scales should be pretty open, from ordinary temperatures to 150° F. or so, but the graduation need not be more accurate than that of the commonest thermometers. Let each be enclosed in an outer bulb, which should extend, if not the whole way, at least some way up the stem, but better the whole way. The outer bulbs should be as nearly as may be alike for all. The bulb should have a tube for exhaustion leading off.

Let the bulbs (or cases) be exhausted to a common pressure of, say, 15 millims., and sealed off. When at the temperature of the apartment let them be plunged together into a water-bath at a temperature of 100° or 150° or so, and observe the rates at which they respectively rise. This would, of course, be the same if they were all alike, but this cannot be assumed, and it will be necessary to allow for the differences, if differences there be. Then keeping one for comparison, open the other two, and continue exhaustion, sealing off one, say, when the barometer and gauge are sensibly level, and the other at the highest vacuum which can be attained. Then compare them again as to the rates of rise in the water-bath, and see whether it has been changed by the more complete exhaustion.

There are two ways in which heat can get from the case to the thermometer—(1) By radiation across the intervening space; (2) by communicating an increase of motion to the molecules of the gas, which carry it to the thermometer. The latter, I suspect, is the way by which chiefly heat is conveyed from the negative electrode of a Geissler's tube to the glass walls.

Now it is quite conceivable that a considerable part, especially in the case of heat of low refrangibility, may be transferred by "carriage," as I will call it to distinguish it from convection, which is different, and yet that we should not perceive a diminution of transference, and consequently a diminution of rate of rise with increased exhaustion, so long as we work with such vacua as can be obtained by good air-pumps. For if, on the one hand, there are fewer molecules impinging on the warm body, which is adverse to the carriage of heat, yet on the other the mean length of path between collisions is increased, so that the augmented motion is carried further; the number of steps by which the temperature passes from that of the warmer to that of the cooler body is diminished, and accordingly the value of one step is increased.

Hence the increase in the difference of velocity before and after impact may make up for the diminution in the number of molecules impinging.

Hence it is conceivable that it may not be till such high exhaustions are reached that the mean length of path between collisions becomes comparable with the diameter of the case, that further exhaustion produces a notable fall in the rate at which heat is conveyed from the case to the thermometer.

Should experiment show that there *is* a notable fall, the inference would be that if the vacuum were absolute the transfer of heat, which in that case would take place only by radiation, would be but small compared with that actually observed in ordinary circumstances.

And now for the applications. You know that Huggins found that the heads of comets and the rest of the tail are partly self-luminous, in the condition of a glowing gas. But how is it that the sun is able to volatilise matter in the nucleus, and render the vapour of it glowing, while the comet is still at a distance from the sun comparable with that of the earth ?

Now if it be true that the greater part of the heat from a heated body surrounded by a gas passes off by communication to the gas in the first instance, and comparatively little by direct radiation, at least till the temperature becomes high and the radiation copious, we can conceive how the heat derived from absorption of the sun's radiation might accumulate in a small body destitute of atmosphere, and without gravitation enough, perhaps, to retain one, till at last it got high enough to volatilise some of its constituents.

Though the cooling of bodies in a vacuum has been tried, it has been only, so far as I know, with such vacua as can be produced by air-pumps.

P.S.—At first I was disposed to blacken the bulb of the thermometer, but I think it best *not*. *If* a positive result should be obtained, then in experiments with an ulterior purport it might be good to blacken the bulb.

23rd *January*, 1877.

Now that I have read Kundt and Warburg's paper I must retract my proposal, for I find it is exactly what they have already done. There would be no use in going over the same ground. But there is one point in which your work might usefully supplement theirs. Your vacua are probably more nearly perfect than what they worked with, and it would be interesting to examine more closely whether, when once the vacuum is high, no further retardation of rate of heating or cooling is reached by further improvement of the vacuum. If *that* can be reached the inference would be that the *whole* of what is then observed is due to radiation. But this could probably be better studied by a single thermometer, the case of which remains attached to the pump, than by two or three. The only object of detaching them would be to facilitate the observation of the rise or fall of temperature. But the thing could be more conveniently done by following K. and W.'s plan of heating a single thermometer— heating it higher than the highest point you mean to observe,— and then leaving it to itself, and noting the time it takes to fall from so-and-so to so-and-so. It does not appear whether K. and W. used a water-bath to keep the temperature of the case first to one temperature and then to another. If not, I don't exactly see how they proceeded, for unless the case were kept to two constant temperatures the rate of fall or rise of the thermometer would be a compound effect, depending, in fact, on the rate of fall or rise of temperature of the case.

8th March, 1878.

Some considerable time ago you were so good as to send me some of your results as to successive swings. I considered them mathematically, and mentioned some conclusions to which they led. I observed that if the force had begun at its full value as soon as the light was let on, the starting-point would have been in series with the successive turning points. Such was very far from being the case, and I remarked that this experiment alone would suffice to show that the action of light was not a direct one. This conclusion had already been otherwise made out by the researches of Reynolds, Stoney, and Schuster. The results of the swinging experiments showed that the force had not quite reached its full value by the end of the first swing, though not far from it.

Had the law of the increase of the force with the time been known, your observations of the turning points, combined with those in which the light was shut off after a second or half second or so, would have sufficed to determine the unknown constant or constants in the expression for the force, and so determine the force completely as a function of the time. As we do not, however, know the law, all we could do would be to assume an empirical law something like the actual law, and determine the constants in it. We should thus get a law which would be not very far from the truth. This, however, is not a satisfactory way of proceeding, and would besides lead to calculations of considerable length. I felt that it would be far more satisfactory to *observe* the deflection as a function of the time; but the way I thought of setting about it was by means of a chronograph something similar to what they have in observatories, and an instrument of the kind is rather an expensive affair.

But in a paper by Messrs Ayrton and Perry, recently read before the Royal Society, they mentioned having employed a piece of clockwork which cost very little to drive a cylinder covered with paper, which was used as a screen on which to receive the spot of light reflected from the mirror attached to the swinging body, on which they found it easy to follow with a pencil the path of the light. They give no data from which to judge of the amount of accuracy of which the method was susceptible; they only speak of the observation as easy to make and as giving useful results.

If the pencil could be followed well enough the result would give the deflection as a function of the time, from which the force could be afterwards obtained as a function of the time by a graphical process, which I need not now describe.

Possibly the registration might be rendered automatic without much trouble by substituting for the lamp a biggish Leyden jar with rather close terminals, working the jar by an induction coil, and using sensitive paper to cover the cylinder, which would then be cut off, developed, and fixed by the usual methods of photography. I think there is a field of research in this direction if the registration could be managed.

The above would of course require the interposition of a lens to form an image of the minute but intense spark.

30th April, 1878.

What I told you about assuming a law of force for trial related to the former experiments, in which the readings of rest at the end of each swing were alone given. When we have the whole curve representing the motion, as in your last experiments, we can get the force without making any hypothesis. It remains to be seen how far errors of observation will tell in the result. The easiest way of getting the result would be by a graphical construction, or perhaps partly by this and partly by calculation. I mean to try my hand at it. It seems to me that the time scale is somewhat too close for accurate results, but I can judge better when I have tried.

In the event of any future measures it would, I think, be convenient, if the apparatus adapts itself to the observation, to stop and screen off the plate after two or three swings, then manipulate with a screen so that the bar shall be nearly at rest when the candle is full on, then start the plate again and take two or three swings. Or if more convenient to draw back the plate and then take the swings for the position of nearly rest. The object of this is to get more accurately the ordinate for rest with candle on.

17th May, 1878.

As I can hardly attack your curves while my lectures last, I will describe what I should be disposed to do in the first instance with the curves.

A good part of the curve is described while the force is sensibly equal to what it finally becomes. I should be disposed, therefore, in the first instance to start from a point sufficiently on for the influence of the change of force to have subsided, and then calculate backwards to see what the previous positions of the tracing point would have been if the force had always been what it now is, and the original motion had been such as to give the motion after some time what it actually is.

For this I want to know the log dec. and the time of vibration, or rather the scale value of the time of vibration. This you regularly register; at any rate they can be deduced from the records themselves. I will suppose them then known.

Let ABC be the traced curve, DE a line parallel to the motion of the plate. Choosing two of the turning-points, B, C, draw tangents there parallel to DE. Let ρ be the number corresponding to the log dec., and divide the distance between the parallel tangents so that the parts are in the ratio of 1 to ρ, and measuring from the tangent on the decreasing side, draw a line GH parallel to DE, then GH will be the equilibrium line for the force equal to its final value. If it cuts the curve in O, we may take O for the origin from which to measure the abcissae in our calculated numbers. Let time be measured along OH, the distance the plate travels in one complete oscillation, or two vibrations from rest to rest, being called 360°. Then if y be the ordinate for a time equal to n oscillations from rest to rest, where n may be integral or fractional, before the spot was at O, we have, supposing the force constant,

$$y = C \times 10^{nD} \sin n . 180° \tag{1}$$

where C has still to be determined. In this expression D denotes the log dec.

The determination of C in the general case involves rather too much mathematics to go into at present. But if D be small, as is practically the case in your experiments, C may be determined very simply. In this case the maximum ordinate at B will without sensible error be the value of y for $n = 0.5$, and therefore if y be this maximum ordinate,

$$y = C \times 10^{\frac{1}{2}D},$$

which gives C. The formula (1) will now give y, calculating, say, for every $60°$, *i.e.* putting $3n = 1, 2, 3, 4, 5, \&c.$, and again to get the curve to the left of O, putting $3n = -1, -2, -3, -4, -5, \&c.$ By laying down the calculated curve we should see how far it agreed with the traced curve, and accordingly where the force differed sensibly from what it finally became. There are graphical modes of dealing with the difference between the two curves, so as to get the force, but it would take too long to explain this fully at present.

Meanwhile I would observe that the first treatment in the manner I indicated would already give us an insight into the manner in which the force approached its final value.

21st May, 1878.

The "Force" of which I wrote, and mentioned its dying away, was the actual mechanical force, exclusive of the force of torsion and the resistance of the air, acting on the suspended body. This force begins from nothing, when the candle is let on, rapidly increases, and would attain nearly its full value in perhaps a second or two after the candle was let on, supposing it were not cut off again; but again begins to change rapidly when the candle is cut off. It would remain sensible for perhaps a second or so after the candle was cut off. The force, as we now know, depends on the heating of the surface of the swinging body by the previous action of the candle.

In the example you gave me in your last the fact of the first arc, from the first to the second stopping-point, being greater than the calculated arc from applying the log dec. to the subsequent swings, proves that the effect of the candle had not subsided,—I mean that the repulsive force due to it had not subsided, when the body was at its first turning-point. But the agreement of all the rest with the calculated numbers proves that it was insensible by the time the body had got to its second turning-point. I find, from measuring on your figure, that the first turn took place 4·5 seconds after the candle was cut off. There was therefore, even so long as four seconds and a half after the candle was cut off, a sensible repulsion due to its action. This was longer than I had anticipated that the action should have been sensible.

9th June, 1878.

With respect to the greater part of the explanation of the movements in the case of the bulb uniformly heated, I would refer you to my former letter.

There was, however, one case which struck me as peculiar as I read the sheets of the MS. which are now in your hands. I had not time to mention to you the explanation which occurred to me. The case is that of the moderately thin and the excessively thin flies of mica when the case was plunged into hot water. Why should the radiometer with the very thin mica manifest a behaviour in most respects the reverse of the other?

The explanation you have given is, I think, right *as far as it goes*. The difference depends on the greater stowage-room for heat of the thicker mica. But how does this operate? It seems to me the state of the case is as follows:—

The thicker mica, having considerable stowage for heat, remains colder when the bulb and the gas within are warmed. The fly therefore goes negatively for a good while. When the bulb is taken out it cools before the fly, and the fly goes positively.

With the thin mica the single turn negatively indicates an action of the same kind. But whence the subsequent positive rotation?

Assuming the correctness of your supposition, that the difference depends in some way on the stowage-room of the thicker as compared with the thinner mica, we are led to regard the positive motion of the thin as depending upon the present actions going on rather than the past history of the fly. The positive motion indicates that the fly is warmer—than what? than the bulb? We must go closer than that into the state of things. Then what part of the bulb, for the whole will not be quite at the same temperature? It is the *inner surface* of the bulb, and the gas, which will be very nearly at the same temperature as this *inner surface*, that we have to deal with; we have to explain how the fly is warmer *than this*.

While heat is being conducted through the bulb the temperature of the glass will not be uniform. Thin as the glass is, we must imagine it divided into strata, of which the temperatures will decrease from the outside to the inside. The skin of water in contact with the glass will also not be quite so hot as the water generally. I will neglect this latter, however, the effect of which will probably hardly be sensible, and attend only to difference of temperature of the different strata.

Poisson speculated that conduction was only internal radiation and absorption. I believe, however, the two are distinct, and go on together, and I will choose my language accordingly. Though the theoretical explanation will not be affected by the theoretical alternative we adopt.

The heat will then be distributed, partly by conduction, partly by internal radiation. The heat radiated inwards by the outer and warmer strata of the glass will partly be absorbed in attempting to pass through the inner strata; but, as the strata are very thin, part will get through. This will fall upon the very thin mica. If for this heat, or a portion of it, mica is "darker,"—that is, more absorbing—than glass, this will warm the mica, and it is besides warmed by the gas, by actual contact. The stowage-room is so small, from its excessive thinness, that the warming by the latter mode can by no means be neglected. Between the two it is quite conceivable that the temperature may be sensibly higher than that of the *inner surface* of the glass, in which case it ought to go positively. The reverse movement in cooling is explained in the same way.

If this be so, we may expect the radiometer to show *some* indication of positive movement to hot glass. I should propose this experiment :—

Cover the radiometer with a very thin beaker, preferably of *flint* glass, so as to resemble the bulb. Heat another glass vessel considerably, and invert it over the whole. Will not the thin mica fly begin to rotate positively, and that with a promptitude which shows that it is by direct radiation from the hot glass, so that we have not got to wait till the covers are sensibly heated ?

12*th June*, 1878.

We had best keep perfectly distinct the two different cases of—(I) different condition of the two faces of a fly ; and (II) different presentation. I will therefore suppose that in (I) the vanes are diametral to the bulb, and in (II) the two faces are alike.

In (I), or difference of faces, I did not suppose that the normal condition when the bulb was heated alike all round was for the fly to revolve negatively. On the contrary, I looked on the normal condition in this case to be for it to go positively. This appears from my paper in the *Proceedings* of December 20th. Thus, for example, to explain the negative rotation of a particular radiometer in these circumstances—see Sect. 13—I had to suppose a relation of darkness with respect to dark heat, the reverse of what we should have expected from the relative darkness for light.

Suppose a fly with black and white faces, and the order of absorption for dark heat is the same as for light. When the radiometer is exposed to radiation from without which gets through without heating the glass, the blacker face is the more warmed, and therefore repulsion takes place. When the whole fly has been heated, and the radiation is then cut off, the fly cools, the white face is left the warmer, and the fly goes negatively.

When, however, instead of heating the fly, we heat the bulb uniformly all round, the fly is colder than the bulb, and therefore there is—or would be if the two faces could move independently of each other—apparent repulsion. The black being more warmed than the white, there is less apparent repulsion from it, and therefore the fly goes positively.

Observe, the relation of positive to negative holds in this way. When we have an envelope which in all its walls, including the surface of any fly within, is at a uniform temperature, there is no motion. When an exceptional portion of the surface is at a higher temperature there is attraction between it and the opposed surface, if not too far off. So, by the usual relation of positive and negative, when a small portion of the surface—wall or fly, no matter—is at a lower temperature than the general space, there is attraction.

Observe from the circumstances of the case the order of temperatures in the second of the two cases just mentioned, is *not* the reverse of what it is in the first. If it were we should have had a reversion of direction of rotation. In the first case the order of increasing temperature is wall, white face, black face, whereas in the second it is not the reverse, but white face, black face, wall.

In the case of difference of character of faces, then, I look on the normal condition to be that the rotation is the same for heating of bulb as for heating through bulb; or positive rotation in the former case, as well as in the latter.

13th June, 1878.

I come now to the case of favourable presentation. In the normal case the temperature of the fly may be taken to be uniform throughout. If the fly be heated, suppose by a red-hot poker, the bulb having been covered with a glass shade, and the poker moved round to guard against the heating of one side of the bulb, the fly goes positively. The object of the shade is to sift the radiation proceeding from the poker, and so keep the glass from being heated,—the glass, that is, of the bulb.

When, instead of this, the bulb is heated by hot water or by inverting over it a glass shade, the exceptional temperature, that of the fly, is lower than the general temperature, and the fly goes negatively. This is what takes place with the cupped aluminium radiometer you gave me; with the thicker of the two mica radiometers you mention in your paper; and, as regards the very first motion, even with the thinner one.

To sum up, I look on a positive motion as the normal one for a heated bulb when the effect depends on a blackening of one face, but a negative motion as the normal one when the effect depends on favourable presentation.

That being the case, I was puzzled by the positive motion of the thin favourably presented mica fly after the single negative turn to begin with. I assume the two faces of the fly to be just alike. I take for granted that it was so. I still do not see how to account for it, except it be by the explanation I gave in my last. I am much obliged to you for trying the experiment I suggested. Had the result been otherwise it would have confirmed that explanation. As it is, though it does not confirm it, yet neither does it upset it. It only shows that, if that be the explanation, glass is highly opaque, though not so intensely opaque as mica, for the kind of radiant heat on which I suppose the heating of the mica to depend. This is what *à priori* seemed more probable than not, so that all we can say is that *either* experiment disproves the explanation, *or* the alternative (which *à priori* seemed perhaps the more likely) is true, in which case the experiment proves nothing either one way or the other.

I don't feel confident by any means that the explanation I threw out is the true one, though I don't at present see any other. Perhaps I may call on you some day when I happen to be in London, and see the behaviour of the radiometer ; for often something is suggested by the sight of the behaviour of an instrument which can hardly be conveyed by words.

13th June, 1878.

On looking over your letter again I see one expression which does not seem to me to convey quite the right idea. You say " black—*or favourable presentation*—is the best absorber of heat." It is the words underlined that I would take exception to. In the case, for instance, of the cupped aluminium radiometer, the two faces absorb heat equally, and are at sensibly the same temperature. The difference is that one faces the bulb and the other faces the open. The impacts are equally intense at the two faces, but, as Stoney has explained, the impacts are more numerous on the side facing the bulb, because on the other side the molecules driven back from the fly, being as it were projected into the open, tend to beat back others which were crowding towards the fly. In the other case, *i.e.* as regards the molecules that were driven back from the fly on the other side, they are projected in good measure against the bulb, and rebound with only the velocity corresponding to the temperature of the bulb. They are not therefore so efficacious in keeping back other molecules as are those that rebound from the side facing the open.

I should have explained that I had in view the case of the fly warmer than the bulb.

It would be a very interesting subject of investigation to take up some time or other, to endeavour to make out how much of the transfer of heat to or from a body, enclosed in a good vacuum, is due to the residual gas, and how much to radiation. Of course as regards what goes on through the bulb there is no question ; the question is, how much of what is conveyed from the bulb to the body, or the other way, passes by radiation from the bulb, and how much is due to transfer by the gas, *i.e.*, to the fly—in the case in which the fly is the warmer—being struck harder than it strikes back, so that energy is transferred from the molecule to the fly?

I mentioned, I think, in a former letter that this has an important astronomical bearing. It seems difficult at first sight to account for the high temperature of comets at such a distance from the sun. It may be that, even in a pretty high vacuum, a good part of the heat lost by a body within it is lost by way of transfer to the residual gas, so that, if it could be perfectly removed, the body going on receiving by radiation as much heat as before, the body's temperature might become much higher than it could while there was still present even a small quantity of gas.

21st June, 1878.

When I saw you last night I did not think of mentioning to you what had occurred to me, that your beautiful contrivance of the minute fly which you can put where you please would answer admirably for investigating the effect of a difference in the state of the surface.

The plan which seems to me to promise the best is the following :—

Take a plane, or hollow cylinder of some metal,—say copper, as being a good conductor of heat. I will suppose a cylinder chosen. The substance had better be a good deal thicker than the very thin sheets of metal you employ; in fact it might be a solid cylinder, except that the time it would take to heat up a solid cylinder would make the experiment rather tedious. The substance ought, however, to be thick enough to ensure a practically constant temperature all round. I should think a hollow cylinder or end of tube, say an inch in diameter and three-quarters of an inch high, might be about a convenient size. I suppose this cylinder to be mounted as a fixture, with its axis vertical, in the middle of a bulb of good size, so that the sides may be deemed to be out of range of direct influence. The globe contains a minute fly, which can be moved about into any position. I contemplate its being held at the height of the middle of the cylinder.

Before enclosure in the bulb, I suppose the surface of the cylinder treated differently on different compartments outside, the compartments being separated from each other by vertical divisions, *i.e.*, the lines of junction being vertical; thus, for example, there might be three compartments dividing the circumference into three angles of 120°: one compartment might be covered, say, with electro-deposited copper, another with lamp-black, and the third left plain and polished.

The experiment consists in this:—The cylinder is first warmed; it might be by candles placed round; it might be by placing the whole for some considerable time in a heated enclosure, and afterwards cooling the bulb with water. The warmed cylinder then projects the molecules faster, *i.e.*, with a higher velocity, than it receives them. The exploring fly is now brought round the cylinder outside to different parts of the circumference in succession. Opposite the middle of a compartment it will probably remain at rest; but according to what you found in your other experiments, near the edge separating one compartment from the other, it might be expected to go round; and the direction of rotation would show from which surface, the various surfaces being at the same temperature, the bombardment is the strongest. In the case supposed we could thus compare smooth copper with rough, smooth copper with lamp-black, rough copper with lamp-black.

In this experiment it is not to be expected that the facility of absorption and emission will affect the result, but rather the state of the surface as to roughness; provided, at least, that any bad-conducting substance laid on is not laid on thick enough to prevent the temperature of the surface from being sensibly equal to that of the metal inside.

If a plain flat plate were used instead of a hollow cylinder, we might compare two pairs of different surfaces, one pair being on one face, with the line of demarcation down the middle, and the other pair on the other face. Possibly experience might show that there would be room for more than three surfaces in the circumference of the cylinder, or two on one face of a plate. But it would seem best to try first the smaller number; and if experience showed there was room for more, a larger number might be used in another experiment, and thereby the multiplication of radiometers, in case experience showed that the mode of investigation was good, might be avoided.

25th June, 1878.

Your experiments with the six disks attached to the same arm seemed to me to differ in one important respect from those I suggested in my last. In your experiments a powder was laid on a bad conductor, and was exposed to radiation from a candle. Hence the temperature of the surface depended on several things; on the absorbing power of the powder, its radiating power, and the conducting power of the substance it was laid on. I should suppose that the differences of temperature produced by these different causes operated more strongly than the difference of state of the surfaces in regard to bombardment.

The experiment I suggested was intended to render the result independent of differences of radiation, differences of emission, and differences of temperature, so as to bring out the effect of roughness, or powdering equivalent to roughness, by itself alone.

Whether we may rely on the equality of temperature in the case of a lamp-blacked surface I am by no means sure. It is possible that the particles of soot, being but feebly connected with the metal below, may cool by radiation to a temperature sensibly below that of the metal beneath. If so the experiment, so far as regards soot, would be illusory.

I really cannot say what ought to be the result to be expected, and that being the case I don't think the experiment would be superfluous. Of this, however, I feel pretty confident, that the results in the cases you refer to, so far at least as I can gather them from the published abstract, depend chiefly on differences of temperature, and accordingly on actions here sought to be eliminated.

In the multitude of experiments you have tried there *may* be some which give the same information here sought to be obtained; but if so, I cannot recall them to my mind.

19th July, 1878.

It was, I think, known before that a thermometer cooled a good deal faster in a so-called vacuum than in air of atmospheric pressure. In air so dense as this, convection currents would no doubt come in for a good deal. The question which I don't think has yet been investigated is whether there will be much difference between different vacua all pretty high, so that we may neglect convection.

The question is, How much is due to radiation and how much to bombardment, when we are working with vacua sufficiently high to render the results uncomplicated by convection ?

Of course you are on your way to investigate this; but I thought it might be no harm to observe afresh that this is the most interesting part of the enquiry.

30th September, 1878.

I found your letter on returning from Scotland late on Saturday night. This morning I verified your numbers.

The formula I gave you was founded on certain suppositions which appeared to me to be very probable, but which have need of being confronted with observation.   It is supposed :—

1.   That the friction is independent of the velocity. This has been established experimentally, at least as a near approximation, for friction of rubbing; but I don't know whether it has been tried as regards friction of turning on the nearest approximation we can make to a point.   This, however, probably is of the nature of rubbing, so long at least as the velocity is not sufficient to make the turning body jog in its support.

2.   That the resistance of the air is proportional to the velocity, other circumstances being the same.   Of this I think with a fair vacuum there is little doubt.

3.   That the principle of superposition holds good as regards the force.   This is probably sensibly true, but has need of being tested by experiment.

In every experimental investigation due allowance must be made for errors of observation.   That is to say, besides the causes of variation which we recognise, or the *methodical* variations which we discover in the course of our researches, there are others which we regard as casual, such as those arising from errors of reading, variations in the rate of burning of a candle which is treated as constant, &c.   One of the very best ways of distinguishing the methodical from the casual is by plotting our results.

If the formulæ I gave you were absolutely true, and the observations absolutely free from error, then, if we were to plot our results, taking for the two co-ordinates lengths proportional to the numbers of turns per second for the one, and to the squared reciprocals of the distances of the candle for the other, we should find that the dots representing the observations lay in a straight line.

I have plotted the observations you sent me rather roughly on the enclosed paper. The dots are far from lying in a straight line; and the question arises, Is the deviation methodical or casual; or rather, since there is sure to be a casual element, can we discover a methodical deviation from a straight line underlying the deviation which we are obliged to regard as casual?

This important question I have not the means of answering, having got the results of one series only, and that containing only four observations. It does not seem probable, à priori, that the true curve, if it be not a straight line, would be so twisty as to take in as a very near approximation all the four dots, and therefore I am disposed to think it probable that there are considerable errors of observation, i.e., variations which at present we can only regard as casual.

I should recommend you to plot in this manner, on any convenient scale, a good number of your observations. You will then be in a condition to judge whether or no there is any methodical deviation from a straight line.

If no methodical deviation can be discovered, the best thing we can do is to draw a straight line by the eye, and obtain our results by measuring from that. Some persons don't like leaving anything whatsoever to the judgment of the operator, and would deduce the constants of the line from least squares, or some analogous method. In such cases as those we are now dealing with I consider myself the advantage, if any, of least squares much more than compensated by the expenditure of time it involves and the loss of the immediate control over what you are about that plotting affords....

1st October, 1878.

I think a good measure of the friction might be obtained directly by ascertaining the distance of a candle when the fly would just not turn round continuously, when started in rotation by applying a vibrating tuning-fork to the support or by some similar method.

But in this way, instead of using a single candle, it would, I think, be much better to use three at angles of 120°, or five at angles 72°, the three or five candles being sufficiently nearly at the same distance from the radiometer to allow us to regard them all as at the mean distance....3 or 5 candles would be a better number than 4, assuming as usual the fly to have 4 arms, because with 3 the system would go through its period 12 times, and with 5, 20 times in one revolution, whereas with 4 it would go through its period only 4 times; so that with 3 or 5 there would be comparatively no dead points, which with the very slow rotations we are now considering might vitiate the result of an experiment, though with a quicker rotation it would not matter, as we should then practically have to deal with the mean force all the way round....

It would be interesting to see whether the measure of the friction thus obtained would agree with that got by the graphical construction applied to the results obtained in the way I pointed out. If it did, that would give confidence in the correctness of the measure.

In the measures you tried I should have thought it good to interpose a glass screen midway, or thereabouts, between the candle and the radiometer, so as to prevent, as far as may be, the heating of the bulb, which possibly was the chief cause of the discrepancies in your experiments. I think it probable that the formula I gave would apply, even if the bulb were heated, *provided* time enough were allowed in any change of distance of the apparatus to come to its permanent state; but so much time would be required for this as to make the experiments extremely tedious. To avoid the necessity of all this waiting, I should have thought it was best to employ a glass screen.

Did you notice whether any difference was made in the results according as you observed with increasing or with diminishing distances? I can imagine its making a difference, unless an inconveniently long time were allowed between one measure and the next.

Of course the formula I gave could not be expected to apply if the observed rotations were affected by the previous history of the fly.

I shall feel curious to know whether the plotted results indicate a distinct curvature, or whether they are so irregular as to indicate that the rotations are affected by some circumstance other than merely the distance of the candle,—such, for example, as what had been done to the radiometer a little while before the measure was taken.

2nd October, 1878.

I have looked at your table on p. 342 of the *Phil. Trans.* for 1876. An inspection of the numbers shows that, even with the smallest force there employed, the minute friction is almost entirely swallowed up. To get a good measure of the friction, at least one observation requires to be made with the force as small as may be. With this condition laid down, I think the method I mentioned ought to give pretty fair results.

There is another method of getting it which is perhaps more accurate, but which involves rather troublesome calculations. It is to set the fly spinning, suppose by presentation of a candle near for a very brief time, and having waited till the force arising from the difference of temperature of the vanes and case may be deemed to have gone, so that the fly may be deemed to have been merely pushed round mechanically, to take three observations of the total angle turned through and the corresponding times, or else two such observations and the total angle turned through when the fly comes to rest. In the first way one observation may be made pretty soon, a second when the fly is revolving very considerably slower, and the third when it has almost stopped. In the second way the first observation may be made pretty soon, the second when the fly has nearly stopped; the third is an observation of a different character.

Probably the second of these methods would be the more accurate, but, on the other hand, it would be the more laborious to calculate....

In every case one of the observations should be taken when the fly is almost at rest, in order to make the influence of the friction more telling.

On second thoughts I am disposed to think that the first of the two methods just mentioned would give the more accurate result, beside being easier to calculate.

In the graphical construction I gave, the friction is determined by the point of intersection of the drawn line with the axis of y, the intercept of the axis being proportional to the friction. The construction shows that, in order to determine this with certainty, the two points must not both be far off from the origin. One should correspond to as small an x—that is, as small a velocity— as conveniently may be.

3rd October, 1878.

I am much obliged by the plotted curve, which I will return after a little. The curvature seems unmistakable. Further, the low part seems to tend almost exactly to the origin, indicating that in those experiments the friction was practically insensible. We must look, therefore, elsewhere for the cause of the curvature. Curvature in the direction indicated by the figure would be produced by a term in the resistance varying as the square of the velocity. Yet at such high vacua other experiments seem to indicate that such a force is insensible. Another way of accounting for it would be by supposing that in the slow rotations, the force arising from radiation acts to greater advantage ; that is, that the longer exposure gives greater advantage than the proportionately longer period of freedom from exposure to radiation gives disadvantage.

17th October, 1878.

As I was returning in the train, next day after seeing your most interesting experiments, I thought of possible ways of confronting further with observation your view that—in the instances you showed me, at least as regards the objects within the dark space around the negative—the fluorescence was produced by the actual impact of the individual molecules against the glass wall or other object.

I think it very probable that you are right; but the thing, if it be so, is so new and so important that I think it very desirable to check the result by any means that seems promising.

I will mention one which occurred to me in the train, but I doubted whether it was worth communicating to you. I am inclined, however, to think it is, and the experiment does not seem to involve any particular difficulty.

I must premise that both theory and experiment agree in the conclusion that there is no sensible creeping of phosphorescence, or fluorescence, which is but a brief, almost momentary, phosphorescence, from the part of the body originally affected to a neighbouring part.

That being the case, if the fluorescence is really produced by impact, it must be confined to a stratum of almost infinitesimal thickness; of a thickness, in fact, comparable with the mean interval between adjacent molecules. But if it be produced by radiation from the incandescent gaseous molecules, which may be supposed to be excessively sparse in the dark space, but to be crowded together at the glass where their previously enormous velocity is checked by the impact, as the medium would not be likely to have an almost metallic opacity for the radiation, the fluorescence ought to be excited further down. At least if two or three different substances were tried, it is not likely they *all* would have an almost metallic opacity for these radiations; and if we may assume that one at least has not, then the experiment I am going to propose ought to give a result.

I think I have seen it stated that glass can be blown so thin as to show in parts the colours of thin plates. It is, I believe, done by blowing out a bulb till it bursts. I don't know whether it can be done with a blowpipe or whether it requires a regular glass-house furnace.

But supposing it can be done, that affords evidence of excessive thinness. In that case I should propose to try, say, the German glass, common English flint-glass which give a blue fluorescence, and uranium glass. Take a fragment of each—it matters little how small—the colours of which demonstrate its extreme thinness, and mount it so as to be exposed as favourably as may be to the molecular stream from the negative, and, either side by side or just behind it, place morsels of the same kind of glass of ordinary thickness.

Now, if the fluorescence be due to actual impact, the fluorescent light ought to be as strong on the very thin glass as on the thick piece of the same kind, and, moreover, the thin pieces ought to cast a perfect shadow. But if it be due to radiation we should expect that the full amount of fluorescence would not be developed on such excessively thin pieces, and, moreover, that some of the rays would be able to get through and affect what lay beneath.

I thought of another way, depending on a particular variety of fluor-spar. I think I will write to Mr Sopwith, who I have little doubt could put me in the way of getting some. I have some; but what I have is partly borrowed, partly in too small fragments to be useful.

I send you a copy of a paper of mine, containing some things bearing on what I mentioned.

I found mica so excessively opaque, with respect to the rays of highest refrangibility, that I don't consider the experiment would be satisfactory with it. That is, the second test, depending on the formation of a perfect shadow, behind which there is no effect. For mica, of course, not being sensibly fluorescent, the first test could not be tried.

21st October, 1878.

You mentioned to me, the night I was at your house, the result of the experiment with the excessively highly exhausted tube according as electrodes close together or at the two ends of the tube were used. You also called my attention to the sharpness of the shadows, which I saw and was very much struck with.

This sharpness, I felt all along, threw a great difficulty in the way of any explanation of the phosphorescence that made it depend on radiation from the gaseous molecules. The only way I saw of getting over this difficulty was, as I endeavoured to explain in my last, by imagining that the molecules rushing from the negative had an enormous velocity till they struck against the wall, and that consequently in the dark space they are comparatively speaking excessively few, those attempting to crowd in being beaten back ; whereas at the wall itself—I mean close to it— the molecules might be supposed to be comparatively numerous. Hence the comparatively crowded molecules confined to the immediate neighbourhood of the glass might act sensibly as lamps, while those in the dark space would be too few to have any sensible lamp-like action. The sharpness of the shadows, on this supposition, would be accounted for by the extreme narrowness of the stratum to which the crowding was confined. The shadows, in fact, would not be optical shadows, but rather, so to speak, molecular shadows, only they would be revealed by an ordinary illuminating effect.

This did not seem to me a probable explanation. Your own view seemed to me the more probable ; only, the effect, if such be the nature of it, is so new that it seemed desirable to put the explanation to whatever tests one could think of.

I quite agree with you as to what you say as to the matter within those very highly exhausted tubes being in a non-gaseous state ; only *I* should say that it was a state in which properties of matter which exist even in the gaseous state are shown *directly*, whereas in the gaseous state they are only shown indirectly, by viscosity and so forth, which are referred to them—referred to them, I mean, in their *theoretical* explanation.

I look on the ordinary gaseous laws as a simplification of the effects arising from the very same cause as in the other case, which simplification is permissible when—and only when—the mean length of path is small compared with the dimensions of the vessel.

You say "Had it been light which radiated from the pole there must have been a penumbra to the shadow." I did not suppose that the phosphorescence was in any case produced by light which radiated from the pole itself: were the pole sufficiently heated for this, alas! for the tube. It is the incandescent gas that I take to be in ordinary cases the source of the light. And it was on account of the sharpness of the shadow that in these peculiar cases I was driven—*if* the phosphorescence were due to a lamp-like action—to suppose that the dark space possessed but very few molecules, and that they were crowded where their velocity was checked by the wall.

22nd October, 1878.

It has just occurred to me that it would be curious to vary the experiment with the discharge in tubes in which the dark space around the negative fills the whole tube, by leaving an air-space nearly as great as the coil can strike over, between the positive electrode and the wire going to the coil....

Under these circumstances would the positive reveal its presence by any luminous phenomenon inside the tube ?

What would take place in a small U-shaped tube, so that a good region around the positive should be well out of the line of fire from the negative, when the exhaustion was of this very high order ? Would the positive branch be quite dark ? Perhaps you have tubes which enable you to answer this question at once.

24th October, 1878.

I never had any other idea than that the greenish light was due to the phosphorescence of the glass. The alternative which I contemplated was this : Was the phosphorescence excited by the actual impact of the individual molecules on the molecules of the glass, which, however probable it might seem from the features of the light, such as shadows, &c., was a conclusion so novel that it demanded to be sifted in any way we could think of; or was the phosphorescence excited, as we have hitherto known phosphorescence to be excited, by radiation?

The difficulty of the last supposition was, Where could the radiation come from? The sharpness of the shadows precluded the idea that it came from the immediate neighbourhood of the negative electrode, which was far from being a mere point, besides which the negative glow did not seem specially intense in the immediate neighbourhood of the negative electrode. The only way I saw for accounting for the sharpness of the shadows on the supposition that the phosphorescence arose from radiation was by supposing that the incandescent molecules were projected with such enormous velocity from the negative that the space around it was kept comparatively clear; that there were comparatively few molecules within it, but that when the molecules struck the glass they were comparatively speaking arrested; that there was a very thin layer of them adjacent to the glass, where they were comparatively speaking densely packed. Thus, if the molecules were thought of as lamps, giving out, be it observed, *not* the green light observed, but the invisible radiation which is capable of exciting the green phosphorescence, it is conceivable that the absence of phosphorescence within the sharply bounded shadows, the sharpness of which proved to me that they were in some way or other truly molecular shadows, might be accounted for by the absence of the thin overlaying phosphorogenic stratum. I grant that this explanation seemed decidedly improbable; but, on the other hand, the excitement of phosphorescence by direct molecular impact was a thing so new that any explanation which referred the phosphorescence to the previously known mode of excitement, if it seemed fairly possible though not probable, deserved to be considered.

Observe I speak of the production of phosphorescence by the direct impact of molecules, not as a thing at all improbable in the present state of our knowledge respecting molecules, but merely as a thing totally new, and as such needing to be defended against all reasonable opponents. Indeed, after what I saw in your house it seemed to me most probable that the phosphorescence *was* produced in the manner you suppose; but anything that held out a chance of referring it to the previously known mode of production seemed to me to demand examination.

My object in proposing to use different kinds of glass in the experiment with thin glass that I proposed was to give greater probability that one glass at least out of the lot was not intensely opaque with regard to the supposed phosphorogenic rays. Were a glass intensely opaque with regard to the supposed phosphorogenic rays, then the full tale of phosphorescence would be produced within so excessively thin a stratum that a thick glass exposed to the rays would have no advantage over a thin one as regards the total phosphorescence produced.

The improbability of the existence of such intense opacity which if it did exist would render the experiment inconclusive, would be increased by using different kinds of glass.

28th October, 1878.

I cannot help thinking it would be worth while to try, for negative electrode in one of the very highly exhausted vessels, a small spherical speculum with a regular optical polish. It would be of course necessary that it should be provided with a platinum stem, for the sake of soldering to the glass. Unfortunately the speculum metal would blacken the glass in time, but this cannot be helped, for aluminium would not take a good polish. It might be in size and shape something like this— [*omitted*].

The object is this:—It appeared from the experiments you showed me that the molecules of gas were projected from the negative in a direction nearly at right angles to the surface. *How* nearly could hardly be judged of by pieces merely shaped by hand. But supposing the projection to be accurately normal then there would be a strict focus, free from any aberration, at the centre of curvature. In arranging for the placing of the speculum in the vessel, this point can easily be found by seeing where a small object must be placed to coincide with its image.

If such a focus free from aberration *were* found, it would show that the projection was very accurately normal. Moreover, from the concentration of the faint deep blue light in a space of insensible width just at the focus, it might be possible to learn something about the spectrum of this faint light, which has never yet, I think, been done; for, though the negative light has been observed, the light so observed was that of the negative light where it is bright, that is, after the commencement of the halo which is formed at some distance from the negative.

It is not, perhaps, likely that the projection would be so accurately normal as this would imply; still, if it were, some interesting things might come out of it.

An experiment occurred to me I think it might be worth while to try, seeing that it is only to make an observation with apparatus you already have.

If there were such a narrow stratum as occurred to me as the only way of explaining the sharpness of the shadows on the supposition that the phosphorescence was excited by radiation, this stratum might be expected to emit a little feebly illuminating light of very high refrangibility, as well as strictly invisible rays. It might therefore be detected if specially looked for, though it might escape notice otherwise.

Now, I should propose to take a tube which has got a flat object, such as a piece of mica, intended to cast a shadow, and holding the eye in the plane of the screen, or rather the least shade to the side of it towards the negative, to look along the surface for a very narrow stratum of faint deep blue light, which, if existing, might be expected to be increased in brightness by the foreshortening. I do not expect that such will be found; still, as you have the tube there it may be as well to look for it. I think it most likely that the phosphorescence in the cases you showed me is really produced as you suppose; but, as this is a new mode of action, it is well to try any other possible explanation, even though not very probable, that presents itself.

I remember a year or two ago, when you showed me a tube that you had got, I think, from Geissler, that would not let a discharge pass, being surprised at the rings of green light. I did not doubt that they were due to the phosphorescence of the glass, but, supposing them to be due in some way to radiation, I did not see why they should be so sharp. I suppose you yourself at that time did not think of molecular impact as being the cause of the phosphorescence.

29th October, 1878.

I am greatly obliged by your letter dated the 28th, but posted to-day, in which you describe the result of the experiment I suggested with the film of the uranium glass.

I am very glad that I made the suggestion, for it has been the means of bringing out some new and very important facts.

The cause that occurred to me for the brief duration of the

phosphorescence of the film was precisely that which on reading to the end of the letter I found was what had occurred to you, namely, the heating of the film. In my paper on the change of refrangibility of light, in the *Phil. Trans.* for 1852, I have mentioned, at p. 532, that glasses and certain sulphides which were sensitive—*i.e.* phosphorescent—while cold, lost their sensitiveness on heating, which returned when they got cold again. There can, I think, be no manner of doubt that this is the cause of the cessation of the phosphorescence of the film almost immediately.

So far, we have had only a new manifestation of what was already known. But now we come to a new and very important result, the superiority in certain cases of the film to the plate. This, in the first place, affords a perfect demonstration that the phosphorescence is not due to radiation. The experiment, which I thought might bring a high additional improbability against the stratum-of-lamps theory—already unlikely enough—yields a result which I had not at all anticipated, and which is perfectly conclusive against that supposition.

But it does more. It demonstrates for the first time the transference to a finite distance of even that short-period vibration on which the phosphorescence depends.

29 October, 1878.

I wrote my last rather in a hurry, being anxious to catch the post. I said, therefore, nothing about the experiment designed to show, as I imagined, the positive light by withdrawing the dazzle, if such a term may be applied to so feeble lights, of the negative light and the phosphorescence resulting from it. I understood you to say, when I was at your house, that all the luminosity came from the negative. I daresay I misunderstood you; but it was with a view to give the positive every chance that I proposed the experiment, thinking that at any rate when well screened the positive light would show itself, especially if a spark were used. The result is in accordance with my expectation.

The screening of the uranium glass plate by the film can only be reconciled with the supposition that the phosphorescence is due to radiation, by the supposition that even a plate, or rather film, that is thin enough to show the colours of thin plates is yet thick enough to absorb all the rays which produce the phosphorescence. This alone would be not absolutely impossible, for I know that that kind of glass is extremely opaque to the rays of very high refrangibility, though *how* opaque exactly I am not prepared to say. But the superiority, under *any* circumstances, of the film to the plate—save only when both are heated together and the plate is the first to cool, which does not apply to the observations we are dealing with—seems to me to be perfectly decisive against the phosphorescence being caused by radiation.

It is very interesting to see the immediate falling off of the phosphorescence of the film, though it falls in completely with our previous knowledge. But it is intensely interesting to see the superiority, in certain circumstances, of the film, and this contributes a material extension to our previous knowledge.

7, KENSINGTON PARK GARDENS, LONDON, W.
30 *October*, 1878 (?).

DEAR SIR GEORGE STOKES,

Your second letter arrived after I had finished the experiment suggested by your first letter.

A tube was made having a thick piece of uranium glass sealed and blown on to the end. The uranium glass was about $1\frac{1}{2}$ millim. thick in centre and tapered off at the sides to the ordinary thickness of the blown tube. In the middle of the tube was a perforated disc of aluminium, and the electrodes were so arranged that a spot of phosphorescence would fall on the centre of the uranium shield.

The tube was exhausted to the best point for phosphorescence, and on passing the current a sharp spot of phosphorescent light formed on the uranium glass. There is no possibility of doubt that the phosphorescence is entirely on the inner surface of the glass, the outer surface not phosphorescing at all.

A confirmatory observation is the following:—The uranium glass contained a little lead, and the blowpipe flame reduced a little lead on the outer surface, so that it was not uniformly clear but patchy. The inside on the contrary was quite clear. When the spot of light was moved about by a magnet it was uniformly bright on the inside, on whatever part of the uranium shield it happened to fall, but looked at from the outside it was obscured or bright as it passed under the dark spots or the clear places.

The effect of parallax is very evident, and leaves no doubt as to which side of the glass phosphoresces.

You say you think uranium glass is not specially suited for testing this point. I have sent the tube by parcel post for your acceptance, and if, on trying it, you think it worth while to try one made of ordinary glass I will willingly try the experiment. But from what you say I don't suppose it will be necessary.

5 *November*, 1878.

I was not aware that you had got so good a figure to your aluminium cup in the highly exhausted vessel. It seems nearly, if not quite, to supersede the necessity for a regular speculum.

I would just, however, call your attention to one thing. You say the focus, such as it was, with the electric bombardment was much further from the metal than the solar focus. This is only what was to have been expected. If the cup were truly spherical, and if the molecules were projected in a strictly normal direction, the focus for the bombardment would be at the centre of curvature. The principal focus, or focus for parallel rays, is at only half this distance from the metal. The centre of curvature is the optical focus, not for parallel rays, such as those from the sun, but for an object itself at the centre of curvature. It is easily found by holding an object, such as the point of a penknife, so as to coincide with its image. If the distance of this point from the mirror be wanted accurately, the two should be viewed through a lens, and the distance of the object from the mirror altered till object and image are seen through the lens in focus together.

It depends upon the degree of goodness of the image in your aluminium mirror whether it would be worth while to try one of speculum metal regularly polished. If the image was so good, as would appear from your description, perhaps it hardly would.

Yet, on the other hand, it might be that regularity of bombardment, if it be regular, would be interfered with by a degree of irregularity of surface which would but little affect the reflection of light. So that, on the whole, I am disposed to think that a regular speculum would be worth trying. To make the effect stronger, the section of it should be a good substantial part of a semi-circle, say an arc of 60° or 90°. There would not, I imagine, be any difficulty about grinding so large an arc. With a mirror of such great an arc as this, there would be greater aberration for the solar image, but there would be no aberration for an object at the centre of curvature; and if the molecular projection were strictly normal, which I think it is more likely not to be than to be, then there would be no aberration in the bombardment focus.

There is another experiment which it would be well to try, and for which you already have the apparatus. The tube you showed me, where the deflection by a magnet is marked by teeth of white paper, would be extremely well fitted for the purpose, because the deflection is large, and is further magnified by the shadow being received very obliquely on the glass.

The experiment is this :—Instead of holding the electro-magnet in the hand lay it on a fixed support, choosing a position which shall give as great a deflection as may be. For the same reason the electro-magnet should be worked by several cells. Marking the place where the shadow falls—*i.e.* its edge—try various arrangements of the contact-breakers of the current which works the coil, and the effect of intercalating or not a small jar in the secondary circuit. The question is, will the place of the edge of the shadow be affected by any of these changes? You have shown it to be affected, as we might confidently have anticipated, given that there was any effect at all, by the strength of the magnetism. Will it be affected also by varying conditions as to the power of the discharge? If the molecules are projected with different velocities in different circumstances, we should expect that the higher velocities would show a flatter trajectory. For though the deflecting force might be expected to increase as the velocity, it would have to increase as the square of the velocity in order that the deflection should be as great at high as at low velocities.

8 *December*, 1878.

By "medium" I assume you mean the residual gas, not the luminiferous ether. For what is a medium? In a gas at ordinary pressures, and even till we come to high exhaustions, the phenomena are the same, *as if* we had to deal with a *continuous* portion of matter the fundamental properties of which are ascertained by experiment,—for example, the property that the pressure varies as the density, and so forth. For simplicity's sake we accordingly make abstraction of the individual molecules, and feign to our imagination continuous matter, which, when considered as the vehicle of some process going on,—say of the propagation of sound,—is called a medium.

Now observe particularly. The "medium" is nothing more than the assemblage of the molecules contemplated from a simplified point of view. When we are dealing with phenomena in which we are obliged to contemplate the molecules individually, as in your late experiments, we must not speak of the medium as well. The medium, in fact, is nothing but the assemblage of molecules, for which latter the ideal "medium" may be substituted in the simpler cases, but not in the cases you are now dealing with.

The only way in which a molecule impelled from the negative could be retarded by the "medium" would be by being influenced by the other molecules. But if it were so influenced it would be driven out of its rectilinear path, or curvilinear when a magnet is in action, whereas your experiments show that the molecules retain their paths, or very many of them (the molecules) at least.

Extracts from Remarks of Detail on Mr Crookes's Paper, of Dec. 3, 1879, "On the Illumination of Lines of Molecular Pressure, and the Trajectory of Molecules."

Page 21.—The experiment as I originally contemplated it would not necessarily have been quite decisive. Had phosphorescence been found under the film it would, so far as I see, have been decisive against the impacts. But the mere absence of phosphorescence underneath would not necessarily have been *quite* decisive the other way. Substances are very commonly extremely opaque with respect to rays of the very highest refrangibility that we have been able to follow experimentally, and *possibly* this might have been the case with the glasses tried, which therefore possibly might have stopped the phosphorogenic rays, even though the thickness was so small as to show the colours of thin plates. It was the unexpected result of a superiority of the film in certain cases that was quite decisive against the radiation theory.

Page 35.—I don't think the constancy of curvature at different distances along the path, when the magnet is moved so as to have the same position relatively to the point where the curvature is sought, is favourable to the supposition of diminution of speed by resistance.

...It is the hits in comparison with the misses during the time, whatever it may be, that the molecules take to traverse the vessel, that you are concerned with.

23 *December*, 1878.

I should not expect that you would find any alteration of the part of the surface of the glass that was so unmercifully pelted by the molecules, that would reveal itself under the microscope. But there is one very delicate test for detecting alterations on a surface, which I may as well mention. It is to use very bright light, and to endeavour to extinguish it by a Nicol's prism, altering by trial the inclination, the azimuth of the Nicol held in the hand, so as to get the nearest approach that may be to complete extinction.

This requires that the second surface of the glass be covered with some sort of black varnish, to stop the second reflection. If stray light can be sufficiently well excluded, the best source of light is the disk of the sun. The observation is, however, awkward, unless the surface be tolerably flat. Perhaps the flat flame of a paraffin lamp, seen nearly edgeways, might answer.

My notion, as I endeavoured to explain when you told me of the result, is that the molecules capable of vibrating in such a manner as to give the green phosphorescence are, as it were, bowled over. I should expect a film of almost infinitesimal thickness to be in a state somewhat analogous to that of a hard-drawn wire. I think it probable that the phosphorescence would take place again if the glass were heated even short of redness.

28 *December*, 1878.

As you had already succeeded in fusing platinum, it is not to be wondered at that you should have succeeded in fusing German glass by the same method.

The result, however, would enable you to try whether fusion, or heating to the point of fusion, will restore the phosphorescing power of the glass after it has been battered out of it, by a more ready method than that I had contemplated.

Probably you had not this in view at the time when you made the experiment, as you do not mention whether the glass had had the phosphorescence knocked out of it, and, if so, whether on cutting off the induction-discharge when the glass fused, allowing the glass to become quite cold, and then letting on the discharge again, the glass phosphoresced as at first.

The experiment with the gold leaves is very remarkable, and would be well worth communicating as you contemplate.

25 *January*, 1879.

The difference between the ray of emitted molecules and what Newton *supposed* to be a line of corpuscles is that the latter is capable of exciting the sensation of light when received directly into the eye. But nobody now-a-days believes that what Newton supposed to be a line of corpuscles is really such.

So much for your letter of the 20th. In that of the 21st the result as to the deadening of the phosphorescence is very similar to what you found before, as mentioned in a former letter to me. I agree with you in thinking that it probably does not depend on the direction of the impacts.

But the shifting of the line of molecular discharge when the current is first turned on is a result quite new, which may help to throw a good deal of light on the whole phenomenon when it is further investigated.

The conjecture which strikes me as the most probable, so far as my present light goes, is that the effect depends in some way on an alteration, perhaps an electrification, of the wall of the tube by the impact of molecules coming from the positive. Accordingly, that it is undone by a subsequent impact from the negative.

I think this idea might be tested by using a **U**-shaped tube, with an inclined plate electrode, and mica screen with hole similar to that in your figure at one end of the **U**, an electrode of any form (say a plate) at the other end of the **U**, and, if it be feasible, an electrode near the bend in the first leg. I will call these electrodes D, E, F; D being behind the mica screen, E at the end of the other leg, and F in the first leg near the bend.

If the experiment be now repeated with D negative and E positive, the question is—will the same effect be produced now that the place of the spot is out of the direct shot from the positive ? If produced, will the shift of the green spot be more tardy ? If it be shifted, will the state of things be restored now that the region of the green spot is out of direct shot from the negative when E is made negative and D positive ?

The object of the third electrode, F, is to enable the previous experiment to be repeated in the same tube when desired, by making F the positive instead of E.

I should not suppose that it is necessary that the electrodes E and F should be plates. I should suppose that wires would suffice, which I suppose would be more easily fixed in.

One of the referees suggests to try a negative plate of this shape in section:—[*wedge-shaped*].

Would there be two sets of molecular projections, one perpendicular to each facet, which are parallel to one or other of two lines (in section)—*i.e.* two planes,—or would the projection be perpendicular to the electrode as a whole?

The same suggests another experiment, which is this:— Connect two electrodes with the coil, and a third idle electrode with earth, and with an electroscope or electrometer. Now let the coil play, and then break the earth connection of the idle electrode. What tension will the electrometer indicate? The idle electrode is supposed to be so placed that the molecular stream is directed upon it.

The same referee suggests trying a flat electrode set obliquely to the plane of the screens. Would the molecules stream through the hole so as in the figure? After what you have done I think it pretty certain that it would.

This suggested to me another experiment:—Make a good earth connection between the negative pole of the coil and the negative electrode of the tube, which are connected with each other. I presume that the phenomena will be as before, though now the positive electrode alone is in tension. The inner terminal of the coil had best be chosen for the negative.

27 January, 1879.

The interesting results you have mentioned with the idle pole fall in very well with the supposition I made respecting the cause of the shifting of the stream of molecules in your former experiment.

That the result of the earth connection, as I had suggested, would be negative is what I had expected, as I mentioned in a former letter. And now I may mention an argument I was prepared to found on it,—that is, on the negative result,—which I avoided mentioning before, as in case I should be wrong in assuming that the result would be negative, I should have had my labour in vain.

Imagine, then, that you had a perfect vacuum in your tube, and that the negative electrode was put to earth, and the coil let play. In consequence of the earth connection there is no tension at the negative electrode. Neither is there current; for everything indicates that a perfect vacuum would be a perfect non-conductor.

Neither is there any sensible statical electricity at the negative electrode, at least if the positive be a mere wire, and be at the other end of the tube....

The negative electrode is simply idle, as might be a piece of metal lying on your table. It is merely in a position in which it is ready to do its work when the circumstances arise.

Now imagine a very minute quantity of gas let in. The tube will no longer be a perfect non-conductor, and the phenomena which you have been investigating will arise.

But inasmuch as the presence of the molecules is a condition precedent to the passage of a current, unless we assume that the state of things at first when the current is let in is quite different from what afterwards takes place, we cannot suppose that an electric vibration of the negative electrode is the cause of the projection of the molecules. In fact, only for the passage of the molecules the current could not be, and we can hardly therefore regard " the electric vibrations of the negative electrode" as the cause of the observed behaviour of the molecules.

It would be well to test by direct experiment the character, as to positive or negative, of the electricity with which the gold leaves diverged, which can be done in a moment with a stick of sealing-wax or glass rod excited, though there can be little doubt what the result would turn out.

5 February, 1879.

I am much obliged by your communication of the experiment with the idle pole. It seems to me that the effect of the touching is to make the idle pole practically the same as if it were faintly negative. I would suggest the following variation of the experiment:—Instead of touching the idle pole, connect it with the negative pole through a very high resistance, the proper amount of which is to be found by trial; perhaps a moist thread, or a thread rubbed with black-lead, would give about the right degree of resistance; but this will have to be found by trial. I should expect that with a suitable resistance a similar penumbra would be obtained.

Another variation of the experiment that seems to me to be likely to be instructive is this:—Instead of connecting the idle pole with the earth, connect it with the inner terminal of a Leyden jar, uncharged, of which the outer coating is connected with the earth. You will probably find the penumbra at first, but after a time it will disappear, or at any rate get much less, but I expect disappear altogether. Now break the connection with the jar by an insulating handle, and the jar will I expect be found charged with a moderate positive charge.

If the changes with the jar are too quick a larger one should be used; if too slow, a smaller.

20 *May*, 1879.

The experiments you so kindly showed me to-day impressed on me more than ever the great advantage that there would be in obtaining, if possible, some substitute for Mr De la Rue's great battery; and what you said last night, after the lecture was over, about the smallness of the amount of electric current which actually passed through those very highly exhausted tubes, leads me to think that it is quite possible that a substitute could be obtained without much trouble. I forget whether I ever mentioned to you the plan I thought might succeed, but at any rate I may as well mention it now.

I should propose to try as large a condenser as you have at hand—say a Leyden jar, or a series of such jars connected collaterally. To render everything perfectly definite it would be as well to connect the outer coating with earth, as well as the outer terminal of the induction-coil. This is also connected with the electrode of the tube which you wish to be negative. The inner terminal of the coil, which is made positive, leads to a point at such a distance from the inner coating, or a knob connected with the inner coating of the jar, that the positive discharge can strike across, but the negative return discharge from the jar is unable to pass, the distance being too great. The inner coating of the jar is also connected with the positive electrode of the tube through a wet thread, which at first had best be taken too long, for fear of injuring the tube, and afterwards reduced as experience indicates is safe and expedient. There is also a break in the connection,—either a contact-breaker or an interval which can be spanned over by a wire when desired.

The bridge being broken the induction-coil is let to play, and when the condenser is filled as full as the coil can fill it the bridge is made. I think it probable that a not extravagantly large condenser would hold electricity enough for the discharge through the tube to last long enough to allow the eye to follow the series of phenomena. If so, I feel convinced that it would be a great gain in the study of what takes place. At present the changes are so quick that the eye merely perceives that the discharge is not steady, but the series of changes are too quick to follow. By commencing with a sufficiently long thread there would be no danger of injuring the tube by an explosive discharge of the condenser; and indeed, unless the condenser were very large, there would be no danger that even a sudden discharge of it would injure the tube. With too long a thread the discharge would probably be too feeble to be of use, but one would learn how far it is safe to shorten the string.

30 May, 1879.

As you are taking so much trouble in preparing a condenser of large capacity, I will make a suggestion which I think may tend to increase its efficiency, and at any rate is not a serious matter to try.

For charging the jar I propose to place the positive terminal of the jar coil, which should also be the inner one if the coil be so wound that there is a difference in the tensions of the outer and inner, at a sufficient distance from the knob of the jar to prevent the positive discharge which strikes across from returning back again. This, however, will prevent the highest tension to which the condenser can be charged from being nearly equal to what the coil could give. I believe the condenser could be charged higher by a little instrument easily constructed, and which would, I expect, act as an electric valve.

For this I should use a tube with a flat negative electrode,—the form is a matter of indifference, but it should be of fair size, though it need not be large,—and a positive electrode formed of a wire filed or ground to a sharp point, and then coated with glass or enamel fused on. The tip is then cautiously filed or ground so as just to expose the merest speck of the metal, or else, when it is nearly exposed, the glass might be pierced by an electric discharge. I do not know which would answer best, but I should be rather disposed to try the purely mechanical process. The tube should not be exhausted very far; I should think 1 or 2 millimetres of mercury would be about the thing. It should not be exhausted to such a degree that an uncovered positive electrode has the discharge no longer confined to one point on it. At what degree of exhaustion that takes place you, with your experience, will be able to say.

The electric valve being placed in the path of the discharge from the positive terminal of the coil to the knob of the condenser,

I imagine that no other air-break will be required, and that the valve will prevent the return of the current that passed through it on its way into the condenser; and as the tube offers comparatively little resistance, it will be possible to charge the condenser higher than when there was a considerable air-space which the discharge had to leap across.

I think M. Gaugain used an exhausted tube as an electric valve, but I am not aware that he used a protected positive electrode, which I imagine will be found to be an improvement.

2 *June*, 1879.

...It is possible that you may have tried a Leyden jar, and found that the capacity was too small to enable you to make anything of it. I should have thought, however, that even a jar of tolerable size would be enough to show whether there was any likelihood of success.

Mr De la Rue's great condenser gave a discharge which lasted for some minutes. Now, on the one hand, a fraction of a second would be enough to show whether there was an effect, even to allow of its study to a certain degree; and, on the other, the tubes to which that great condenser of Mr De la Rue's was applied were such as gave good conduction, and therefore required a copious supply of electricity. With a view to the employment of a condenser, I asked you once whether much current passed in your highly-exhausted tubes, and you told me that very little passed; at least I fancy that that is what you told me, though now that I think of it I think you said something about the whole of the discharge of the coil as passing.

In any case it would be well worth while, if you have not done so already, to feel your way with the use of such high-tension condensers as you have at hand, before proceeding further.

20, MORNINGTON ROAD, LONDON, N.W.
7 *June*, 1879.

DEAR PROFESSOR STOKES,

I have tried your experiment with condensers made of Leyden jars of different sizes and separate or connected together, but can get no results worth recording. Where little electricity passes (slow discharge through wet string) the phosphorescent light is too faint to see anything distinctly, and if the string is shortened the passage of the current is too quick. If you can suggest a better way I shall be glad to try it.

The valve acts very well*, and seems almost entirely to cut off the little amount of + in the − discharge of the coil, and *vice versa*. Perhaps two valves would make the action certain.

Will not valves on the coil answer the required purpose, as the condenser seems difficult to arrange properly?

I have had a visit from Mr Garnett. He asks for the two experimental tubes (yours and Professor Maxwell's) to take down to Cambridge for you to experiment with. He will come for them in about a week. Besides these tubes would you like any valves or other apparatus made? There is plenty of time.

9 June, 1879.

I certainly feared that the capacity of even two or three Leyden jars put together might not be sufficient to get any result. However, as I knew that great resistance was offered to the passage of the current in those excessively high vacua, I thought it possible that the quantity passing might be such as could be furnished by a good-sized jar,—at least so far as to give a duration which could be observed with a quick eye. From what you tell me of the result with several jars, I fear the quantity would be too great for a high-insulation condenser of reasonable capacity.

In thinking over your letter, however, another mode of making the experiment occurred to me, which would I think allow you to obtain the results, though it would not be so luxuriously comfortable as a high-insulation condenser of very great capacity.

I will first describe the experiment in relation to the pendulum experiment I suggested:—

Screw home the ordinary contact-breaker of the coil, and lead the primary current through a mercurial contact-breaker instead. If you don't happen to have one suitable for the purpose, it would be very easy to rig one up by using a lamina of elastic steel somewhat larger and stiffer than a piece of watch-spring of the common dimensions. A platinum wire is attached perpendicularly at one end of the lamina, to make contact by dipping into mercury. The lamina is gripped in a vice, and allowed to vibrate. The time of

---

* On May 31 he had written, " I will make a valve for the induction spark. I don't know Gaugain's, but the one you describe is not unlike Reiss's which seems to answer well. Reiss used a disk and a point as poles in a tube which could be exhausted to different degrees. The point was not however protected. This will be a great improvement. Prof. Edlund (*Phil. Mag.*, Ser. 4, XL. 30) has tried many experiments with this valve."

vibration can be regulated by the length left outside the vice. The time should be regulated so as nearly, if not quite, to synchronise with the time of vibration of the pendulum. The height of the mercury surface should be regulated so as to come to the point of the platinum in the position of rest of the lamina, in order that the intervals between successive breaks of contact may remain the same—assuming the vibrations isochronous, which will practically be the case nearly enough—as the amplitude of vibration dies away. It would be convenient that the vibrations of the lamina should be a shade slower, rather than a shade faster, than those of the pendulum, in order that the phenomena may be seen in their natural order rather than the inverted order, with reference to the direction of motion of the pendulum. The primary current is led through the vice and the mercury. The mercury is covered, as usual, with alcohol or turpentine—I mean oil of turpentine. Alcohol, I think, is what is generally used. It might be well to introduce a valve into the secondary circuit, to prevent the attention from being distracted by the phenomena attending make contact.

As a practical matter the direction of the primary current is not a matter of indifference, as, if it be wrong, explosive discharges occur at the surface of the mercury and alcohol, which scatter the fluids about. I forget which way the current should pass, but I think the mercury should be negative. It is easy to try, and then make it right. Hence if, for convenience, a current reverser should be placed in the primary circuit, the mercury break ought not to be included in the part of the circuit which is under its command, but in the remaining part.

Everything being arranged, and the pendulum tube being in the secondary circuit, the current is let play. At each break the discharge passes for a very brief time, but at each successive break the pendulum is a very little further in advance in its course, since a very little longer time than that of one vibration of the pendulum has elapsed since the last break. In this way I think it would be possible to follow the successive phases of the phenomenon depending on the position of the pendulum, somewhat in the manner of the toy called "the wheel of life."

I do not recollect for certain the details of Maxwell's experiment, and I have not his report to refer to, as it is still in your hands[*]. But I think it is to try the effect of an idle pole which is temporarily

* See *infra*, p. 433.

placed in tension by an independent coil. If so, it would be well to introduce mercurial breaks into both primary circuits, arranging their laminæ for nearly, but not quite, synchronous vibration. Both vibrations might here with advantage be a good deal quicker than in the former case, where the time of vibration was conditioned by the swinging time of the pendulum.

18 *June*, 1879.

When I wrote about the vibrating slip of steel I had in my mind another plan which I did not mention, thinking that the slip would be easier. But as you apprehend some difficulty about the slip, I may as well mention the other, as it may be well you should have the two plans before you. It is simply to employ a very small extemporised pendulum, one arm carrying a bent arm of platinum wire dipping into mercury covered with alcohol, and so adjusted that in the position of rest the wire just touches the surface of the mercury. I think a sufficient pendulum for the purpose could be made out of a couple of fragments of a knitting needle and a cork. A leaden bullet would serve for a bob.

18 *August*, 1879.

As to the influence of an inserted piece of metal, which may be made an electrode at pleasure, on a molecular discharge that brushes by, the idea is very tempting that it acts by the attractions and repulsions of the statical charges of itself and the molecules which rush past. Such was your view, and such at first was mine. But Maxwell, in his report on a paper of yours, gave a calculation from which he inferred that, under no reasonable suppositions as to magnitudes involved, could the influence actually found be accounted for by the supposition that it depended on statical attractions. To put this point to a further test I suggested to you an experiment with a Leyden jar, which you were good enough to try with some additions of your own, especially discharging the jar while the experiment was going on.

The aggregate results, and especially the absolute indifference of discharging the jar, proved to my mind conclusively that the effect did not depend upon statical charge acting merely as such, but depended on a current from the electrode when made such.

Mr De la Rue's battery shows the thing beautifully. When the electrode is unconnected it produces no effect, notwithstanding that it quickly becomes strongly charged with positive electricity.

When connected with earth it becomes a feeble negative electrode, to the extent—considering only the permanent state of things, and not any merely temporary charging up of the battery, &c., with a statical charge—that the leakage of the battery permits. When it is connected with the negative electrode a similar effect, but far more violent, is produced.

20 *August*, 1879.

It struck me in reading the paper that there was a very important, as it seemed to me, observation that you mentioned to me, and of which nothing is said in the paper. I am not sure whether you are quite certain of the result.

You dwell almost needlessly on the non-exhibition of the spectral lines of the gas in the green light, which none of us attributed to anything but phosphorescence, but have omitted to say anything of their non-exhibition in a case in which *à priori* we should have expected them.

The focus-shaped luminosity appears to be formed under two totally different conditions. In the first place, with electrodes of certain shapes,—such as a ring, or a cup, or a hemi-cylinder,—as the exhaustion proceeds from that which first shows anything of a good negative glow, which at first closely invests the electrode, the glow as it expands gets the part which is convex towards the electrode, becoming more and more curved at the tip, till at last it presents the form of something like a cusp rounded off. This, however, is merely the ordinary glow tending, in one point, to turn itself inside out. This form of luminosity somewhat focus-shaped was known before, and it naturally presents the spectral characters of an ordinary glow. It belongs to the *outer boundary* of the dark space.

As the exhaustion proceeds this passes away with the rest of the glow, which becomes extravagantly enlarged. We now get the new focus-shaped blue glow which you have discovered, and which belongs to the *interior* of the dark space. I take it these two luminosities are quite distinct in their nature, notwithstanding that when the first disappears and the second appears they occupy nearly the same place. Now you told me that the *second* blue glow showed nothing of the spectral lines of the negative glow of the gas; yet we must attribute it in some way to the presence of the gaseous molecules.

I think this is a very important point to bring out, if you feel quite satisfied that the failure to see the ordinary spectral lines of the gas did not merely arise from the feebleness of the light. I am not quite certain of your meaning in Sec. 520, as printed. You say "blue light has disappeared, and no lines are detected...." Detected in what? I took you to mean in the green light, which we know is due to phosphorescence. To judge by the words, you can hardly mean in the other blue light which has taken the place of the former blue light that has disappeared, inasmuch as you say nothing about this other blue light.

I attached so much importance to the spectral character of the *new* blue light that I often thought of calling at your house to see it through a prism with my own eyes, and try if I could satisfy myself that the failure to see any lines did not merely arise from the extreme feebleness of this very feeble light.

APPENDIX A. Prof. CLERK MAXWELL'S Report on Mr Crookes's Paper "*On Repulsion resulting from Radiation*," Parts III. and IV. (Read February 10th, 1876.)

THE experiments recorded in these papers relate to bodies placed in a vessel as completely exhausted as possible. These phenomena are more regular than those observed by Mr Crookes in less highly rarefied air, and the forces are more powerful.

Mr Crookes has now succeeded in making the phenomena so regular that they can be subjected to the test of measurement and comparison. He has therefore made several different kinds of instruments for the measurement of repulsion force.

Sect. (128). A differential radiometer, consisting of two pith disks, one white and one black, at the ends of a horizontal rod suspended by a silk fibre.

(135) The same with a bar of pith, one end white and one end black, with magnet and mirror, suspended by a silk fibre and controlled by a magnet.

In both these instruments light is incident on both surfaces, but the black surface is repelled more than the white.

In the first instrument there is equilibrium when the stronger specific action of the black surface is counteracted by its greater distance from the source of radiation and the greater obliquity of the rays. This can take place only when the source is very near the instrument.

In the second, the directive force of the magnet tends to bring the instrument back to zero, and enables it to be used when the source is at a considerable distance, so that the rays may be considered as parallel.

(143) The rotating radiometer, or "Light-Mill." This is also a differential instrument, the sides of the vanes being alternately black and white, so that the effect of radiation is to drive the mill round. The final velocity of rotation depends on the friction of the pivot, and whatever other resistances may act on the exhausted chamber.

It appears from the experiments that the velocity is inversely as the square of the distance of the source; and as it has been already shown that this is the law of the pressure on a surface at rest, it follows either that the resistance to the motion of the "light-mill" is proportional to its velocity, or that the pressure resulting from radiation is different for surfaces at rest and in motion.

The most important experiments with the light-mill are those which show the effect of the change of temperature of the instrument.

When the instrument is cold, and luminous radiation falls on it, each vane goes with its white surface foremost. This Mr Crookes calls the normal or positive direction.

Under other circumstances the black surface goes first. This may be called the negative rotation.

(166) An aluminium light-mill, while being heated, went +

(167) Afterwards, while cooling, it went −

When the bulb was cooled by evaporation of ether +

„   surrounded by a heated glass shade −

„   surrounded by hot air from Bunsen's burner −

„   heated by experimenter's body −

[At (165) the effect of the hot glass shade on the brass light-mill seemed reversed, but the statement is not quite clear.]

These results lead me to think, what indeed Mr Crookes at Section (195) seems to hint at, namely, that the effect depends on the difference of temperature between the white and the black surface.

If we admitted the corpuscular theory of light, we might suppose that the force was due to the impact of the corpuscles. This force would be greater at the white surface, from which the corpuscles recoil, than at the black surface, where they are merely stopped. According to the electro-magnetic theory of light there is a pressure in the line along which light is propagated. This would be greater if (as near a reflecting surface) light is propagated both ways, than near a black surface, where there is no reflected light. Hence neither of these hypotheses explains the phenomenon of the black surface being most repelled.

But if we suppose that the force depends on the temperature of the surface (that is, not on radiation, but on "thermometric heat"), I think we can explain most of the facts.

For a black surface is known to absorb and to radiate more than a white one, even when the radiation is not luminous.

Hence when the radiometer is exposed to radiation from sources of high temperature, the black surfaces absorb more, become warmer, and are "repelled."

When the radiometer is hot the black surfaces radiate most, and become cooler than the white ones, and the motion is negative. When a hot glass shade is put over the instrument, or when heated air rises round it, the heating effect acts on both white and black surfaces, but probably most on black; but the radiation of the black surfaces through the transparent hot body cools them more than the white ones, and the motion is negative.

Cooling the instrument by ether renders it less able to radiate, and the black surfaces are now most warmed by radiation from without, so that the rotation is positive *.

Mr Crookes's view, as expressed in (195), agrees with this up to a certain point. He admits that in normal rotation the black surface is hotter than the white, but he attributes the motion to the *radiation from* the black surface, owing to its higher temperature. Now if the same radiation falls on a white and a black surface, the total radiation from the white surface will be greater than that from the black, because some of the radiation is reflected directly by the white surface, whereas what falls on the black is absorbed, and the radiation of the black is only that due to its own temperature.

It is not, therefore, because the black surface radiates more, but because it is hotter, that it is more repelled.

With respect to the cause of this repulsion Mr Crookes (130) has shown clearly that it is not evaporation of water, as in Prof. Reynolds's *illustrative* experiment (*Proc. R. S.*). Indeed Prof. Reynolds himself shows that at the low pressure obtained by the mercury pump, and at ordinary temperatures, liquid water cannot permanently exist.

But the experiments of Kundt and Warburg (*Pogg.*, 1874—5) show that in the best exhausted vessel there is still enough matter to offer considerable resistance to motion and to conduct heat. (They seem, however, to have succeeded in exhausting a vessel of small capacity till the passage of heat was not affected by the kind of gas with which it was originally filled.)

Hence any inequality in the pressure, arising from evaporation or any other cause, may continue to exist in an exhausted vessel as long as it would in a less rarefied gas.

But whatever may be the cause of the phenomenon, the determination of its magnitude, as compared with the radiation which occasions it, is of the greatest value as a means of testing

[* On these points, cf. Prof. Stokes' remarks, *supra*, pp. 377, 386 *seq.*, and his paper reprinted in *Math. and Phys. Papers*, Vol. v. pp. 24–35.]

the various hypotheses. Mr Crookes has made great progress in the construction of measuring apparatus. He has employed three methods—the "Torsion Rod" of Michell and Coulomb, the Horizontal Pendulum of Zöllner*, and the Torsion Balance of Ritchie.

Of these the best adapted for quantitative measurement is the Torsion Rod proper, in which the force to be measured is balanced against the torsional elasticity of the suspension fibre. The instrument described at the beginning of Part IV is of this kind.

[Some calculation follows, showing that the data of this kind in the paper do not suffice to deduce the pressure on the disks.]

The horizontal pendulum, in which gravity is the principal force, is liable to great alterations of zero if the instrument is not rigidly fixed. Hence Mr Crookes abandoned it.

The torsion balance described at the end of Part IV is a common gravitation balance, in which the final adjustment is made by the torsion of a fibre instead of by a rider. The consistency of Mr Crookes's results is a proof of his wonderful skill in the construction of such apparatus.

He has ascertained that the pressure due to weak sunlight, on a certain day, was 0·009 grain per square inch. This is $\frac{1}{381000}$ of an inch of mercury, or 0·089 dynes per square centimetre.

This is an important measurement, but what would give it more value would be a simultaneous measurement of the heating power of the sun's light by an actinometer, the light being subjected to the same absorbing media in its path to the actinometer as in its path *to the pith*.

The experiments in which the instrument itself was brought to various temperatures are important, and should be extended.

But if the disks themselves could be constructed of two substances separated by a small interval, the temperatures of the two surfaces might be separately altered †.

Thus if a pith disk with a hole in it had a copper disk at its back, the back of which is blackened,—then if the copper disk were heated by means of a burning glass, the rays passing through the hole in the pith, the copper would be hotter than the pith, and the whole might move towards the light; and other experiments might be devised with compound disks to determine whether temperature or radiation is the cause of repulsion.

These remarks are not intended to depreciate the value of Mr Crookes's papers, which I consider worthy to be printed in the *Transactions*, and likely to become important documents for the history of Science.

No explanation is given in the paper of the mode in which the Morse instrument was made to register the revolutions of the radiometer as shown in the drawings.

---

* I think it was first used in America.   [† Cf. *supra*, p. 388.]

APPENDIX B. PROF. CLERK MAXWELL'S REPORT* ON
MR CROOKES'S PAPER, "*Contributions to Molecular Physics
in High Vacua.*" (Dated March 27, 1879.)

THIS paper contains a further investigation of the discharge from
the negative electrode, some of the properties of which were
described in the paper read December 5th.

The author has tested two suggestions† made by one of the
referees of the former paper, to see whether the direction of the
discharge is really as it seemed to be, normal to the surface of
the negative electrode at the point from which it is projected, or
whether this appearance depends on the general lie of the surface.

To test this Mr Crookes made an electrode of corrugated
metal, every part of which was inclined one way or the other
to the plane containing the bottoms of the furrows. The discharge
from this was in two series of plane strata normal to the sides
of the individual furrows, and no sensible part of the discharge
was normal to the plane formed by the ridges. The other test
was by placing a plane negative electrode obliquely behind a mica
screen with a hole in it. In this case also the discharge through
the hole appeared to be normal to the surface of the electrode.

These experiments confirm the result of the former paper—
that the path of the discharge, when undisturbed, is normal to
the surface of that part of the negative electrode from which
it proceeds, and is independent of the position of the positive
electrode. The experiment with the corrugated electrode shows
that several sheets of the discharge may cross each other without
interfering much with each other.

Another experiment (607) shows that two parallel and con-
current streams of the discharge repel one another.

The action of magnetism on the discharge is examined more
fully. The results agree with those of the former paper in
showing that the force which acts on the molecules is in the same
direction as that which would act on a conductor placed in the
line of the discharge, and carrying a current of positive electricity
towards the negative electrode.

In other words, the molecules are deflected in a direction
normal to the line of discharge and to the line of magnetic force;
and if we consider the molecules as moving in part of a circle, the
direction in which they are moving round the circle is opposite to
the direction of a positive current which by flowing in a circuit
embracing the whole field would produce the actual magnetic
force. This is made very clear in paragraph 609, but in pars. 616
and 617 it is shown that in certain cases a discharge in a low

[* This Report was communicated by Prof. Maxwell's permission to the author,
who expressed his thanks and his desire to have an opportunity of showing the
phenomena to Prof. Maxwell.]
[† Cf. *supra*.]

vacuum, which certainly moves in the same direction as a conductor would move, travels sometimes the same way as the molecular discharges, and sometimes the opposite way. Now the low vacuum discharge was between a point above and a ring below, a magnet being placed vertically below the point. The lines of force proceeding from the magnet diverge as they ascend, but the case is that of a current forming the generating line of a cone, and the rotation of the discharge is the same as if it had been carried in a copper-wire.

But in the high vacuum both poles were in the axis, so that the discharge proceeding from the negative pole was always *from* the axis, whether it proceeded from the upper or the lower pole, and therefore the direction of rotation depended on the vertical part of the negative force at the negative pole. As this is weaker at the upper pole, perhaps the rotation would not be so fast when the upper pole is negative. The phenomenon depends entirely on the manner in which the discharge leaves the negative pole, and not on the manner in which it gets beaten to the positive pole. Hence the rotation should be in the same direction whether the positive pole is above or below the negative pole.

In par. 617 it appeared to me that the rotation in the high vacuum went the opposite way, *in all cases*, to that described in par. 609, but in this I may be wrong.

In par. 596 two idle poles were introduced between the electrodes, one being connected to earth and the other to an electrometer. When the electrode next the electrometer was positive there was an instantaneous deflection, probably due to induction. When this was discharged to earth the electrometer became again charged positively, but in a gradual manner. When the electrode next the electrometer pole was made negative the effect was more doubtful, but the author seems to think the induction positive.

May not the charge of the idle pole be mainly due to positively charged molecules which are working their way from the + to the − electrode, and which take no part in the brilliant display of the negative discharge? This might be tested by trying whether the positive electrification of the idle pole is strengthened by being placed near the positive electrode, and weakened by an idle pole connected to earth placed between it and the + electrode; and secondly, how is it affected by the negative discharge being made to play on it or being deflected from it, say by a magnet....

To test the action of an electrified pole on the molecular discharge, a metal plate might be placed with its plane parallel to the line of discharge, or, still better, two idle poles might be placed one on each side of the line of discharge, and these might be electrified one + and the other − by Holtz's machine.

This would give a powerful electric field with the force everywhere in one known direction, and by the use of magnets the direction of the discharge between the two plates could be varied at pleasure, so as to make it play against one or other when they are connected to an electrometer.

I have confined my remarks to the dynamics of the discharge, because I think that magnetic and electric tests are the best known means of investigating the discharge while in transit. It is proved to be in some respects like a current as regards magnetism, and in others, though less clearly, like a charged body as regards electric force.

I have only now to recommend that this paper be printed in the *Philosophical Transactions*.

PROF. STOKES ON MR CROOKES'S PAPER (dated March 27, 1879).

Page 2, par. 590.—That the phosphorescence is due to a molecular stream seems satisfactorily established; but whether the cause of the molecular stream is simply the ordinary repulsion of statically electrified bodies is far more doubtful. It *may* be so, but it cannot by any means be regarded as established. The theory that the molecules act simply as carriers of electricity, and are repelled from the negative simply by the ordinary repulsions due to statical electricity, is very tempting, but does not explain all the phenomena. I mentioned already that Prof. Maxwell had calculated the deflecting force of a magnet as compared with that of a statically-charged idle electrode, on the supposition that a charged molecule behaved like an element of a current, and found that the former was as nothing compared with the latter; whereas experiment reveals the former, and shows apparently nothing of the latter.

Again, in the experiment I suggested of connecting Maxwell's idle pole with a Leyden jar instead of an electrometer, you told me that no difference was perceived as to the phosphorescence in the first action from the action after some time, and no difference made by discharging the jar by the finger. The presence of the jar, however, in either case *did* make a difference, the shadow showing a penumbra, which you consequently discovered to be a fluctuation. I remarked, I think, at the time that this seemed to show that the effect of the jar depended not upon its charge, as one might expect on the statical repulsion theory, but upon the current out of it*.

[* It was only in vacua so high that the conductivity no longer screened off the electric field that electrostatic deflexion has since been observed and measured.]

The relation of the charge as ordinate to the time as abscissa may be expressed rudely by a figure something like this:—[*omitted*]. Experiment shows that the jar after a time is positively charged. Therefore the current in, represented by the ascending line, is not at first balanced by the current out, represented by the descending line. As the charge mounts up the current out increases, till after a time the mean charge is sensibly constant. Discharging the jar by a touch reduces the state of things at once to the initial state. The curve would more properly be like this...

The effect, I think, shows very clearly that the observed result is not due to charge, but to the in-going currents by which the idle pole acts as a negative electrode, and is referable to the class of phenomena which are otherwise manifested in the mutual repulsion of molecular streams.

P. 9, par. 603.—I cannot think this explanation correct. The experiment with the Leyden jar seems to me to prove that the deflection of the shadow depends upon current, not charge. I mean upon current, not upon potential.

But the thing seems to me to admit of being put to the test of a crucial experiment. If the deflection depends upon current the shadow of the idle insulated pole is what I may call the normal shadow. When the pole is connected with earth it is no longer idle; currents pass through it to and fro, and the negative current deflects the molecules, enlarging thereby the shadow.

According to your explanation, when the pole is unconnected the molecules are inflected, and the shadow consequently contracted. When the pole is connected with earth the tension is eased, and the shadow regains its natural width.

The experiment I would suggest is this:—...

P. 23, par. 631.—I think Becquerel's observation of the spectrum of rubies rendered phosphorescent by radiation should here be referred to, not so much as a matter of priority (for Becquerel's phosphoroscope is pretty well known, and the description of a particular spectrum calls for no special notice in relation merely to priority, unless it is something particularly remarkable) as because it affords the best proof yet obtained of the identity of the phosphorescent light, whether the phosphorescence be produced by radiation, as in Becquerel's experiment, or by the molecular stream process, which the film experiment proved conclusively to be due to something different from radiation.

April 22, 1879.

Dec. 24, 1880.
My dear Professor Stokes,

I have the pleasure of sending herewith my long promised paper on "the Viscosity of Gates at High Exhaustions."...

The investigation was commenced at your suggestion nearly five years ago and it has occupied much of my time since. During this Autumn I have repeated a good many of the determinations, finding my laboratory convenience much better than hitherto.

I think I have thrown a little new light on Maxwell's Law; and it is a great loss to Science, and a matter of deep regret to me personally that he is not with us to give his sound criticism on the paper.

I am rather nervous as to the results of your scrutiny. I have indulged somewhat in theory towards the end, but not I think to a greater extent than the facts warrant.

Jan. 3, 1881.

Your letter received this evening relieves me of some anxiety. It would certainly be by far the best for you to write a few pages discussing the facts given in my paper*. I felt this all along but I did not like to suggest it to you knowing how much your time is occupied. Now you propose to do this, my plan of action is very simple; it would be best to leave the paper as it stands, making the few verbal alterations and adding a paragraph giving the times of oscillations for different pressures in the gases worked with.

I will start to-morrow getting some of the times of oscillation.

2 Feb., 1881.

Dear Mr Crookes,

Before your apparatus is finally dismounted, though whether before or after the experiments on the effect of temperature, does not matter, there is another investigation which might be borne in mind.

I think it may be concluded from your experiments that a gas is *not* physically defined merely by its density under a standard pressure and by its coefficient of viscosity. Up to high exhaustions these two constants appear to be sufficient for its definition, so that if they are given the chemical nature of the gas is a matter of indifference. But something else must apparently come in, in

* This formed the Appendix reprinted in *Math. and Phys. Papers*, Vol. v. pp. 100–117.

order to define the gas even as regards its mechanical—let alone its chemical—properties when we come to very high exhaustions.

The question arises, is this something a function of the molecular weight alone, of the density under a standard pressure; or is this not the case?

7 *Feb.*, 1891.

I am extremely obliged to you for all the trouble you have taken about the question which I asked.

I think you have not quite understood the notions I had in my head. I had no idea that the metallic dust projected from an electrode would be made to glow so as to show the metallic lines by a subsequent discharge, or even by a subsequent part of the same discharge, when of such a kind as to show a finite duration when viewed in a small mirror turned by hand. My notion was that if we could get an instantaneous discharge, and the vacuum were a decidedly high one, then at the moment of the discharge the lines might be well shown.

I recollect Mr De la Rue a long time ago showing a discharge at the Royal Society the appearance of which I well recollect, though I have forgotten some of the details of the arrangements.

He had an egg-shaped glass vessel, I should suppose 6 or 8 inches in the longer diameter, with electrodes at the ends, the negative electrode being very large. The vacuum was a high one, I think a carbonic acid vacuum, that is, carbonic acid well pumped out and then most of the residue absorbed by caustic potash. I do not recollect what the electro-motor was, but from my recollection of the intervals between the discharges I think it must have been a battery (see later); I do not think it can have been an induction coil. Yet a small number of cells would hardly, I should suppose, have been sufficient to force the discharge through, and I do not recollect that there was any very large battery there. Be that as it may, the appearance was this. Immediately after a discharge all was dark; then minute tufts of faint bluish light were seen at various points of the large negative electrode. These increased in intensity, being however but feeble at the strongest, till a sudden discharge took place. The chief visible feature of the discharge was a tuft of green light, of dazzling brilliancy, from one point of the large electrode, the particular point varying casually from one discharge to another. The electrode was probably copper. There can be little doubt I think that the dazzling tuft would have shown well the metallic lines.

I have been thinking since that the electro-motor must have been an electrifying machine, used to charge up a Leyden jar, the coatings of which were connected with the two electrodes.

I do not recollect for certain that the large electrode was the negative one; only, from the nature of the case I think it must have been.

The faint spots of bluish light denoted of course electric leakage going on across the glass vessel while the jar was being charged up to a tension that enabled the discharge to take place.

From some of Mr De la Rue's later work, it would seem that when the exhaustion becomes very high, the main part of the energy of the discharge is expended in getting the electricity from the negative electrode into the gas, whereas when it is more moderate it is expended in getting the electricity across the gas. Hence to concentrate the energy as much as possible on the first obstacle, which we might expect to be favourable for bringing out the lines of the metal, we should apparently aim at two things; one, a decidedly high exhaustion, the other, confinement of the discharge to or nearly to a single instantaneous discharge of the Leyden jar, or other condenser used. With the improvements in the Sprengel pump that you have introduced, the first condition can be secured with certainty. The exhaustion of course should not be so very high that we are unable to force a discharge across. The second would seem to be attained with greatest certainty by charging the jar by means of a frictional or influence machine; for we could thus go on charging the jar till the tension was sufficient to drive the electricity across, without almost immediately filling the jar up again. If an induction coil were used instead of an electrifying machine, it would seem to be necessary to choose a jar of suitable capacity; for if the capacity of the jar or other condenser were too great, when it was charged by the break-contact current the tension at its maximum would be too low to strike across, and if it were too small there might be two or more discharges of the jar for one break of contact, with a current between, like what you have without any jar at all intercalated; or even if there were but one discharge of the jar it might be followed by a current through the exhausted space having a duration rendered appreciable by viewing it by reflection in a moving mirror.

I do not think it would be necessary to examine the spectrum of the metal. The eye could tell whether there was any such brilliant metallic tuft as was shown in that experiment of Mr De la Rue's. If there were, it would be sure, I think, to show the spectral lines of the metal strongly. And the conditions for obtaining it appear to be a high exhaustion and a jar which has to be charged up to a rather high tension before it will discharge itself across the vacuum, and which is large enough to give a strong discharge when the discharge does take place.

P.S. Pray do not trouble further about these experiments. I am greatly obliged to you for the trouble you have taken. I do not recollect whether the electrode where the brilliant tufts were shown in Mr De la Rue's experiment was positive or negative. I merely guessed that it was negative from the fact of its being large.

8 April, 1891.

Many thanks for your very interesting inaugural address*, which I have just finished reading. You have brought together the different parts of your great subject in an extremely attractive way.

I had no idea till I read your paper (see p. 9) that in a stratified discharge with compound strata, differently coloured for different component members of a group, the differently coloured strata gave spectrum indications of different substances, mixed together in the rarefied gas.

7 May, 1891.

It occurred to me this morning that what you showed me of the effect of breathing on the Lippmann's coloured photographs of the spectrum completely falls in with, and in fact confirms, Lippmann's explanation.

Supposing that the rays which impressed the plate were incident perpendicularly on it, the interval between the plane strata of reduced silver would be half a wave's length in the medium (gelatin I think) which was the vehicle of the silver salt, or about 3/4ths of the half wave-length in vacuo. When you breathe on the plate it imbibes moisture, almost immediately through its whole thickness as it is so thin, and consequently swells. Hence the interval between the silver planes is increased; and therefore to find the interval which suits the red light you must travel to a part of the plate where the interval had been less, such suppose as previously suited the green. Hence the reflected colours will march onwards in the direction from the red to the blue, and

* As President of the Institution of Electrical Engineers.

will go back again when the moisture evaporates, and consequently the film contracts.

I feel little doubt in my own mind that Lippmann's explanation is the true one.

4, WINDSOR TERRACE, MALAHIDE, Co. of DUBLIN,
14 *Sept.*, 1894.

I was lunching to-day with the Provost of Trinity College, Dublin (Dr Salmon), and I met among others Professor FitzGerald. I asked him about Lord Rayleigh's discovery of a new gas. I had been travelling about in Germany, and when I returned from the Continent (on Aug. 14) I went immediately to the Isle of Man. So I have been out of the way of getting scientific information. I got some further information about the new gas from Professor FitzGerald. From a brief allusion in some newspaper I imagined that in the electric discharge it showed only a single line, but FitzGerald told me there were a lot of lines coinciding with the lines of the line-spectrum of nitrogen. This might be due to an impurity of nitrogen in the new gas, or it might be due to the fact that what was deemed to be nitrogen contained the new gas, or it might be due to the new gas being allotropic nitrogen and being decomposed by the electric discharge. With reference to the second possible supposition, I asked him how the nitrogen with which the comparison was made was prepared. He did not know, but said that the nitrogen spectrum used for comparison was prepared by you, or rather (which is not necessarily the same thing) that the photograph was taken by you. I should be glad to know therefore whether the nitrogen used was prepared directly, by the decomposition of some compound of nitrogen, or was prepared from atmospheric air by removing the oxygen, vapour of water, and carbonic acid, the residue being assumed to be nitrogen.

I am to leave this on Monday for Cambridge. I suppose you have by this time made a good deal of progress in the photography of the spectra of the phosphorescent light of the rare earths you have been studying.

Sept. 17, 1894.

DEAR SIR GEORGE STOKES,

I have only just returned home after a stay with my family in Wales, and therefore have done no work with the spectra of the rare earths for some time.

Just before I left London Ramsay brought me two vacuum tubes, one containing the new gas and the other pure nitrogen.

I took some photographs of each of them and sent them to him. He was in a hurry to have them at the B.A. so I had no time to take any measurements of wave-lengths. The nitrogen tube gave the well-known nitrogen spectrum and the new gas tube gave some of the strongest nitrogen lines rather faintly, and some other strong and sharp lines in places where no nitrogen lines occurred on the other plate. The inference is that the faint nitrogen spectrum was caused by a little impurity of nitrogen remaining in the new gas, and that the other lines belonged to the new gas. Ramsay wrote to me afterwards that he had made the measurements. The new gas was prepared from the atmosphere by absorbing the bulk of the oxygen with hot copper, and then passing the residue over heated magnesium which absorbs the nitrogen. The residue was then purified from oxygen, carbonic acid, excess of nitrogen, &c., by the best available means. Ramsay told me before I took any spectra that there might be a little nitrogen remaining.

I intend to repeat the experiments and compare the new spectrum with that of oxygen, carbonic acid, and other gases which might possibly be present. Also I want to get very accurate measurements of the wave-lengths of the new lines.

Lensfield Cottage, Cambridge,
25 Sept., 1894.

Dear Mr Crookes,

    Thanks for your kind reply to my letter. I assumed leave to forward it to Professor FitzGerald for his perusal. The account of the spectrum of the new gas which I had read in some newspapers led me to imagine that it showed a very definite line, in, I think, the blue, which does not occur in nitrogen. But the account I got from Professor FitzGerald in conversation gave me a different idea. He told me that on comparing the spectrum of the new gas with that of nitrogen it seemed to him that the same set of lines were seen in the one as in the other. He mentioned that Ramsay pointed out to him one line in particular, but he (Prof. F.) could not satisfy himself that it was absent in the spectrum of nitrogen. He said it was in the same place as a band in the spectrum of nitrogen....In returning your letter he says that he has hardly at all worked with the spectra of gases, or indeed photographic spectra in general, so that you are a far better judge of the evidence than he could be. It seems to me that the evidence in favour of one or other of the two rival suppositions, that the new gas is (1) a new element, or (2) allotropic nitrogen, turns mainly on the interpretation of the spectra obtained. The account which you give of the spectra seems to favour (1), that which FitzGerald gave me to favour (2). The density which Lord Rayleigh gives for the new gas is just what we should have thought the most probable on the supposition that the gas is allotropic nitrogen. It seems most probable that oxygen being denoted by O_2 and its density being 16, ozone is O_3 with a density 24. So nitrogen having a density 14, we should expect as most likely a density of 21 in allotropic nitrogen, if such a thing there be.

Sept. 26, 1894.

DEAR SIR GEORGE STOKES,

I think the preponderance of evidence is rather on the side of the elementary character of the new gas, but there are strong arguments in favour of the "allotropic" view. My spectra were sent off to Ramsay before I had had time to examine them carefully. That FitzGerald found many of the N lines in the new gas lines I can well imagine, as it was contaminated with a little N. I am now fitting up the apparatus to make some of the gas for myself, and will then take careful photographs of it and N.

June 24*th*, 1895.

MY DEAR SIR GEORGE STOKES,

I want to trespass on your kindness to ask if you can give me an improved formula for interpolating lines in my photographed spectra.

The formula I am using is

$$\lambda_2{}^2 = \frac{(n_3 - n_1)\, \lambda_3{}^2 \lambda_1{}^2}{(n_2 - n_1)\, \lambda_1{}^2 + (n_3 - n_2)\, \lambda_3{}^2},$$

where λ_3 and λ_1 are the wave-lengths of two known lines, n_3 and n_1 are the distances of the two known lines from datum, n_2 is the distance of the one unknown line from datum, and λ_2 is the wave-length of the unknown line.

This formula was tested by W. Gibbs, and the results recorded by him in the *Am. Journ. Sci.* 1870, vol. 2, p. 45. He found the error in the calculated wave-lengths to be about $\frac{1}{2}$ millionth of a millim., while I myself have found a discrepancy of 1 millionth of a millim. between the calculated and known wave-lengths of certain cadmium lines.

In the photographs of the spectra I am now observing I have many easily separated lines whose wave-lengths differ from those of their neighbours by less than 1/4th of a millionth of a millim., consequently it appears to me that to get the greatest value out of my work I must employ a formula which will give me a result correct to at least one more decimal point than the above.

LENSFIELD COTTAGE, CAMBRIDGE,
28 *June*, 1895.

Pray excuse an earlier answer. On Monday I went to London, and slept there. Tuesday through Cambridge to Ely, and slept there. Wednesday home, latish. Thursday to London, and back late.

I have looked into Gibbs's paper, where he gives numerical examples of the applicability of the approximate method I proposed. Of course I made numerical trials myself, but I did not print the results, and I did not care to keep the trials, so that Gibbs's figures were useful.

Since I proposed the method at a meeting of the British Association*, the determinations of wave-lengths have increased in accuracy, and increased exactitude is required. Still the Gibbs numbers show that the approximate method I proposed is *almost* sufficient, so that a very small correction will suffice. If a curve be plotted with the squared reciprocals of wave-length for abcissae, and the refractive index for ordinates, it is nearly a straight line, and my method was founded upon taking it, for a moderate interval in the spectrum, as a straight line. If this be not accurate enough, as it is not quite, it will doubtless be sufficient to take one term more in the ordinary interpolation.

If I had done the numerical reduction myself, I think I should have started by taking the squared reciprocals of the wave-lengths, or of so many of them as I was going to employ, I would then have found by interpolation the squared reciprocals of the wave-lengths for the new lines, and having thus formed a table of squared reciprocals would have formed the table of wave-lengths

---

* On the whole subject see *Math. and Phys. Papers*, Vol. v. pp. 293–5.

from that. If x be the squared reciprocal, and if we confined ourselves to two terms in the interpolation, we should thus get

$$x_2 = \frac{n_3 - n_2}{n_3 - n_1}\, x_1 + \frac{n_2 - n_1}{n_3 - n_1}\, x_3 \quad\ldots\ldots\ldots\ldots(1),$$

which is identical with the formula you used, only the numerical
calculation is carried on by turning the formula you used upside
down.

The result of this you find to be not quite exact enough,
though very nearly so. To make the small correction we must
take in a third line of known wave-length. This should be chosen
well removed from the selected known lines 1 and 3. If chosen
in the interval 1—3, it had best not be *greatly* distant from the
middle. There is, however, very wide latitude of choice in this
respect. Thus if the line chosen for correction divided the interval
1—3 in even so unequal a ratio as 1 to 3, the correction which I
am going to name E, which serves as the foundation for the
corrections for the new lines, would be $\frac{3}{4}$ of what it would have
been if the selected line had been in the middle of the interval.
But it will be better I think to take the third known line outside
the interval 1—3, and some way from the nearest end line 1 or 3.
Thus if it lay at a distance outside half as great as the interval
1—3, the correction E (which is the fundamental small quantity
that we wish to emerge from errors of observation) would have
been thrice as great as if the line had been chosen in the middle
of the interval 1—3. I will denote by an accent what relates to
the third line chosen for making the correction.

Now for the correction. Apply the formula (1) to calculate
the squared reciprocal of the wave-length for line ′ from those for
lines 1 and 3, and let E be the error, or difference between the
numbers got by calculation and observation. Then to get the
correction (E_2) for the new line intermediate between 1 and 3,
we have

$$E_2 = \frac{(n_2 - n_1)\,(n_3 - n_2)}{(n' - n_1)\,(n_3 - n')}\, E.$$

If you prefer working with the formula you have hitherto
employed, and wish to get the correction $(\delta\lambda_2)$ to the wave-length
of the intermediate line 2, apply your formula to calculate the
wave-length of the known line ′ chosen for making the correction,

and let $\delta\lambda'$ be the error of the interpolated wave-length, then

$$\delta\lambda_2 = \frac{\lambda_2{}^3}{\lambda'^3} \frac{(n_2 - n_1)(n_3 - n_2)}{(n' - n_1)(n_3 - n')} \delta\lambda',$$

where at the second side the first approximate value of the wave-length of line 2 is to be taken.

It is really a toss up of advantage which way you work. When you have made a trial of the correction on some of the known cadmium lines, I should be glad to know how it works.

P.S. You will only want to work with 2 or at the most 3 significant figures in making the correction.

LENSFIELD COTTAGE, CAMBRIDGE,
29 *June*, 1895.

As I was carrying my last letter, begun last night but finished this morning, to the post, I recollected that the refractive indices of the new lines were not known. I had thought of the n's as refractive indices, according to the notation usually employed on the Continent, and then it occurred to me that as I have just said the refractive indices of the new lines were not known.

But I suppose I may take for granted that the differences of the n's with suffixes in your letter are not the differences of refractive indices exactly, but differences of the measured lengths on your scales. So my momentary qualm had no foundation.

Though I did not mention it in my letter, I had tried Gibbs's first set of numbers against my correction. The result I think shows that the correction I gave you, if the wave-lengths of the cadmium lines can be trusted to such a degree of accuracy, will give the wave-lengths correct to 0·0000001 metre. I do not know what the limit of accuracy of the wave-lengths of the cadmium lines which you assumed as known may be.

7, KENSINGTON PARK GARDENS, W.,
July 4th, 1895.

MY DEAR SIR GEORGE STOKES,

Your letters of the 28th and 29th arrived in due time, and very fortunately I have my mathematical son at home for a short time, and he has been of great assistance to me in working out a sufficient number of instances to enable me to test

the new formula you have sent me.   I append a couple of cases :
[*omitted*].

I have other examples, but they all come within the limits you
said.   The new formula is a great improvement over the old one,
and it increases the accuracy more than ten-fold.   Whereas, with
the old one I could not rely on more than three figures accurate
and on the fourth as approximative, now I can feel certain about
the fourth figure, and get more or less close to the fifth.

Having supplied me with so valuable a means of getting
accurate results, I have no right to ask for more, but if it were
possible to simplify the operations it would be a great comfort.
At present it takes from 20 minutes to half an hour to work out
the corrected wave-length of one line, and when each plate contains from 20 to 100 lines, and I like to correct the results by
working out the figures from separate measurements on two or
three different photographs, the time occupied mounts up in an
alarming way.

The new formula has done good service in another way.   It
has enabled me to put my finger on two or three errors I made
in measuring the wrong line.   Two lines of a metal lying close
together, it is not always easy to decide which is the measured
one.   Calculation now has disclosed the fact that I had taken the
wrong line, and putting the measurements to the other line at
once removed a tiresome discrepancy.

You are aware that my photographs are taken on a curved
film, while the measurements are taken when the film is flat.
The radius of curvature is about 7 inches.   The actual distances
ought to be measured across the chord of the arc, it seems to me,
instead of along the arc itself, as I do.

The measurements are in decimals of an inch.   The fifth
figure is not to be trusted, but the fourth may be.

The standard wave-lengths are those of Kayser and Runge,
based on the latest wave-lengths of Rowland.

I remain, very truly yours,

LENSFIELD COTTAGE, CAMBRIDGE,
6 *July*, 1895.

I found your letter on my return from London last night. As to the length of the process for giving the correction to the simple form for approximate interpolation, I contemplated, after the method had been tried numerically, in case it was found useful to substitute a graphical for a numerical process, as being quite sufficient for determining the small correction.

If you had been using diffraction spectra, obtained by a grating, then wave-lengths are what should have been used in the interpolations required to get the wave-lengths of unknown lines from the measurements of known and unknown lines taken on your plates, combined with the known wave-lengths of as many lines of known wave-lengths as you please to take for standards. But as your spectra are obtained by refraction, it is convenient to work throughout, not with wave-lengths, but with squared reciprocals of wave-lengths. I find by experience that in numerical calculation where you have several processes to be carried on on several numbers, given say by observation, it is far better, and involves less risk of error, instead of taking the numbers one by one, and carrying out the processes on each in succession, to take the processes one by one, and carry out each in succession on all the numbers. It is much like the making of needles, where one man cuts the wire into the proper lengths, another, if need be, straightens it, another stamps one end to flatten it, another drills it, another sharpens the ends. So in numerical calculation the less you have to think of a change of process, the more you work by a humdrum repetition of the same process, the better.

Thus if I were interpolating for a great lot of wave-lengths from measurements taken on refraction spectra, I would first write down in a vertical column the known wave-lengths that I want in any way to deal with, whether it be to use as standards, or as furnishing means of testing the success of the method of interpolation employed, or of comparing the measurements on which the assumed wave-lengths were founded with your own measurements.

Suppose there were, say, 50 such wave-lengths. I would write these down in a vertical column. I will suppose that you wish to keep in 6 significant figures. As Barlow's table of squares, square roots, reciprocals &c., goes only from 0 to 10000, though the results are given to 7 figures, it would be shorter to use logarithms than to interpolate from Barlow's table. Well. I would write down in a second column the 50 logarithms, in a third the 50 complements, in a fourth the 50 doubles, in a fifth the 50 antilogarithms. Having thus got the squared reciprocals, I would throw overboard wave-lengths altogether, and work by squared reciprocals. These would show just as well as the others the degree of accordance between the results of one observer and another, or between different measurements of the same observer. It would be convenient if all tables of wave-lengths were accompanied by squared reciprocals, as the two sets of numbers would be ready for public use, the first for those who are working with diffraction, the second for those who are working with refraction spectra. It goes without saying that the final numbers for the wave-lengths of unknown wave-length lines may be got from the squared reciprocals by a calculation which is anything but formidable, even for numerous lines, provided it be carried out process by process for a good large batch of lines together.

The interpolations may be done from your plates measured after being made straight. There would be no advantage in substituting chords for arcs. But the interpolation should be done separately for different plates unless the part of the spectrum for which the prisms were set at minimum deviation, and the part of the spectrum for which the camera was pointed, were the same in the two. And even on the same plate it remains to be found by experience for how large a range of spectrum the same interpolation will suffice, that is to say, how much we can take in without having recourse to another trio of standard lines.

Now for the interpolation. Let q be the squared reciprocal of the wave-length, x the measured distance of any line. Then the first approximation, which is equivalent to the formula you used, gives

$$q = q_1 + \frac{q_2 - q_1}{x_2 - x_1}(x - x_1) \qquad (1)$$

and we have to add the small correction, which I will call δq.

For this, apply the approximate formula (1) to a sufficient number of known lines to serve as a test, and find the errors δq. Now plot the results, taking the measured distances x for abscissas, and the errors δq for ordinates. The heads of the ordinates will form a curve which with a suitable choice of scale will have a gentle curvature. The scale of ordinates should be large enough to prevent errors of drawing from exceeding errors of observation. The scale of abscissas should be large enough to make the curve of only gentle curvature, in which case it may be taken as a circular arc, and drawn accordingly, unless the range of the spectrum included to be too large to make the correction sensibly represented by the ordinate of a circle even though the scale of abscissas was not so small as to induce a deviation from circularity due to this cause. It is an object to use a circle because the curve is so easily drawn, but a smooth curve drawn with a flexible rod will do as well.

The places of the unknown lines on the axis of abscissas are now laid down from the measured values x, and the small corrections δq for these lines are got by measuring the ordinates of the curve.

If you change the reference lines 1, 2, the values of δq must be calculated afresh for as many standard lines as you want for drawing the curve. If the circumstances of two plates be not quite the same, the curve drawn for the first will serve for the second provided we refer the places of the unknown lines to known lines near them, as the change or ordinate corresponding to the small error of abscissa we might thus be liable to would be quite insensible in its effect on wave-length.

7, KENSINGTON PARK GARDENS, LONDON, W.,
4 March, 1894.

DEAR SIR GEORGE STOKES,

Allow me to thank you for the reply you have so kindly sent to my query as to the curvature of the field of my spectrograph. I regret there is no instrumental means of getting the field flat, but it is well to know it cannot be done as thereby time will be saved. I must try photographing on curved plates.

I find the coma you speak of on one side of the lines, but have succeeded in keeping it down by putting a small diaphragm in front of the collimating lens.

26 *November*, 1896.

I am building a large spectrograph, similar to the one of which I have shown you some of the results, but employing four double prisms of quartz (Cornu) instead of two. Preliminary trials show that I shall get very fine results, the sharpness being as good as in the present instrument while the length of the spectrum will be about 15 inches.

I have met with a little difficulty which I think a few lines from you will enable me to get over. The four prisms bend the light round to such an extent that the blue and ultra-violet rays overlap the path of the incident ray from the slit; so that the slit and the photographic plate will in one position occupy the same place.

I propose to get over this difficulty by introducing a right-angled prism in the path of the ray of light between the slit and the collimating prism, so as to put the slit quite out of the way of the camera. All my quartz prisms are made each of two halves— one right handed and one left handed, and what I want you to tell me is whether I must have the right-angled prism made in the same way, and if so how are the half prisms to be cut in relation to the axis of the crystal. The Cornu plan is so perfect in removing all trace of duplication of the lines even under the highest magnification I can apply to them that I do not wish to spoil the beauty of the spectra by having an imperfect right-angled prism if a little extra trouble will give me a perfect one.

You have been so good on many occasions in helping me over awkward places that I hope you will not think me too troublesome in asking for advice in this instance.

LENSFIELD COTTAGE, CAMBRIDGE,
30 *November*, 1896.

It may be useful to you to know to what degree of accuracy the two angles at the base of the reflecting prism must be made equal to each other.

Supposing there was a difference of one minute in the angles which ought to be equal, that would introduce a duplication of the image of the slit on the plate of only the one 400,000th part of the focal length of the objective of the camera. Supposing that to be, say, 20 inches, the separation would be the 1/20,000th of an inch. If the error were of 5' the separation on the plate would still be only 1/4,000 inch, which I greatly doubt if you would find sensible. Hence though accurate cutting would no doubt be required, the accuracy does not appear to be by any means superhuman.

The thing is rendered much easier by the fact that one of the angles is arbitrary within pretty wide limits. The optician would cut out a prism from a crystal free from defects, cutting the block for a prism of 90, 45, and 45 degrees, and would finish the hypotenuse and one side. He would then measure the angle between the finished faces, and fine-grind the other face to as nearly as might be the same angle. When the face was fine ground, I should think that by rubbing on a little oil the face would give a sufficiently good reflection to permit of measurement by reflection ; if not a piece of parallel plate could be made to stick . by capillary attraction by putting a small drop of oil between. The measurement would show whether any correction was required. If it were, it could then be done before polishing, so as not to have to spoil a finished face in order to correct the angle. I should think he could get the angles nearly enough equal without any great difficulty.

An error of 5′ deviation from equality in the angles would produce an angular error in the image of only 2‴·7, about the 20th part of the smallest angle which two well defined points can subtend so as still to be seen by the naked eye as two.

P.S. You did well to ask me, for you would hardly naturally have thought of making the rays travel in a direction perpendicular to the axis, where the difference between the refraction from or into air of the ordinary and extraordinary is greatest.

3 *Dec.*, 1896.

The length of the spectrum would increase in the same proportion as the number of the prisms, the prisms being supposed alike.

A single prism would, I take for granted, cost distinctly less than a compound prism. The difference perhaps would not be great, and yet I think it would not be altogether trifling.

But I think the plan of a reflecting prism half way between the slit and the objective of the collimator ought not to be rejected without consideration. The prism would be half size, linear, of what would be required the other way. An error in the equality of the angles would be only half as serious, or to express it other-wise, the limit of toleration of an inequality in the angles would be doubled. The compensation for focus, if sufficiently serious to be worth considering, would be easily effected by a very simple kind of block, which would not consume much material, and would only require two faces to be polished, and those quite narrow, and little taller than the slit. The block would be placed close to the slit (it might be anywhere between, but the nearer it is placed the less size is demanded). The length in a fore and aft direction would equal one of the short sides of the reflecting prism, the thickness in a right and left direction need only be enough to allow the rays on their way to the reflecting prism to clear the corners. The block would be cut perpendicular to the axis of the crystal, so that the axis is horizontal in the block and vertical in the prism. Only the two vertical faces, fore and aft, need be polished.

2 *Dec.*, 1896.

As you make no objection to the position of the reflecting prism being *after* the rays from the slit have passed through the objective of the collimator, I presume that this is either what you had contemplated or what you acquiesce in, probably the former. If however you wish it to come sooner, which would make a smaller prism suffice, please let me know that I may consider about the best mode of compensation for a slight difference of focus in the two images whose centres are coincident.

4 *Dec.*, 1896.

Would you kindly tell me approximately the focal lengths and apertures of your collimating and camera objectives, that I may be better able to judge of the relative advantages of different plans?

I suppose you have some reason for not liking the use of a plane mirror of speculum metal. It would cost I should think much less than the quartz, and so far as I know it would serve your purpose. Silver would not do, as it reflects feebly a portion lying about the extreme end of the solar spectrum.

8 *Dec.*, 1896.

Certainly if an additional compound prism would not cost much more than a reflecting prism of comparable size, or a reflecting prism of half size with the addition (if thought necessary) of a small transmitting block, it would seem best to have the advantage of the additional dispersion, which would increase the length of the existing spectrum in the ratio of 4 to 5.

Taking the index at 1·57, I find the loss by reflection at one surface of quartz at a perpendicular incidence would be 4·9, say 5, per cent. At the angle for minimum deviation for a prism of 60° it would for a single reflection be about 14 per cent. for light polarized in the plane of incidence, and only about 1/3 per cent. for light polarized perpendicularly to the plane of incidence. For a compound R and L prism the modification of the state of polarization in passing through one half at minimum deviation would be neutralised in passing through the other half, so that such a prism would behave like one cut from a doubly refracting crystal exempt from rotatory polarization, and therefore light polarized in and light polarized perpendicularly to the plane of incidence would be transmitted independently, so that after passing through four such prisms (supposed in perfect adjustment) the light would be mostly polarized perpendicularly to the plane of incidence, and there would therefore be but little loss by reflection at the two oblique incidences. You would have considerably less loss by reflection in the compound prism than you would have by reflection at a speculum, even if its reflecting power were as good (as it probably would be) for the rays of high refrangibility with which you chiefly work as for the visible rays.

At a single right-angled reflecting prism you would have the loss due to two reflections at a perpendicular incidence. In a compound Cornu prism you would have this and in addition the loss due to two reflections at an oblique incidence; but for the reason I have just mentioned the latter would not be very serious for light that has already passed through four such compound prisms.

I do not think it will signify much your collimator not being quite in focus for all the rays. It puts as it were your slit at a great distance instead of at infinity, and very good spectra can be got with a distant slit.

13 *Jan.*, 1897.

I have not waited to hear from you, but have tried the effect of substituting the index 1·57, which is about that of quartz for the high rays, for 1·5, the result of which is given in Coddington. The advantage of the second arrangement (curved faces both forward) over the first (flat faces together) comes out very nearly, though not quite, as great, 11 to 3 instead of 4 to 1. The substitution of the second arrangement for the first would therefore reduce the spherical aberration to very nearly one quarter. As you have a very wide field, the effect on the highly oblique pencils is important to consider. The calculation for these is troublesome; but by a calculation having some relation to it I have been led to think that the very oblique pencils would also be improved rather than otherwise. Very probably the second arrangement is the one adopted, but if it should be the first I think it would be well worth while to try the effect of so very simple a change.

I do not think it would matter practically which lens of either pair were put outside; but if it be as easy to put one as the other, which is probably the case, I would take for the outside lens in each pair quartzes of opposite name, that is one right-handed and the other left-handed.

7, KENSINGTON PARK GARDENS, LONDON, W.,
13 *Jan.*, 1897.

DEAR SIR GEORGE STOKES,

The lenses are not compound ones. They are plano-convex, the collimating lens being cut from right-handed quartz and the camera lens from left-handed quartz. They are mounted with their convex sides next the prisms. Thanks for your explanation of what puzzled me as to the apparent duplication of lines when out of focus.

With the five prisms and longer focus lenses the curvature of the field is becoming very serious. At the red end when I started the curvature of the plate had to be 470 mms. radius. Then it had to be altered to 173 mms. radius, and now I have had to curve it to a radius of 145 mms. and I have at least six more positions beyond to get before I am at the highest point in the ultra-violet. In addition to the curvature, the plate, or rather, the tangent to the centre of the plate, has to be at an angle with the incident light, so the result is that the higher rays will strike the surface at a glancing angle. My plates are 162 mms. long, so the further end is getting nearly in a line with the incident rays. No sharpness is possible under these conditions. I fear there is no way out of the difficulty but taking a very narrow part of the spectrum at a time, say a few mms. and putting the plate normal to the rays.

LENSFIELD COTTAGE, CAMBRIDGE,
14 *Jan.*, 1897.

For what part of the spectrum do you focus the collimator, and for what part do you set the prisms for minimum deviation, when you attempt (if so it may be) to photograph the whole spectrum at once?

14 *January*, 1897.

After I had written to you last night I seemed to have a glimmering doubtful recollection that I had asked you the same question once before, and that you had told me the lenses were placed curved faces forward in both halves. I am however by no means sure whether I ever asked the question. If I did, I must beg you to excuse the troubling you. Be the existing arrangement what it may, the lines are beautifully sharp.

7, KENSINGTON PARK GARDENS, LONDON, W.,
20 *January*, 1897.

MY DEAR SIR GEORGE STOKES,

I have received and studied your letters. First I will answer your questions.

The focal length for the *D* ray of the lenses in my smaller spectrograph is 315 mms. The focal length for the same ray, of the lenses in my present instrument, is 673 mms. All measured from the convex side. In each case the convex side is next the prisms. Both collimator and camera lens are of the same focus. The apertures of the 315 mm. lenses are 52 mms., those of the 673 mm. lenses are the same. The recent lenses are new ones. I have the old ones by me. They are in the old apparatus which I do not wish entirely to dismantle, as it will be very useful in taking spectra which are too faint for the large instrument.

Before getting your last letter I had tried the experiment of shortening the distance between the slit and the collimating lens. It was formerly set at 673 mms. and now I have pushed the slit in 45 mms. reducing the distance between slit and lens to 628 mms. I am working high up in the ultra-violet rays, and the plate takes in wave-lengths between 3403·74 cadmium, and 2573·12 cadmium. The prisms are set for the min. dev. of a line very near 2573.

When the collimator was 673 mms. from the slit the angle

of the plate had to be about 42°, and the curvature of the plate 150 mms. radius. Pushing the slit in 45 mms. has had the effect of reducing the angle to 9°, and the curvature to a radius of 235 mms. These are very great advantages, and to-morrow I am going to try the effect of altering the positions of the prisms so as to get a line near wave-length 3403 at min. dev. to see if the curvature of the plate will be still further reduced. Pushing in the slit has had another advantage. It has brought the lines at the blue side closer together, while at the same time the lines at the red end are not much altered. I thought I would tell you the effect of these alterations before I do much more, as I do not care to alter the adjustments blindly. It takes about two days hard working to get the focus, angle, and curvature right after I have made any alteration.

In my old instrument I could get all the spectrum on one plate. Now I am using longer plates and it takes about six to take the whole. My least refrangible line that will impress itself on the plate is the double sodium line. But I want to allow space for the red lines in case a method of sensitising the plates for the red end is subsequently discovered. Methods are now known, but they make the plate so slow as to be useless.

LENSFIELD COTTAGE, CAMBRIDGE,
26 *March*, 1897.

The spectra you showed me were really wonderful for sharpness of definition, considering that the whole would be 10 feet long. Indeed the definition would have been most remarkable even had the slip you showed me represented the whole of the spectrum, instead of only a small fraction of it; though of course it is to be said that from the small variations of index within the range, the conditions of getting sharply a small fraction of a long spectrum are in some respects a little simpler than those of getting the whole spectrum within the same length.

What you showed me as to the presence or absence of certain lines in two iron spectra taken under what appeared to be identical conditions bears out your remark in the discussion, that there is still a great deal to be done in the laboratory beyond what has already been done, in relation to the important subject which Lockyer is discussing.

8 *Jan.*, 1898.

In the existing arrangement of your prisms, the right and left-handed halves of each compound prism are traversed at almost exactly the same inclination to the axis, not alone in the position of minimum deviation, but also by the rays of the spectrum right and left of that for which the deviation is a minimum. Hence the compensation. But in the arrangement you are thinking of (slightly inclined instead of Cornu's method, back to back) one half is traversed perpendicularly to the axis and the other at an inclination depending on the distance of the point of the spectrum under consideration from that point for which the direction of traversing is normal. I assume that you contemplate placing each half prism, as regards azimuth, in the position which had its fellow been with it would have been that of minimum deviation for the compound prism so formed, not in the position of minimum deviation for the half alone, which is quite a different thing.

Before venturing on the contemplated alteration, it would be well to test by eye observation (or if this should be inconvenient by photography) whether the difference of inclination will introduce a sensible double refraction. I rather expect it may.

For this purpose take a single compound prism, or if more convenient your whole train. I suppose you have an eye-piece that you can use in lieu of the camera plate in the observing telescope, or if not that you could temporarily substitute some other telescope for it for the purpose of eye observations. Choose any point of the spectrum for your standard, say the *D* line. I will suppose that you use a single compound prism, but the method will be the same if you find it more convenient to use the whole set. Focus the collimator for the standard, and adjust the prism for minimum deviation of the same. Having observed the standard line at its best, leave the front half undisturbed, and turn the rear half prism in either direction through an angle equal the largest angle you want to have in your half field, and notice on your test object whether the imperfection of the compensation of the two halves is sensible. I rather expect there will be a sensible double refraction. If so, the new arrangement must not be adopted.

8 *Jan.*, 1898.

... I said (as regards the last arrangement I proposed) that the inclination of the second prism to the first, i.e. the inclination of the flat faces to each other, should be made in the trial as great as the angle subtended at the centre of the camera objective by lines drawn, one to the part of the field for which the rays emerge perpendicularly, the other to the edge of the field which you want to have good. I was not so explicit, but that was what I meant.

In truth that would exaggerate the angle to be encountered, as for the sake of being on the safe side I did not allow for the refraction towards the normal of the emergent axis supposed reversed in direction to so as to enter instead of emerging from the last prism.

If you care to rig up an arrangement in which the halves only are used, being placed in the relative positions you have drawn, you must remember that it will *not* test the doubtful point unless you turn the prism next the telescope *from* the position in which the rays emerge perpendicularly, through an angle which (after making allowance for refraction of the axes of the pencils) is about 5/9 of the greatest angle subtended at the centre of the observing telescope by the half field as above mentioned.

7, KENSINGTON PARK GARDENS, LONDON, W.,
10 *January*, 1898.

DEAR SIR GEORGE,

I have been trying a series of experiments this morning, and I now send you the results. I fitted up the spectroscope with a much more powerful eye-piece, and longer focussed quartz lenses than I usually employ, so as to get a good magnified image of the slit. I first put the prisms in the following position (I need not draw the lenses, they were in the proper positions to receive and send on the rays along their axes). The slit was as fine as I could use, and the light was a sodium flame. No amount of magnification that I could apply would show me the two components of D. This was only to be expected with one prism....

Briefly, the experiments show this:—When the light passes through the prism-components parallel to the optical axis there is no double refraction, no matter how the components are situated. But when the light does not pass along the axis double refraction shows itself, increasing in amount with the divergence of the axis and path of light.

LENSFIELD COTTAGE, CAMBRIDGE,
11 *Jan.*, 1898.

I was confident beforehand that there was nothing to
fear from the change of arrangement so long as the pencils pass
along the axes in the two half prisms, as in your expt. 4. I
mean, so long as, being placed in the intended position, you are
dealing with light which passes through both in the direction
of the axis. The question was, supposing your second half prism
turned round from that position through a small angle, as great
however as that which your half field subtends at the centre of the
camera objective, would the amount of double refraction which
would then come in be perceptible, so as sensibly to impair the
image of a line ?

I think the result of your trials is fatal to the new arrangement
unless that angle through which you have turned the second prism
was decidedly greater than the angle actually demanded.

Let O be the centre of the camera objective, ABC an arc
of the focal band, or of the photographic slip, for which you
require to have the image good, C being the point for which the
rays passed through the second half prism in the direction of the
axis. Settle what is the greatest angle AOC or COB that you
demand to have the image good for. Cut a piece of paper or card
to an angle 5/9 or say 2/3 of this, and use it for a template for
turning the prism. Having placed the prisms as in fig. 1 back to
back, or fig. 4 of your letter (no matter which, but fig. 1 will
be rather the simpler) so as to get your test image quite good,
turn the second prism through the angle given by the template,
and see whether the image is practically good up to the angle of
turning. If it be good, you have turned it in your trials through
a needlessly large angle ; you have put the prism to an unreason-
ably severe test. But if even when turned no more than through
the angle given by the template the image is distinctly inferior,
such as you would not tolerate, your proposed plan must be given
up unless you are content to take the photographs more piece-
meal, using a larger number of photographs, each taking in a
smaller arc.

While you are about it I should be glad if you would try this
experiment. Place the prisms, as in fig. 1 back to back, in the
normal position. Notice the goodness of the image. Now turn
the second half separating them through the angle A, and notice
the goodness. Now leaving the second half in the last position

bring the first half up to it, which comes of course to the same thing as turning the whole prism together from the first position through the angle A. Notice the goodness. I should like to know whether there is much difference in goodness of image between the second and third positions.

The angle A may be that of the template, or in case the images were still (in the previous trials) found good so far, a somewhat larger angle.

28 *Nov.*, 1898.

In the first place I should like to know what I am to assume as to the breadth of the object to be photographed. If it were a flame, you would not want a condenser at all. Your collimator subtends at the slit an angle of about 1 in 12, so that if you brought your flame near the slit, say 2 inches off, it would require a breadth of flame of only one-third of an inch to fill the collimator. But I feel pretty sure that it is not with flames that you are mainly working.

I suppose you are working chiefly with induction discharges which show the metallic lines due to the electrodes. In that case the luminosity is very narrow, at least the most intense part of it, and I suppose I shall not be far wrong in assuming that you want to work with a luminosity that may be regarded as infinitely narrow, so that the image formed by the condenser, assuming it perfect, would be practically as narrow as you please.

As the refractions produced by the two cylindrical lenses are independent of each other, taking place in perpendicular planes, there is no reason why the two lenses should be placed near together. The place of No. 2 (the one with its axis horizontal, and which therefore refracts in a vertical plane) may be chosen at such a position as to give a sufficient height to the illuminated portion of the slit to give the height of the lines on the photographic plate what you think sufficient, without making them needlessly high, so as to waste light. It is the other lens we are mainly concerned with.

Now if the luminosity itself has no sensible breadth, so that we may regard it as a point or vertical line as the case may be, we gain light by bringing the lens nearer to the luminous point. For the angular width of the refracted pencil is given—that being equal to the angle subtended at the slit by the collimator—

the incident pencil which feeds this becomes wider. There is the obstacle of spherical aberration, which however, when not too extravagant, can be corrected by a lens, which must be concave.

I must close this now to catch the post, but I will write again.

2 Dec., 1898.

Pray excuse the delay. I am at present engaged in lectures, besides some other matters. I don't expect to lecture beyond Tuesday, and I shall then be free comparatively speaking. Meanwhile I should be glad to know whether the focal length of the collimator is roughly the same as that of the camera objective, or say half of it or whatever it may be. I only care to know that I may know whether I may take the height of the image as approximately that of the slit, or double that of the slit. But the shortest plan would be to give me roughly the height of the slit directly, and then you need not mind the other.

I *think* I shall come to the conclusion that it would be an improvement to move the condenser somewhat nearer to the spark.

Would it suit you to do me the honour of dining with me as my guest at the anniversary dinner of the Cambridge Philosophical Society on Saturday, Dec. 10?

4 Dec., 1898.

I had previously made some simple calculations (arithmetical) about places of foci &c., and last night I plotted, or rather drew, the course of the rays in the horizontal and vertical planes. The figure showed how little the condensing lenses come really into play. With the distances you have got, the only parts of the lenses which come really into use are bands of breadth about 1/4 inch for the one of shorter focus and 1/3 inch for the other; or, as either band would only come into play over a portion of the other, squares of say 0·35 inch a side would be the only part really in use.

The work of the condenser is much lightened by the substitution of the new collimator of (say) the same aperture and double the focal length of the old, as the cone of rays which the condenser has to supply is now of only half the angle it was.

If I may assume that the source of light is infinitely narrow, a considerable improvement may be made by bringing the condensing lenses nearer to the spark. If the breadth of the luminosity is somewhere about that of the slit, the change in position would I think make little change in efficiency one way or other. But you can get rid of the curvature, which may as well be done.

I have not yet drawn the course of the rays for one or two other positions of the lenses.

One advantage of cylindrical over spherical lenses is that the work of each is independent of that of the other, so that in considering where you had best place the one you need not trouble yourself with the place of the other; only of course one must not invade the space occupied by the other. They need not however be nearly together if it should be found well to separate them.

7 *Feb.*, 1898.

Baily repeated Cavendish's experiment on the attraction of leaden balls with a view to determine the mean density of the earth. Baily was an extremely accurate experimentalist, and had expended a great deal of time and trouble on the experiments when he showed his apparatus to Professor James Forbes. Forbes, who had worked a great deal at radiant heat, suggested to Baily, with a view to mitigating or removing certain irregularities referable to changes of temperature, that he should gild his balls. Baily tried this, and found his results so much improved that he rejected a great lot of work, and began his observations afresh. As you mentioned disturbance by temperature changes, it is possible that gilding the balls might greatly reduce the disturbances.

I don't know your object; but in case it should be to try whether there is such a thing as specific gravitation, I may as well mention that Bessel repeated, with all the refinements of modern accuracy, the fundamental experiment by which Newton proved (what could not have been foreseen) that the attraction of gravitation exercised by a body is simply proportional to its mass (as measured by inertia) irrespective of its chemical nature.

7, KENSINGTON PARK GARDENS, LONDON, W.,
Feb. 8, 1898.

MY DEAR SIR GEORGE,

...To begin with I took gold and gold, and worked for a long time with these so as to find out all the peculiarities of the apparatus. It was made and remade a dozen times before I settled down to the one now in use. Having obtained very good and concordant results with two spheres of gold, I replaced one of them by a sphere of magnesium of exactly the same weight in air, and repeated the same experiments. The apparatus is so arranged that at any time I can tell the exact distance from centre to centre. I have not finished the magnesium-gold experiments, but I may say that in every case the attraction between Mg and Au is far greater than that between Au and Au.

I know Baily's difficulties well, for over 40 years ago I copied out nearly his whole papers and subsequently printed privately all the portions which referred to the difficulties he met with. I do not think gilding would be of any use in my case, for the whole thing is in glass, sealed up and embedded all round in about eight inches of wool, feather pillows, and wooden caseing. It is in perfect darkness and in a room that scarcely changes in temperature. A narrow glass tube connects it with the pump, and small windows, stuffed with eight inches of wool, can be removed when I want to get to the inside of the case. A small tube leads the ray of light to the mirror on the moving beam. I observe from a telescope some distance off. The room is entirely set apart for this purpose and is not entered by anyone but myself or assistant when experiments are going on.

I was not aware of Bessel's experiments, but will look them up at once. If he has anticipated me I may as well stop and do something else. Can you tell me where I shall find an account of them?

After magnesium I was going to try all the other metals against the same sphere of gold, and then try them against each other and other metals. The permutations would be very great and I can only hope to try a few typical cases and leave the rest to younger men.

I am still in hopes there may be something left for me, for Bessel had no quartz fibres and did not work in a vacuum on masses of only 57 grains in weight.

LENSFIELD COTTAGE, CAMBRIDGE,
9 *Feb.*, 1898.

DEAR SIR WILLIAM,

From my remembrance of the look of the German book in which I read Bessel's paper to which I referred in my letter, I did not doubt that it was the *Berlin Transactions.* The title is given in the *R. S. Catalogue,* Vol. I. p. 345, No. 108. There are two abstracts referred to as well. "Versuche über der Kraft mit welcher die Erde Körper von verschiedener Beschaffenheit anzieht. Berlin, *Abhand.* 1830, pp. 41—102; *Pogg. Ann.* XXV., 1832, pp. 401—417; *Astr. Nach.* X., 1833, col. 97—108."

It would be useless to try your experiment for the object for which you designed it. Bessel has proved the independence of the attraction on the nature of the attracted body with an accuracy incomparably superior to anything you could do in your way. It is true that in Bessel's work the attracting body was always the earth. But attraction is a thing mutual; if bodies *A* and *B* attract each other, you may consider *A* the attractor and *B* the attracted, or *B* the attractor and *A* the attracted, just as you like. Hence to test whether or no there is such a thing as specific gravitation there is no need to change the nature of both bodies; one will suffice; and when the earth and a ball attract each other, you may consider the earth as the attractor and the ball as the attracted, and may limit yourself to changing the nature of the attracted.

I have no doubt the difference you found between gold-gold and magnesium-gold arose from electrical attraction. The rarefied air acted slowly as an electrolyte, and the different metals got slowly charged, positively and negatively, up to a certain difference of potential. I feel little doubt that if the aluminium ball were gilt the result would be very different; for then you would have two balls alike as to surface, though very different in the interior.

The amount of attraction would very probably depend on the nature of the residual gas in the exhausted chamber. You have worked, I understand, with air. I have no doubt that your apparatus is very delicate, and it may perhaps do good service in electricity; but of the likelihood of that Lord Kelvin would be a far better judge than I.

I supposed the gradual alteration, towards one side, of the

middle point of the swings, as the amount of swing decreased, to be due to the unequal effects of the viscosity at equal distances on the two sides of the resting point, and that may be the cause of it. But it may also be due to a very gradual accumulation of electric charge on the balls.

12 *Feb.*, 1898.

I have not finished the numerical calculation, but as the last post which would be delivered in London is now close, I write to say that your beautifully delicate apparatus may yet do good service, though in a way quite different from that for which it was constructed. We can calculate very well the deflection due to gravitation, and on deducting that we shall get the residual deflection due to some other cause. That I should think is probably the deflection due to a minute electrical charge depending on the dissimilarity of the two metals.

For a given potential, the charge on a sphere varies as the radius, whereas the attraction of gravitation exerted by the sphere at a given distance varies as the cube of the radius, *i.e.* given distance from the centre. Hence if you take a radius 10 times as great as before, the proportion of attraction of gravitation exerted by the sphere to the electrical attraction is 100 times as great as it was for the smaller sphere. Hence with your small apparatus the electrical attractions come out into evidence.

15 *Feb.*, 1898.

I had occasion to go to London to-day shortly after breakfast. I was at the R. S. for a short time, and took occasion to look into the notice, in Poggendorff, of Bessel's work to which I referred you. I see that not only was there no specific gravitation to be discovered, but the accuracy of the observations was such that we could say that if there were any it was not more than the 10^{-6} or so of the whole. This is vastly closer than what you could possibly detect with your instrument.

Let A, B be two mutually gravitating masses. Leaving A unchanged, let the nature of B be changed (gold, aluminium, quartz &c.) and appeal to experiment to show whether the attraction is or is not proportional to the mass of B. Nothing hinders us from taking for A the earth itself, which makes the experiment comparatively easy and accurate.

Though I think the question of specific gravitation is already

settled, your apparatus is available for the repetition, with improved conditions, of Cavendish's experiment. Also the cause of
the difference of result according as the "weight" was of gold or
aluminium has yet to be investigated. You have already been
richly rewarded by the results you arrived at by hunting out
unexplained anomalies.

22 Feb., 1898.

I have verified a few of your numbers, just to see that your
mode of calculation and plotting is right, which I find it to be.
I found however some numerical errors or errors of copying, whichever it may be. I think there may be errors of both kinds. The
first is pretty clearly a mis-copy; the second looks like an arithmetical error. I think it would be well to check your additions
and subtractions. The numbers taken out of the table are not
likely to be wrong.

It has I think now been shown that the way in which uranium
acts as a discharger is by rendering the air (or other gas) intervening between the charged and uncharged, or (as the case may
be) differently charged metallic bodies, a conductor. In a perfect
vacuum there would be no discharge, and in a very high vacuum
the discharge is imperfect, but whether only much slower, or
whether even if time be given less complete, I do not recollect.
I think it is in Nature that a paper by Lord Kelvin on the subject
appeared a few months ago. A similar conclusion follows from
some recent work of Röntgen's. As you want a good vacuum, the
protection afforded by uranium would be but imperfect.

But the weight and ball can be quite sheltered from each
other's electrical influence by a cage or screen of metal or metallic
gauze in metallic connexion with the earth. I do not know
whether your "weight" is insulated or in metallic connexion with
the earth. Anyhow the suspended ball is insulated by the quartz
fibre, so that if it got charged the charge would influence the
apparent attraction of the weight and ball.

But if the ball and weight be screened from each other by a
metallic screen in connexion with the earth, then if the ball gets
charged it will induce towards the cage, but the external effect
will be nil. The cage and ball will electrically attract each other,
and the resting point will be different from what it would be if
there were no charge. But the external effect of the charges on
the ball and cage will be nil, so that, provided the charge does

not alter, the difference of the resting points weight up and weight down will be almost unaffected by the charge. I say "almost," because the small difference of half a millimetre or so between the resting points with and without the weight will make a little difference in the electrical attraction between the ball and the screen in case it is the ball that is charged; not much however if the screen be kept as far from the ball as the weight will allow.

If you consider the idea of a screen to prevent electrical action between the ball and weight worth thinking about, I can enter into further particulars.

7, KENSINGTON PARK GARDENS, LONDON, W.,
Feb. 28*th*, 1898.

DEAR SIR GEORGE,

...I meet with a radiometer effect when the apparatus is not properly screened from external influences by non-conductors of heat. I think I can recognise these disturbances, and they have almost entirely disappeared now. I think the movements are due to some other cause. I cannot think it possible that the upper part of the bulb is of a different temperature from the lower part.

When the apparatus is left to itself the weight is left where it happened to be at the end of the last experiment—sometimes up and sometimes down. I set up the swing when I commence an observation by altering the position of the weight. This is sufficient.

I have had a hard day's work to-day. I have kept the pump going without stopping all day long, and have been observing swings with the weight alternately up and down at different distances along the scale so as to get as many observations as possible under the same circumstances.

My "weight" is electrically connected with earth, but the ball is insulated. I could silver the quartz fibre and so make it conduct if you think this would be of use.

LENSFIELD COTTAGE, CAMBRIDGE,
27 *Feb.*, 1898.

DEAR SIR WILLIAM,

I meant to have calculated and sent you the effect of the mere attraction of the magnesium ball. In consequence of your letter received to-day, I did the calculation this evening, and find that at the distance of a centimetre between the centres of the gold and magnesium balls the gravitation of the magnesium ball would cause a deflection of only 0·003 mm.

I hope to write to you to-morrow about electrical screening. The facts you mentioned in your last render it all the more interesting to make out the cause of the attraction.

1 *March*, 1898.

The smallness of the gravitational effect in your apparatus depends on the smallness of the scale. In an apparatus similar to yours in every respect, except that the scale of everything was ten times as great linear, the mutual attraction would be 10,000 times as great. The small scale is therefore advantageous as getting rid of effects of gravitation, which has been already well investigated.

P.S. If the object be to determine whether or no the attraction be electrical, it will be well to gild the fibre. But if the object be to make what can be made out of it on the supposition that the attraction *is* electrical, it will be better to have the fibre insulating.

If it were gilt, I suppose there would be no difficulty in removing the gold by a solution of cyanide of potassium, for enough of the fibre at any rate to make the fibre insulating.

7, KENSINGTON PARK GARDENS, LONDON, W.,
March 4th, 1898.

DEAR SIR GEORGE,

I do not think the movements are radiometric for this reason. I have on more than one occasion kept the weight down (near the ball) all night, and on looking in the morning the position was the same as it had been 16 hours before. The temperature of both must have been the same. On now raising the weight, the ball immediately swung back, showing that it had remained for 16 hours under a constant pull, which could not have been due to difference of temperature.

DEAR MR CROOKES,

It is stated by Eilhard Wiedemann that when phosphorescence is produced in the wall of a Crookes tube by the electric
bombardment (he would say by the Kathodenstrahlen, which I
don't believe) it is at the *outer* surface that the glass phosphoresces,
and that the reason why this was not perceived is that the
wall is usually made thin, so that you can't tell at which
surface the phosphorescence takes place; but when you try
with a thick wall you then see that the phosphorescence takes
place at the outer wall.

This seems to me strange if true*....I can imagine that it might
be difficult to distinguish between an internal reflection at the
outer surface, of phosphorescence at the inner, and phosphorescence
at the outer surface itself, unless the discharge were limited by an
internal screen with a hole or long aperture forming a broad slit,
or else allowed to fall on a body casting a shadow. In either case
I think it would not be difficult with care to distinguish by means
of parallax between the two.

According to my knowledge so far, I am very strongly disposed
to believe that with respect to the nature of the so-called "Kathodenstrahlen" you are right, and Wiedemann, Lenard and Röntgen
are wrong.

I should be glad to know whether you have critically examined
the question, Is the seat of the phosphorescence the immediate
neighbourhood of the inner or outer surface of the glass wall?

29 *Oct.*, 1896.

Many thanks for your letter. Pray do not trouble yourself to
make a tube such as you proposed. I do not think uranium glass
would be specially suited to the object I had in view; rather
the contrary; for it is so very sensitive to phosphorogenic action
that it might rather obscure the issue to be tried, which is whether
the phosphorescence produced by what the Germans call the
Kathodenstrahlen is ever shown on what we should be disposed
from our theory to call the "wrong" side of the glass.

Two tubes have been constructed in the Cavendish Laboratory,
in the absence of Prof. J. J. Thomson, who has not yet returned

* This point was settled in Oct. 1878, *supra* pp. 410 *seq.*

from America, which, combined with a re-perusal of what E. Wiedemann has written, settle the question in my mind, and show that Wiedemann's conclusion is based on a wrong interpretation of the facts he had observed.

2 *Nov.*, 1896.

Most hearty thanks for your very kind present of the beautiful "Crookes" which reached me safely on Saturday. I have not yet tried it, as I had the cares of a lecture on my shoulders to-day, but hope to do so without delay. I shall probably take it to the Cavendish Laboratory; for though I have an induction coil of my own (in fact I have two) I have not worked them for a great many years, and it would take too long to wait for trying the tube till I should have got everything into order. But the tube is a strong inducement to me to set to work again at some experimental work.

It is curious how all the Germans, so far as I have seen, oppose you, and say that the cathodic rays are processes in the ether, the propulsion of matter being merely a secondary, and non-essential, phenomenon, while so far as I know all your countrymen support you. The French, so far as I can gather opinions from papers in the *Comptes Rendus*, where the issue is not directly raised, are divided in opinion as to whether the cathodic rays are streams of molecules, or are something going on in the ether.

10 *Nov.*, 1896.

I think I told you that I tried the tube you so kindly sent me both in air, and with the lower part of the tube immersed in a beaker of water so as to diminish the range within which total internal reflection could take place. I could not feel sure that there was not, in addition to phosphorescence at the inner surface, also a phosphorescence extending from the outer surface some way into the glass. It was very difficult to say whether this was merely the effect of total internal reflection of the phosphorescence which took place at the inner surface.

I afterwards bethought me of trying bisulphide of carbon, with which there could be no total internal reflection, as it is more refractive than at any rate crown glass, and probably than the uranium glass. I tried that to-day, and with this certainly there was no phosphorescence on the outside.

If therefore there were any when water was used, it must have been connected with the conducting power of water, rendering conceivably possible something of a Leyden jar effect, like what takes place when a discharge is passed through a moderately exhausted tube and the wall is touched with the finger. But I think it was very probably only an internal reflection of the phosphorescence at the inner wall.

P.S. It would of course be very remarkable, would in fact amount to a discovery, if there were any phosphorescence produced at the outside of the glass when it was immersed in water. The thing seems so unlikely that I think it must have been the total internal reflection that gave the apparently external reflection I saw when the tube was immersed in water (*i.e.* external to the *inner* surface of the wall though internal to the outer surface).

31 *March*, 1897.

The *Rendiconti* of the meeting on March 7 of the Italian Lynxes, which reached me this morning, contained (p. 183) an interesting paper by Majorana "Sulla deviazione electrostatica dei raggi catodici" which is in full accord with your (and I may say the British generally) view of the nature of the so-called cathodic rays, though some (*entre nous* wrong-headed) Germans will have it that the Kathodenstrahlen are a process going on in the ether.

I have come to pretty definite views regarding the nature of the Röntgen rays, which I have given in germ and moderate development, indeed pretty far developed, to the Victoria Institute, the British Association, and the Cambridge Philosophical Society. The last two were oral communications, and indeed so was the first, but it was taken down. I hope shortly to draw up a full paper for the Cambridge Philosophical. The whole thing is based on the view that the "Kathodenstrahlen" are not rays at all, but streams of molecules.

I am just off to London.

12 *April*, 1900.

I have long had it in my mind to return you the photographs you sent me a good while ago to show the results obtained with your improved quartz spectroscope. I am afraid I am very much given to the bad habit of procrastination. I did not recollect exactly where I put them, but the other day in looking for something else I came across them. I now return them with many apologies. I hope they will reach you undamaged.

I was looking the other day into your paper in the *Phil. Trans.* in which you show the curious effect of samaria in modifying the phosphorescent spectrum of lime and other oxides. I could not help speculating on the rationale of the dulling effect of a small quantity of samaria on the phosphorescent spectrum of lime. It seemed to me that the samaria must act as a sort of transformer, taking up vibrations from the lime and passing on the molecular agitation to a wider molecular group, which might be made to give out light of a somewhat lower refrangibility.

This leads me to ask whether, when the phosphorescent spectrum of a mixture of lime and samaria, the quantity of samaria being small though not the very smallest, is examined with low dispersion, there is any extra brightness in the lime spectrum on the red-end side of the dull band which samaria produces in the lime spectrum.

There is also another question I should like to ask, which is whether you have examined the absorption spectrum of the earths themselves and mixtures of earths. Most probably you have; but as a powder does nor lend itself for observation of absorption as well as a solution, I presume that the absorption spectra you have given are those of solutions of salts of the earths. Still it would be worth while if you have not done so already (as most likely you have) to take a look at the absorption-spectra of the earths themselves. I need not say that when the specimen is accessible, and you don't object to contact with a substance which can afterwards be washed out, a drop of oil or glycerine would somewhat facilitate the observation.

I suppose you have by you the tubes you used for the observations mentioned in your paper in the *Phil. Trans.*, which could be examined without more ado, except of course as to making the powder more transparent, which is not however essential and which you would hardly care to do unless the preliminary observations should show something worth going into further.

The spectrum of *anhydrous* oxides might be very different from that of solutions.

I am not quite certain whether the absorption spectra you give in your paper are those of the powders, anhydrous oxides or mixtures of them, or of solutions of the salts; but I rather suppose the latter.

26, Princes Square, W.,
19 *April*, 1900.

I think the experiment we talked of would be important in the way of clearing the ground. According to the wagtail notion I should expect the same luminosity in the half in air and in the half in vacuum; according to the idea of kicks by molecules of air I should expect the luminosity to be, not perhaps destroyed, since no vacuum we can make is perfect, but at any rate greatly reduced.

7, Kensington Park Gardens, London, W.,
April 21*st*, 1900.

Dear Sir George,

"Radium" was equally divided and put into two glass cells. One was exhausted to the highest attainable point and the other was kept open to the air. No difference could be detected in their luminosity, and the impression on a photographic plate after five minutes' action was of the same intensity in each case.

Cambridge,
23 *April*, 1900.

Dear Sir William,

Thanks for the interesting information as to the result of the experiment about "radium." The result is in accord with what, in speaking to you, I called the "wagtail" theory of the Becquerel rays which I ventured to throw out in my "Wilde" lecture* at Manchester about two years ago. I have some copies, but I cannot at the moment lay my hands on them or I would send you one.

I have not seen the permanent luminosity of radium (oxide?) mentioned. I suppose it must have been noticed, but your experiment should I think be published at such time as may be deemed most suitable.

23 *April*, 1900.

The fact you showed me of the emission of light by radium is extremely interesting. Did you discover it yourself, or was it mentioned before? I do not recollect having seen it mentioned. The experiment you have just tried is I think very important as helping greatly to clear our ideas.

* *Math. and Phys. Papers*, Vol. v. pp. 254–277.

It seems pretty clear to my own mind now that the influence of "radio-active" substances is of two kinds, absolutely different in their nature; one an actual radiation, a disturbance of the ether, and propagated by it; the other an emission of "radiant matter." What you told me of the behaviour of radium shows that in its case a considerable effect is due to a real radiation. Some experiments of Lord Kelvin's on metallic uranium, mentioned in *Nature*, show, as it seems to me, that in that case most of the effect is due to radiant matter.

And now I have an experiment to propose which I think may help further to clear the ground. It is this. Divide your radium into two portions, about equal, which are laid in two heaps as nearly alike as may be, one on quartz, fairly thin if you happen to have such a piece by you, at any rate not very thick, and the other upon glass of somewhere about the same thickness. Lay the quartz and glass side by side on a photographic plate, and give them the same exposure. Develop. Will the two be much alike, or will that under the quartz be distinctly darker?

23 *April*, 1900.

I have come across, and send you, a copy of my Wilde lecture. The little I have said about the Becquerel rays is at pp. 23, 24.

I retract the inference I drew in my letter just posted, from Lord Kelvin's experiments with uranium, for he only examined the effect of pressure on the rate of discharge by uranium of a charged body. Now real radiation would not cause discharge in the absence of gas, therefore feebleness of discharge, which takes place when the air is exhausted, does not prove feebleness of radiation.

7, KENSINGTON PARK GARDENS, LONDON, W.,
April 24th, 1900.

MY DEAR SIR GEORGE,

I hasten to say that the self-luminosity of radium compounds is not my discovery. I believe it was first mentioned by M. and Mdme. Curie in their papers on Radium in the *Comptes Rendus*. It is a recognised fact now, so there is no objection to your speaking about it. Becquerel also has described it.

I have an experiment now going on which is similar to the one you suggest. I have a long flat plate of pitchblende which gives a very uniform blackening all over its surface when laid on a sensitive plate for 48 hours. I have cut strips of differently coloured glass and arranged them between it and the sensitive surface. I am contemplating trying all sorts of things in this way, to see if there are differences of transparency among media. Of course metals will vary greatly, but there must be some law, possibly according to density, or atomic weight. I don't think there will be any difference between glass and quartz, from an experiment of a somewhat different kind I tried a little time ago. But I will try it again.

April 27th, 1900.

Thank you for your "Wilde" lecture. I remember reading a report of it at the time it was delivered. It was very hard to follow, as it probably was reported incorrectly. I shall value the present copy and study it.

Since writing to you about the results of the vacuum and air experiment with the radium compound I have printed a positive from the negative. The image on the negative was very dense, and I could detect no difference in their intensities. When printed, however, you see there is a slight difference in favour of the radium which was not exhausted. This gives the bombardment hypothesis an advantage over the wagtail one.

1 *May*, 1900.

DEAR SIR WILLIAM,

Many thanks for the photographs showing the photographic effect of Radium, as produced through glass, according as it has air about it or is in vacuum*.

I think the result proves that the theory that the radiation is due to kicks will not answer. At the same time the presence of air may very well increase the radiation. To speak metaphorically, if the dog wags his tail so as at each wag to strike against a half opened door, the agitation he produces in the air may be greater than if the wagging tail did not come against anything but air. For air read ether; for the half opened door read a molecule of air; for the dog's tail read a sort of termination of a molecular structure, agitations propagated into which become

---

* The spectrum has been closely investigated by Sir W. and Lady Huggins (*Roy. Soc. Proc.*) and later by Walter, and found to be the band-spectrum of the nitrogen of the air.

more smart, in a manner analogous to the agitation produced about the lower end of a chain suspended from one end, produced by an agitation at the upper end, which travels down the chain.

IVY HOUSE, KINGSLAND, SHREWSBURY,
12 *Dec.*, 1900.

DEAR SIR WILLIAM,

Coming here to-day I put on a fur coat; for though it was very mild I thought it might be hard frost when I returned on the 27th. It was not till my journey was just over that I discovered in the breast pocket a stamped (but not posted) letter addressed to you. I have opened it to see when it was written and what it was about. I find it is dated 25 April, 1900! Doubtless I put it into my pocket to post, and there it slumbered ever since. It is an old story now, but such as it is I send it.

7, KENSINGTON PARK GARDENS, LONDON, W.,
Dec. 14*th*, 1900.

MY DEAR SIR GEORGE,

I was glad to receive your letter of April 25, although, as you say, it has been a long time coming. Reading it by the light of recent research I am struck by the accuracy of your deductions. I suppose by this time you are satisfied of the self emission of light by radium compounds? I have had some in my desk for about a year, and it shines now as brightly as it did originally. The whole of this time it has been practically in darkness. The glass tube in which it is sealed is assuming a rich purple colour, the result, probably, of the incessant bombardment, day and night, by ultra-violet rays.

IVY HOUSE, KINGSLAND, SHREWSBURY,
15 *Dec.*, 1900.

DEAR SIR WILLIAM,

You told me of the purple colour of the wall of the tube containing a radium compound, and I asked if the colour were superficial or penetrated into the glass, and you said it was on the surface. But in your letter of yesterday you speak of its probably being produced by incessant bombardment by the ultra-violet rays.

I should not myself use the term "bombardment" as applied to the ultra-violet light, or to the Röntgen rays, though I should

as applied to the cathodic rays.   I think it would be a convenient
nomenclature to speak of "rays," "jets" and "emanation," "rays"
referring to a disturbance propagated in the ether, "jets" to a
discharge of molecules, and "emanations"* including both.   If
we mean to speak of an isolated portion of the whole emanation
we may use the expression emanation-thread.

If the radium emanation is of a mixed character, as appears to
be the case, we may speak of it as consisting of rays and jets.   A
ray of light is decomposed by a prism into rays of different
refrangibility.   A cathodic jet is decomposed by a magnet into
jets of different deflexibility.   A radium emanation-thread is
decomposed by a magnet into a jet or jets and a ray.   This
was found by Mr Strutt, Lord Rayleigh's son, though I have not
followed the literature closely enough to allow me to say whether
he was the first to find it.

The purple colour you mention may have been produced
by radium jets or radium rays.   I should think the former more
probable.   When I say radium jets I mean merely that the jets
were sent by radium; whether the propelled matter came from
the substance of the radium or from the air (or other gas) in
contact with it, I leave an open question, but I think the latter
more probable.   It is true that very highly concentrated Röntgen
rays act superficially on glass in a manner detectable by certain
phenomena, but cathodic rays act much more strongly in a manner
to show itself.   On the other hand glass containing much
manganese, used as "glass soap" in its manufacture, gets purple
after long exposure to light; but this purple is not superficial but
goes through, and it is conceivable that radium *rays* might have
a similar effect, though it does not seem probable.

There might be some interest in deciding whether the purple
colour in your radium tube *is* superficial or goes through.   If the
wall be very thin this can hardly be done merely by looking at the
tube as you had it up, but may be done by immersing the tube (or
such portion of it as you care about) in water, or if need be (should
the glass wall be *very* thin) in some liquid of higher refractive
index.   If R, R' be the outer and inner radii of the tube (I
suppose the outer and inner surfaces cylindrical and coaxial),
μ the index of the glass, μ' that of the fluid outside, we must have,

---

* This term is however now applied to the active gas discovered by Rutherford,
emitted from radio-active substances.

in order that the boundary of the inner surface may be seen,

$$R' \text{ less than } \mu' R / \mu.$$

This gives for the thinnest wall that will permit it,

$$\text{thickness} = \frac{\mu - \mu'}{\mu} R.$$

If we take for glass $\mu = \frac{3}{2}$ and for water $\mu' = \frac{4}{3}$, we find $\frac{1}{9} R$ for the least thickness of wall that will allow the inner surface to be seen. If the outer fluid be air the thickness of the wall must exceed $\frac{1}{3} R$, or even more if there be some lead in the glass so that $\mu > 1 \cdot 5$.

If in my long-delayed letter I went into the question whether Becquerel had mentioned the *permanent* luminosity of certain radium compounds, it was not, as you seem to have supposed, from doubt as to the reality of the phenomenon after what you had told me, but because I wanted to see whether or not you had been anticipated by Becquerel.

7, KENSINGTON PARK GARDENS, W.,
Dec. 16*th*, 1900.

MY DEAR SIR GEORGE,

I have been thinking long over the radio-activity of the curious substances found in pitchblende, and I am gradually coming to a somewhat different opinion to the one you expressed in the letter, dated April 25th, but, alas! only now received.

You consider the radio-activity of these bodies to be of two kinds, utterly and completely different in their nature. That was my view for a long time, but experiments I have been carrying on for the last two or three months have made me think that one explanation may account for the two kinds of action. I am now inclined to think that all the radio-active actions are to be accounted for by the theory of "bodies smaller than atoms" propelled from R. A. substances. I thus suppress your (R) and throw all the work on your (P). The objections to this are grave, but I think they may be answered, and if this is so we shall not be obliged to have two totally different explanations for sets of phenomena which glide insensibly one into the other.

Your R is a disturbance in the ether and belongs to the category of the Röntgen rays and ordinary light. The chief differences between R and P are that R is not affected by a

31—2

magnet, and that it penetrates opaque bodies with more or less facility. Both R and P affect a photographic plate, and both make air a conductor of electricity. I will therefore give reasons based on the magnetic and penetrating qualities.

I consider that the magnetic deflection of the rays is simply a question of degree. In a paper I published in 1878 (*Phil. Trans.*, Part 1, 1879, pp. 157—161) I showed the deflection by a magnet was in proportion to the velocity of the radiant matter. At a low vacuum the deflection was great, while at the highest vacuum I then could get the deflection was very little....I therefore look on the non-deviability of the rays from Radium only as a proof that they are projected with too great a velocity to be affected by the magnet, or perhaps they carry a too small charge. The radium rays are thus the extreme in one direction. I should use similar arguments to show that the Röntgen rays were projections and not (directly) affections of the ether.

The greater penetrating action of R over P is more easily answered. I take the case of polonium which emits rays or emanations easily deflected by the magnet, and scarcely able to penetrate any screen. The emanation scarcely extends many centimetres round it, and acts more like a sea fog than like a ray of light. Here however we find some few rays that are less deflected than others, and in exposing a photographic plate some distance off, or behind a screen closer to the polonium, after very many days an impression is seen.

It may be urged that Thomson's ultra-atomic particles are only existent theoretically, and no instance is known of such a phenomenon as a gas or projection or emanation passing through matter. It has long been known that certain gases would pass through iron, platinum, palladium &c., at an increased temperature, but these passages were explained by pores or flaws in the metal. Now however it has been shown by Villard (*C. R.* June 25, 1900), that hydrogen will pass through fused quartz at a red heat— a temperature far below the softening point of the quartz. Now if a dense body like hydrogen gas will get through quartz, how much more easily will particles much smaller than the ordinary chemical atom get through glass, aluminium and black paper?

I am trying many experiments to see if I can get anything which can be considered crucial. The following, which I am repeating with many different active substances, is one on which I think an argument may be based.

I took some active radium compound (the rays from which are the most penetrating and least, if at all, deflected magnetically) and put equal quantities in two adjacent cells turned out of brass. Over cell *A* a wide lead tube was put, a few inches high. Nothing was put over cell *B*. Now if the radiations from the radium were of the nature of light the intensity of action at a certain distance off would be the same whether a lead tube were round it or not. On the top of the lead tube, and projecting over the open cell of radium, I put a photographic plate, the sensitive surface being equidistant from both cells. On developing, hardly any action was seen over the cell that was open—the emanations had diffused about and produced a general fog over all the plate it could get at. But over the cell covered by the lead tube the action was strong. Now two sources of light put in the positions of cells *A* and *B* would have affected, say, the square centimetre of plate immediately over each of them equally (allowing for a trifling increase owing to reflection from the sides of the lead tube). But were the emanations of a material character then they would be confined by the lead tube, while where no tube was they would diffuse away and produce a general fog.

One great argument against my view is that radium and barium chloride is self-luminous. I could not get over this for a long time, but it appears that to get self-luminosity we must have some other substance present*. In my case it is barium, and the colour of the luminosity is very like the phosphorescent glow of a barium compound in a radiant matter tube. It is therefore reasonable to say that the self-luminosity is that of the barium, rendered phosphorescent by the close approximation of the emissive body radium.

A mass of experiments tried during the last few months are tending towards the theory I have ventured to put on paper. I should be so thankful if you would give me your opinion, and suggest, if you think well of it, some experiments which can be considered crucial.

Believe me, very sincerely yours,

WILLIAM CROOKES.

* Cf. *supra*, p. 480.

LENSFIELD, CAMBRIDGE,
15 *April*, 1901.

DEAR SIR WILLIAM,

I feel rather presumptuous in what I am about to write, as if (to use a common proverb) I were "carrying coals to Newcastle." Nevertheless if you should not have come across a recently-unearthed paper written by an eminent man long since dead, the thing may be of some interest to you.

What induced me to write was the failure (which I have just been made acquainted with) of an elaborate experimental research by a student, consequent on the presence of a little vapour of water which it was thought had been provided against. It is true the presence was found to be due to a small unexpected leak, and not to moisture coming from glass walls, as I conjectured when I read of the presence, but had not yet read on to the part where it was accounted for; yet the idea I started with led me to think of an interesting paper by Fraunhofer at p. 33 of his Collected Works, a copy of which was recently sent me by the Academy of Sciences at Munich. It is taken from an obscure German serial not indexed by the Royal Society, and is entitled "Versuche über die Ursachen des Anlaufens und Mattwerdens des Glases, und die Mittel, denselben zuvorzukommen."

It occurred to me that possibly the walls of a glass vessel intended for very high exhaustions of air or other gases might be improved as to resistance to condensation of moisture by a very simple process mentioned by Fraunhofer, namely, by filling the vessel (tube, or whatever it might be) intended to be used with concentrated sulphuric or hydrochloric acid, which is left in contact with the glass for 24 hours or so before the glass is cleaned from the acid.

As to resistance to tarnish, much depends on the composition of the glass. But even in using a glass of favourable composition for resistance to tarnish, it struck me·that perhaps the resistance might be further improved by treatment with acid as mentioned by Fraunhofer.

How far improved resistance to alteration by moisture may be accompanied by diminution of power of condensing on the surface other gases, if there is any such diminution at all, is more than I can say.

7, Kensington Park Gardens, London, W.,
May 28, 1901.

My dear Sir George,

Your letter of the 15th April has started me on some experiments which I wanted to complete before replying. I have not been satisfied for a long time with the working of a delicate aluminium-leaf electroscope of M. Curie's design. It would not hold the charge for a sufficient time, and on speaking about it to Professor Ayrton he suggested that there might be a little leakage through the support of the Al leaf. We therefore tested many supposed good insulators in the instrument, and the results are in all cases disappointing.

The edge of the Al leaf is viewed by a telescope containing a ruled glass scale in the eyepiece, and the measures are the movement of the leaf across the divisions per minute.

The original mounting was a mixture of sulphur and paraffin wax, which is said to be an almost perfect insulator. With this the leakage was 0·43 divs.

The sulphur-paraffin was removed and the support was a long, somewhat thin lead-glass rod, very carefully cleaned and artificially dried. Here the leakage rose to 0·60 divs.

Next we used quartz rods prepared by Shenstone, and treated in various ways. To our surprise the leakage in this case was worse than when glass was used, being 0·65 divs.

I now put into practice the hint conveyed in your letter. I took some glass rods of lead glass (this glass is the best insulator) and boiled them for some time in strong acids. They were then thoroughly cleaned and artificially dried and tested. The leakage here was 5·60 divs.

From this experiment it seems as if the boiling in acids, which has always been supposed to increase the insulating properties of glass, has done much harm, and it is better not to interfere with the original fused surface of the glass.

Another thing that upsets our previous ideas is that quartz is no such good insulator as it was thought to be.

In all cases the interior of the apparatus was dried with sulphuric acid.

Professor Boys has suggested that the leakage observed is a sort of invisible brush discharge from the sharp edges of the aluminium leaf and the brass plate. I scarcely think this is the explanation, for were it so no electroscope would be able to hold its charge. I have an idea that as I have been working so long with uranium and other active bodies from pitchblende, the instrument has become soaked with these emanations, and the air thereby is rendered a slight conductor. We know that radio-activity, like some diseases, is catching, but the infected metal or other body soon recovers when the strong radio-active substance is removed from near it.

Believe me, very truly yours,

THE ATHENÆUM, PALL MALL, S.W.,
1 *June*, 1901.

DEAR SIR WILLIAM,

That was a remarkable result which you obtained, that the action of acid should have so *promoted* the leakage of electricity. I notice one difference between your procedure and that of Fraunhofer, viz. that he did not use heat with the acid. But I do not think it likely that your result was due to this.

The most probable explanation seems to me to be that the acid left the surface hyper-microscopically pitted, and that the roughness, though too fine to be visible, facilitated the propulsion of electrified molecules of the air or other gas. There is no reasonable doubt, I think, that a perfect vacuum would be a perfect insulator, and that the way in which points act in discharging electricity is by...(as may be seen from the forms of the equipotential surfaces) driving the electrified molecules away from the charged body. I have been writing with a view to the *difference* between the acid-treated and the non-treated support.

LENSFIELD, CAMBRIDGE,
4 *June*, 1901.

I feel rather curious to know what acids they were that you used in the treatment of your glass support.

It is very remarkable how the leakage was increased by boiling the glass with acid. Your immediate object was to get good insulation. But the cause of the great leakage consequent on the treatment with acids, which was quite unexpected, seems to be worth trying to make out when leisure permits. Your wonderful radiometer was the outcome of an investigation designed to follow out the cause of obscure and minute indications of an attractive or repulsive force, and you reaped a rich harvest of discovery. Who can say what may lie hid under this unexpected action of acids?

As to the cause of the leakages 0·43, 0·60, 0·65, the fact that the first is only about two-thirds of either of the others looks as if it took place not through the air but along the stem that supports the leaves. If however it took place, not by creeping along the stem, but by convection through the air, it is conceivable that it might depend, not only on the nature of the gas surrounding the leaves, but upon the nature of the leaves themselves. It is conceivable that there might be more from aluminium than from gold, the metal usually employed for a leaf electroscope. I do not know whether experience has shown that aluminium under ordinary circumstances holds as well as gold.

However, there is no use in a tyro like myself speculating in these things.

14 June, 1901.

This week's *Nature* contains (p. 157) an interesting paper by Professor Rutherford on the emanation from radium. In it he refers to an interesting paper by Mm. Curie and Debierne, which somehow I had missed reading though the number was in my house. It is *C. R.* for March 25, 1901 (Vol. 132, p. 768). It seems they got the active substance as a gas, which when thus obtained partly separated was wonderfully active. From its diffusibility Rutherford estimates the molecular weight as somewhat high, say between 40 and 100.

If it can thus be got mixed with air as a gas, there seems hope of making out something more about it. But probably you have seen these papers.

When I see the numbers you mentioned, I may be better able to guess what the experimental behaviour as to the additional electric leakage due to "infection" may be, and therefore better able to see or guess how best to correct for it.

June 15th, 1901.

Dear Sir George,

Like you I missed the paper of Curie and Debierne in the *C. R.* I will look at it. Rutherford's paper is very interesting, but I cannot agree that the chief radio-active body in pitchblende is a gas, in the ordinary meaning of the word. I have long had an idea that the radio-activity is produced by the separation of material particles (probably Thomsonian ultra-atomic) from a body or bodies contained in pitchblende. These particles are "ionised" or charged with negative electricity, and temporarily may possess the properties of a gas, or vapour. The great difference in action between radium and polonium shows that there is more than one body capable of so liberating small corpuscles, but what is the special peculiarity which enables these bodies to allow particles to be chipped off we do not know. I think my work on *Electrical Evaporation* (copy enclosed) throws some light on these obscure actions. I can imagine if a body has some of its constituent particles lightly held, they will be apt to fly off under the influence of an extra negative charge. At the surface of any volatile liquid molecules are continually flying off into the air, and I have shown that the surface of a (so-called) fixed solid, under electrical stress, behaves in the same way, giving off particles and actually distilling onto adjacent surfaces. These particles, in the act of distilling, would very likely act as radio-active bodies. The ordinary substances, such as gold, cadmium, &c., I experimented with, only behave in this manner when under added electrical stress; but suppose there be bodies which are capable of temporarily fixing additional atoms of electricity— unstable perelectrides—these might have in themselves a sufficient surcharge of electricity to become temporarily volatile, and the gradual dissipation of the extra atoms of electricity would produce all the phenomena of radio-activity.

Forgive my crude speculations. I feel as it were groping in an unknown laboratory in the dark. If by any means we can evolve a spark of light in any direction it is one's duty to follow it up. I hope you will not think what I have said utter nonsense!

I have left no time to copy out and send you the figures connected with the electric leakage experiments, but I will try and send them this afternoon.

Believe me, very sincerely yours,

June 15, 1901.

I wished to see if I could carry down radio-activity by precipitating barium sulphate in the presence of a radio-active body. Thoria was the radio-active body used, and I precipitated successive lots of barium sulphate in the presence of the same lot of thoria. My idea was to see if I could exhaust the activity of the thoria in this way, and eventually get a thoria inactive. I precipitated ten successive lots of barium sulphate, with the following results:

Empty electroscope, time required to sink (leak) 5° 240″
Electroscope + pure thoria   „   40″
„   + pure barium sulphate „   240″
Electroscope empty   „   600″
„   + barium sulphate, No. 1 ppt.   190″
„   empty   600″
„   + barium sulphate, No. 2 ppt.   405″
„   empty   550″
„   + barium sulphate, No. 3 ppt.   42″
„   empty   600″
„   + barium sulphate, No. 4 ppt.   80″
„   empty   570″
„   + barium sulphate, No. 5 ppt.   300″
„   empty   600″
„   + barium sulphate, No. 6 ppt.   105″
„   empty   600″
„   + barium, No. 7 ppt.   240″
„   empty   465″
„   + barium, No. 8 ppt.   315″
„   empty   510″
„   + barium, No. 9 ppt.   285″
„   empty   510″
„   + barium, No. 10 ppt.   410″
„   empty   430″
„   + barium, No. 11 ppt.   350″

These obervations of the activity of the barium sulphate precipitates were taken a short time ago (Feb. 18). On Sep. 14 I took the same precipitates just after they had been precipitated and dried. They were much more active. Their action varied between 2″ and 20″. This shows that the radio-activity conferred on barium sulphate by its being precipitated in the presence of thoria is only a temporary quality. A few experiments at that time showed that heating the active barium sulphate to redness did not appreciably impair its activity.

I have worked out the electric radio-activity by the method

you gave me, and from the look of the figures I should think they were accurate, as far as possible when the results are so varying and sometimes contradictory. But there is one thing I should like to manage, if possible. The figures as got by the "reciprocal" method come out upside down. That is, the strongest stuff gives the lowest figure, and the weakest stuff the highest figure. I want to express the results so that I can plot the two curves superposed for comparison.

LENSFIELD, CAMBRIDGE,
17 *June*, 1901.

DEAR SIR WILLIAM,

Just for curiosity, I tried if I could see any relation between the amount of activity of your precipitates and their order of sequence, but such I did not find. It was indeed not to be expected that I should, so much as to the thorium carried down with the barium sulphate is likely to depend on the manipulation of the precipitation. It will be more interesting to see whether the electricity-discharging and the photographic effects correspond.

28 *Nov.*, 1901.

In connexion with some speculations about fluorescence and phosphorescence, and about the theory of incandescent mantles, I should like to ask whether you have tried the effect of thoria in bringing out the citron line due to a trace of yttria. I mean under electric bombardment.

My speculations lead me to think that lime, zirconia and thoria might be very good, and perhaps in that order. You specially remarked the thing with reference to lime.

P.S. I have just dipped into your abstract in *Proc. R. S.* Vol. 38. I should expect that thoria would bring out well the spectrum of samaria.

7, KENSINGTON PARK GARDENS, LONDON, W.,
Nov. 30*th*, 1901.

MY DEAR SIR GEORGE,

I have tried the three experiments suggested in your letter of the 28th inst. I made three mixtures in the form of nitrates; they then were evaporated to dryness and ignited with sulphuric acid and put into three separate vacuum tubes connected together in series, so that the exhaustion and treatment might be the same in each case.

Mixture No. 1 contained......$\begin{cases} \text{Samaria} & 100 \\ \text{Lime} & 80. \end{cases}$

Mixture No. 2 contained......$\begin{cases} \text{Samaria} & 100 \\ \text{Zirconia} & 60. \end{cases}$

Mixture No. 3 contained......$\begin{cases} \text{Samaria} & 100 \\ \text{Thoria} & 50. \end{cases}$

I took different proportions as the atomic weights of the three added earths differ, thoria being the heaviest and lime the lightest.

The tubes were well exhausted and heated so as to get the most favourable conditions for phosphorescence. They were then sealed off and left all night.

This morning the calcium tube phosphoresced brilliantly. The thoria tube offered much higher resistance to the spark, but after a little time it also phosphoresced well, but not so well as the lime tube. The zirconia tube was non-conducting, and it was only when I warmed it with a gas flame that I could get the spark to pass. Then the phosphorescence was very poor, and the tube seemed full of gas (fog).

I allowed the tubes to rest for some time, and then connected the lime and the thoria tubes together in series, so that the same spark should pass each. At first the thoria tube offered great resistance ; the lime tube phosphoresced well, but the spark struck across the terminals of the thoria tube. I then, with a gas flame, carefully heated both tubes, taking care not to heat one more than the other. The thoria tube then phosphoresced well, and there was little difference to be seen between the two tubes.

I put the lime and the zirconia tubes in series in the same way, but could not get a good phosphorescence of the zirconia tube.

My opinion is that lime is the best of the three to induce phosphorescence. The tube retains its condition and will allow stronger sparks to pass without injury, while at the same time it will phosphoresce with a very weak spark. Thoria has a tendency to increase the vacuum till it will not conduct, although when heated it induces phosphorescence almost as well as the lime. Zirconia is far inferior to the other two. It requires much attention to the vacuum and the heating to get it to phosphoresce, and when sealed off at the best point it rapidly deteriorates in condition, and it cannot by subsequent heating be brought to its original (poor) condition.

I am quite at your service for any further experiments you would like to be tried.

LENSFIELD, CAMBRIDGE,
9 *Dec.*, 1901.

DEAR SIR WILLIAM,

I am afraid you will think me ungrateful for not having written before to thank you for your kindness in trying some experiments I had thought of. About the time your letter reached me I was engaged in winding up my lectures for the term (for though I am in my 83rd year I still go on lecturing), and besides on two days we had friends here; and further I wanted to look into the subject before writing to you.

I must plead guilty to having caused you needless trouble, by trusting too much to my memory instead of having looked into your papers again to see whether my memory was correct. I thought I recollected that you had failed to get the citron line of yttria even when that earth was present in abundance, but obtained it well when that earth was mixed with a relatively large quantity of lime. Trusting to this supposed recollection, I imagined that the failure to get the citron line of yttria even when there was plenty of the earth present might have been due to too great concentration, and that the office of the lime was to dilute the yttria; just as a very strong solution of sulphate of quinine does not show the fluorescence as well as a dilute solution.

I have now read the abstract of your paper in the *Proceedings R. S.*, Vol. 35, p. 262, and I see my memory was wrong, and that your failure to get the citron line was due to the non-employment of sulphuric acid, the excess of which is afterwards driven off by heat. It looks as if the molecular group which gives the citron line is anhydrous sulphate of yttria, or perhaps (when lime is employed) an anhydrous double sulphate of yttria and lime.

Regretting the having caused you needless trouble by neglecting to look into your papers before writing,

I remain, Yours sincerely,

G. G. STOKES.

CORRESPONDENCE WITH PROF. S. P. THOMPSON, F.R.S.

Feb. 28, 1896.

DEAR SIR GEORGE STOKES,

I made yesterday an observation of such curious interest that I am minded to bring it before your notice. I find that if a phosphorescent substance such as sulphide of Barium is exposed to ordinary white light so as to be well insolated, and brought to the shining condition, it emits afterwards (and apparently also during illumination) not only ordinary light that can be cut off with an aluminium sheet, but also something else that is not cut off by aluminium, and is, in this respect at any rate, the same as the X-rays of Röntgen, that it can traverse aluminium and act on a photographic plate. If it is true that there are fluorescent (or phosphorescent) substances that deviate from your law of degradation of frequency (or wave-length), this would seem to present an extreme case of such deviation. But if these be Röntgen's rays, then I have succeeded in manufacturing them out of common light by a sort of reversal of the process of fluorescence. Do you know of any other instance in which fluorescence or phosphorescence has been found to be reversible in its operation?

I am, Yours most sincerely,

SILV. P. THOMPSON.

29 *Feb.* 1896.

DEAR PROFESSOR THOMPSON,

Your discovery is extremely interesting; you will I presume publish it without delay, especially as so many are now working at the X-rays.

For my own part I am not at all disposed to believe that the Röntgen rays are due to normal vibrations, the hypothesis to which Röntgen himself leans. I think it far more probable that they are transversal vibrations of excessive frequency. That being the case, I think what you have discovered belongs to the same class of phenomena as Tyndall's calorescence. For a very long time I have thrown overboard the conjecture I made in my paper of 1852 as to the cause of fluorescence, and have regarded it as a disturbance extending from more limited to more extensive molecular groups. Long ago (I happen to have a memorandum fixing it to 1886) I arrived at a little interesting dynamical problem throwing an interesting light on the way in which the period got lengthened, but I did not publish the result*. I look on calorescence as an agitation passing from wider to more minute molecular groups. In your discovery, I think we have something of the nature of calorescence; only that whereas in Tyndall's work the disturbance was excited in the first instance in wider molecular groups, in yours the wider groups are already something like the chemical molecules of the peculiar substance.

I am in correspondence with Lord Kelvin about the Röntgen rays, and I should like to refer to your discovery, but do not mean to do so till you have published your result. I should be glad therefore of a notice of the publication. Perhaps you may be writing to him yourself. Of course if you have done so I am free to say anything to him. He is very enthusiastic, and might let something slip out without thinking about it.

Yours very truly,

G. G. STOKES.

2 *March*, 1896.

I fear you have already been anticipated. See Becquerel, *Comptes Rendus* for Feb. 24, p. 420, and some papers in two or three meetings preceding that†.

It seems probable that the agent in this and in your case is the X-rays, but their nature has yet to be made out. I think it most probable that they are transverse vibrations of extreme frequency, but that has yet to be made out. The supposition is by no means exempt from difficulty.

[* Cf. addition (1901) in *Math. and Phys. Papers*, Vol. III. p. 410.]

[† This was Becquerel's cardinal discovery of radio-activity in uranium salts. Prof. Thompson's letter describing his experiments, of later date, as given above, and to which Prof. Stokes' letters kindly lent by Prof. Thompson are in reply, specified sulphide of barium, which is not sensibly radio-active; but in answer to an inquiry he now states that he had been trying various substances, including uranium nitrate, and that he had found, conclusively on Feb. 26—7, that the latter was the only one to which the aluminium foil was not opaque, though black paper was transparent to others.]

28 *May*, 1896.

It may be worth while to mention to you a little experiment, though containing nothing new in principle.

I made a solution of Uranine in water, using very little of the substance, so as to get a solution showing well the fluorescence. Tying a thread round the brick of Calcium Sulphide you gave me, I exposed it to daylight (the sun not shining at the time) and then carried it into a dark room and immersed it in the solution. The blue light emitted was absorbed by the solution, so that the brick was not seen glowing with the usual blue light unless it were so hung in the solution that the blue rays which it admitted had but a small path to traverse in the solution before they emerged from the wall of the glass vessel containing the solution on the side next the eye. But they excited the phosphorescence of the solution so that the brick was seen enveloped with a mantle of green light, fitting it more or less closely according to the strength of the solution.

I tried the experiment yesterday, putting the brick into a narrow glass vessel (I did not happen to have by me a test tube broad enough to admit it) and dipping the containing vessel into the beaker containing the solution. However the brick admits of being put into the solution without being spoiled by the wetting.

There is of course nothing new in principle in the experiment. I may mention a little experiment I tried about I suppose 43 years ago. I took three fluorescent media, A, a solution of sulphate of quinine, B, a slab of uranium glass, C, a madder solution, virtually a solution of purpurine in alum solution. The solutions were contained in vessels with glass sides. The sun's light was reflected horizontally into a darkened room and passed through a lens so as to form a narrow concentrated beam which was passed into one of the solutions, or the glass slab, as the case might be. A little way from the fluorescent beam, in a lateral direction, was placed a lens to form a real image of the fluorescent beam, and this image was received in another of the media. When the first medium was A, the image of the fluorescent beam excited fluorescence in B and C, but not in A. When the first medium was B, the image excited fluorescence in C, but not in A or B. When it was C, the image did not excite fluorescence in any of the three. I don't recollect that I ever mentioned this experiment to anyone, the thing is so obvious. The principle is just the same as in the experiment with the brick.

P.S. I may as well mention, in case you should not have seen it, that in the last number of the *Comptes Rendus* is a paper by Becquerel in which he mentioned that the metallic uranium shows the remarkable phenomenon which you and he discovered*, independently, about four times as strongly as the salts of uranium he had previously used.

26 Sept. 1900.

Thanks for your letter of the 24th in which you mention the best way of using the brick form of the phosphorus. I have informed my friend in Ireland, who told me he had ordered a brick.

I have tried on a larger scale the use of melted paraffin instead of hot water as a vehicle for the slightly waxed powder† which Mr Horne sells in cans, and it seems to promise to be very successful.

I had not noticed the brownishness of the phosphorus while in the phosphorus condition. It is of course well known that the rays of lower refrangibility than those that excite the sulphide of calcium, when allowed to fall on a plate already phosphorescing from previous exposure, cause it to give out light (of a whiter tint, as I have noticed) *until* the result of the previous exposure is exhausted. The rays thus working must doubtless be absorbed. But it had not occurred to me that this absorption might be sufficient to cause the plate to appear to be sensibly less luminous than it would be if it had not been exposed at all, and were subjected to the action of the lower rays as above mentioned. I do not however think that the whole of the brownishness is thus to be accounted for. I have it in my mind to try an experiment with a view to seeing whether any sensible amount is thus to be accounted for. But I think a good deal is accounted for by ordinary absorption, irrespective of previous exposure. For I noticed that with paraffin, and still more with bisulphide of carbon, a very decided brownishness is seen independently, as it would seem, of previous exposure. There is but little of this to

[* See footnote *supra*, p. 496.]

[† A special preparation of sulphide of calcium protected from moisture by a coating of wax.]

be seen when the powder is exposed to the action of hot water, as recommended in a paper on the cans. There is more when paraffin is the vehicle, and still more when bisulphide of carbon is used. If I try the experiment I was thinking of, I will let you know the result.

3 *Oct.* 1900.

I have tried your experiment a few times, but I cannot say that I have succeeded. Sometimes I seemed to have a slight suspicion that perhaps the unexposed substance might have been of a slightly paler brownish colour than the part that had been exposed, but it did not go beyond a possible suspicion. The difference between our results may very probably have been due to the employment of a different material; I used the substance from a can, you very likely used a brick, which I have not at present tried.

I think in all probability almost the whole of the slight brownishness in my trials was due to a real absorption independent of exposure, the effect of which in the case of a powder would come out much more strongly if the powder were impregnated with a liquid of approximately the same refractive index, and which did not act chemically upon it.

Paraffin answered extremely well. I could only try on some of Horne's powder that had been already slightly waxed. I think for chemical reasons that paraffin might very likely be rather superior to wax. I have half a mind to write to Mr Horne. I don't suppose that it would be disagreeable to you that I should.

Yours very truly,

G. G. STOKES.

INDEX TO VOL. II.

Numerical reductions, advantage of system in, 452.

Objective, reversal of components, 194; experiments, 194; resulting features of the aberration, 195; specification of photographic, 196; components examined by phosphorescence, 196.

Oersted, H. C., 168.

Ohm's law, as to limits of, 37.

Optical instruments, British practice, 88, 91.

Optical reflection, experiments by Maxwell on, 33; boundary conditions in theory, 41.

Optical systems, Maxwell's geometrical theory, history, 2.

Organ-pipe, maintaining action in, 184; amplitude, 184.

Oscillations, free tidal, 169; forced, requires concomitant semi-free one, 170; tidal analogy, 172.

Parliamentary duties, 115.

Peirce, C. S., on American pendulum survey, 309—313.

Pencils, oblique, 459.

Pendulum, proposed new standard, 267, 269; design, 270, 253—271, 311; precaution in travelling, 312.

Pendulums, effect of viscosity at different exhaustions traced, 105, 110; mathematical problems, 105; magnetic influence on steel rod, 107; reductions compared with Baily's and Sabine's, 253; methods of eliminating air resistance, 255; formula for reversible, 256; sensitiveness, 257; theory, 276; correction, 277; temperature effect, 279; tracing discrepancies, 280; time lag in temperature, 281; advantages of graphical method, 282, 285; influence of eddies, 284; correction for humidity, 294; viscosity, 295, 317, determinable by light model, 320; correction for buoyancy, 318; flexibility needed, 319; invariable, advantages of, 316; temperature corrections, 321; collected memoirs on, 322; limits of validity of theory, 325.

Phosphorescence, decays by diffusion, 417; tiring of, explained, 422; of mixed rare earths, 493; and absorption, 496.

Photometer, Maxwell's, 6; another on Swan-Brodhun pattern, 25.

Plücker, J. R., reasons for award of Copley medal, 61; magnetocrystallic theory, 64; positive and negative lights in vacuo, 68; cathode rays, 69; on geometric complexes of rays, 69; plans of work, 71; relation to Sylvester, 71.

Polarization by gratings, early experiments, 102.

Poisson, S. D., magnetic theory, 65, 167; on waves in water, 175.

Prism, constitution revealed by its dispersion, focussing for pure spectrum, 329; reflecting quartz, accuracy needed in, 456.

Proceedings of the Royal Society, suggested changes, 166.

Prout's hypothesis of a primordial substance, 78.

Publications, exchange of, 58.

Pulse in air has condensed and rarefied layers, 348.

Quartz films, optical behaviour, beads on, 336.

Quaternions, Maxwell on, 43.

Radiation, theory of exchanges in spectra, 84; pressure of, 434.

Radio-activity, decay of, 491.

Radiometer, Maxwell on, 44; repulsion on the case, 369; explanation of, 373; best type of vanes, 374; discs sensitive to different radiations, 375; effect of heated envelopes, 377; effect of form of vanes and bulb, 388, 401; absorption by bulb evaded, 389; effect of local heating, 391; of concavity, 402; of thermal capacity of vanes, 398; of roughness, 404; reversals of motion, 400; precautions needed, 405; Maxwell on, 433; radiometer disturbance in Cavendish experiment, 473.

Radium, source of its luminosity, 478; both radiation and projection involved, 479; influence of air, 480; coloration of glass by, 481; location of colour in glass, 483; cathode rays from, 482, 484.

Rainband spectroscopy, 225; difficulties, 228.

Rainbows, supernumerary, from quartz fibre, 332.

Ramsay, Sir W., on argon, 445.

Rankine, W. J. M., a theory of double refraction, 100.

Rayleigh, Lord, 99—125; treatise on sound, 102; at Stornoway, 115; on wave-groups, 157.

Rays, Plücker on complexes of, 69.

Refraction used to locate internal colour, 483; slight double, compensation for, 457.

Repulsion, radiometer, on swinging disc, 378; its growth and decay, 381; determination of change in the force, 383; time required for establishment, 386, 397; deduced by chronographic reductions, 394; tangential force, 390; theory, 406; new feature, 410.

CAMBRIDGE : PRINTED BY JOHN CLAY, M.A., AT THE UNIVERSITY PRESS.

Printed in the United States
By Bookmasters

CAMBRIDGE LIBRARY COLLECTION

Books of enduring scholarly value

Lucasian Professor of Mathematics at Cambridge and President of the Royal Society, Sir George Gabriel Stokes (1819–1904) made substantial contributions to the fields of fluid dynamics, optics, physics, and geodesy, in which numerous discoveries still bear his name. The *Memoir and Scientific Correspondence of the Late Sir George Gabriel Stokes*, edited by Joseph Larmor, offers rare insight into this capacious scientific mind, with letters attesting to the careful, engaged experimentation that earned him international acclaim. Volume 2 (1907) includes important professional correspondence with James Clerk Maxwell, James Prescott Joule, and many others, with particular attention given to Stokes' activities with the British Meteorological Society. Many of his foundational innovations in optics are also explicated in these letters, serving in place of the authoritative volume he unfortunately never had the opportunity to complete.

www.cambridge.org/clc

CAMBRIDGE
UNIVERSITY PRESS
www.cambridge.org

ISBN 978-1-108-00892-1

9 781108 008921